TURBULENCE IN FLUIDS

FLUID MECHANICS AND ITS APPLICATIONS
Volume 1

Series Editor: **R. MOREAU**
MADYLAM
Ecole Nationale Supérieure d'Hydraulique de Grenoble
Boîte Postale 95
38402 Saint Martin d'Hères Cedex, France

Aims and Scope of the Series

The purpose of this series is to focus on subjects in which fluid mechanics plays a fundamental role.

As well as the more traditional applications of aeronautics, hydraulics, heat and mass transfer etc., books will be published dealing with topics which are currently in a state of rapid development, such as turbulence, suspensions and multiphase fluids, super and hypersonic flows and numerical modelling techniques.

It is a widely held view that it is the interdisciplinary subjects that will receive intense scientific attention, bringing them to the forefront of technological advancement. Fluids have the ability to transport matter and its properties as well as transmit force, therefore fluid mechanics is a subject that is particulary open to cross fertilisation with other sciences and disciplines of engineering. The subject of fluid mechanics will be highly relevant in domains such as chemical, metallurgical, biological and ecological engineering. This series is particularly open to such new multidisciplinary domains.

The median level of presentation is the first year graduate student. Some texts are monographs defining the current state of a field; others are accessible to final year undergraduates; but essentially the emphasis is on readability and clarity.

For a list of related mechanics titles, see final pages.

Turbulence in Fluids

Stochastic and Numerical Modelling

by

MARCEL LESIEUR

National Polytechnic Institute,
School of Hydraulics and Mechanics,
Grenoble, France

Second revised edition

POLYTECHNIC LIBRARY
WOLVERHAMPTON

Acc. No. 495825

CLASS

CONTROL

532.
0527
LES

DATE 29. MAY 1991

SITE RS

KLUWER ACADEMIC PUBLISHERS

DORDRECHT / BOSTON / LONDON

Library of Congress Cataloging in Publication Data

```
Lesieur, Marcel.
    Turbulence in fluids : Stochastic and numerical modelling / Marcel
  Lesieur. -- 2nd rev. ed.
        p.    cm. -- (Fluid mechanics and its applications ; v. 1)
    Includes bibliographical references.
    Includes index.
    ISBN 0-7923-0645-7 (alk. paper)
    1. Fluid mechanics. 2. Turbulence. 3. Transport theory.
  I. Title. II. Series.
  QC145.2.L47  1991
  532'.0527--dc20                                              90-46046
```

ISBN 0-7923-0645-7

Published by Kluwer Academic Publishers,
P.O. Box 17, 3300 AA Dordrecht, The Netherlands.

Kluwer Academic Publishers incorporates
the publishing programmes of
D. Reidel, Martinus Nijhoff, Dr W. Junk and MTP Press.

Sold and distributed in the U.S.A. and Canada
by Kluwer Academic Publishers,
101 Philip Drive, Norwell, MA 02061, U.S.A.

In all other countries, sold and distributed
by Kluwer Academic Publishers Group,
P.O. Box 322, 3300 AH Dordrecht, The Netherlands.

Printed on acid-free paper

All Rights Reserved
© 1990 by Kluwer Academic Publishers
No part of the material protected by this copyright notice may be reproduced or
utilized in any form or by any means, electronic or mechanical,
including photocopying, recording or by any information storage and
retrieval system, without written permission from the copyright owner.

Printed in the Netherlands

à mes Parents,

à Stéphanie et Juliette,

Preface to the first edition

Turbulence is a dangerous topic which is often at the origin of serious fights in the scientific meetings devoted to it since it represents extremely different points of view, all of which have in common their complexity, as well as an inability to solve the problem. It is even difficult to agree on what exactly is the problem to be solved.

Extremely schematically, two opposing points of view have been advocated during these last ten years: the first one is "statistical", and tries to model the evolution of averaged quantities of the flow. This community, which has followed the glorious trail of Taylor and Kolmogorov, believes in the phenomenology of cascades, and strongly disputes the possibility of any coherence or order associated to turbulence.

On the other bank of the river stands the "coherence among chaos" community, which considers turbulence from a purely deterministic point of view, by studying either the behaviour of dynamical systems, or the stability of flows in various situations. To this community are also associated the experimentalists who seek to identify coherent structures in shear flows.

My personal experience in turbulence was acquired in the first group since I spent several years studying the stochastic models of turbulence, applied to various situations such as helical or two-dimensional turbulence and turbulent diffusion. These techniques were certainly not the ultimate solution to the problem, but they allowed me to get acquainted with various disciplines such as astrophysics, meteorology, oceanography and aeronautics, which were all, for different reasons, interested in turbulence. It is certainly true that I discovered the fascination of Fluid Dynamics through the somewhat abstract studies of turbulence.

This monograph is then an attempt to reconcile the statistical point of view and the basic concepts of fluid mechanics which determine the evolution of flows arising in the various fields envisaged above. It is true that these basic principles, accompanied by the predictions of the instability theory, give valuable information on the behaviour of turbulence and of the structures which compose it. But a statistical analysis of these structures can, at the same time, supply information about strong non-linear energy transfers within the flow.

I have tried to present here a synthesis between two graduate courses given in Grenoble during these last few years, namely a "Turbulence" course and a "Geophysical Fluid Dynamics" course. I would like to thank my colleagues of the Ecole Nationale d'Hydraulique et Mécanique and Université Scientifique et Médicale de Grenoble, who offered me the opportunity of giving these two courses. The students who attended these classes were, through their questions and remarks, of great help. I took advantage of a sabbatical year spent at the Department of Aerospace Engineering of the University of Southern California to write the first draft of this monograph: this was rendered possible by the generous hospitality of John Laufer and his collaborators. Finally, I am grateful to numerous friends around the world who encouraged me to undertake this work.

I am greatly indebted to Frances Métais who corrected the English style of the manuscript. I am uniquely responsible for the remaining mistakes, due to last minute modifications. I ask for the indulgence of the English speaking reader, thinking that he might not have been delighted by a text written in perfect French. I hope also that this monograph will help the diffusion of some French contributions to turbulence research.

Ms Van Thai was of great help for the drawings. I am also extremely grateful to Jean-Pierre Chollet, Yves Gagne and Olivier Métais for their contribution to the contents of the book and their help during its achievement, and to Sherwin Maslowe who edited several Chapters.

This book was written using the TEX system. This would not have been possible without the constant help of Evelyne Tournier, of Grenoble Applied Mathematics Institute, and of Claude Goutorbe, of the University computing center.

Finally I thank Martinus Nijhoff Publishers for offering me the possibility of presenting these ideas.

Grenoble, October 1986   Marcel Lesieur

Foreword to the second edition

Four years seems to be a good period of time to assess one's old points of view in such rapidly evolving field as Turbulence and Fluid Mechanics. The new possibilities offered by direct-numerical simulations have provided a lot of information on vortex dynamics, coherent structures and transition, compressible or rotating flows. The third chapter now gives a basic presentation of the linear-instability theory applied to shear or thermally unstable flows. A substantial part of the phenomenology in Chapter VI is devoted to mixing-length theory applied to turbulent shear flows. Concerning the stochastic models, it seemed necessary to include more information on the D.I.A. and R.N.G. theories. New calculations and experiments on stratified or shear flows have been incorporated, with emphasis put on the three-dimensional structures topology. Recent results on the intermittency of isotropic turbulence, and on passive scalar diffusion are also included. Finally, I rewrote Chapter XII on large-eddy simulation in order to make it more general and accessible to graduate students. This is the general point of view which has been my guideline during the write up for this second edition.

All this makes for a much more substantial book. I hope the original spirit of the first edition has not been lost, but I think it has resisted well to my attacks. Particular thanks are extended to Pierre Comte and the graduate students of our group for their important visual contribution which illustrates so well coherent structures and transition. Many thanks also to Olivier Métais who contributed greatly to certain of the new numerical results shown here, and for his permanent and total support and interest. Jim Riley was unlucky enough to spend his sabbatical with us in Grenoble during these last few months, and influenced many of my conclusions, while pretending to correct the language of a few chapters. Finally, I am indebted to all the sponsoring agencies and companies who showed a continuous interest during all these years in the development of fundamental and numerical research on Turbulence in Grenoble.

Grenoble, June 1990   Marcel Lesieur

Contents

Chapter I

INTRODUCTION TO TURBULENCE
IN FLUID MECHANICS

1 - Is it possible to define turbulence?

Everyday life gives us an intuitive knowledge of turbulence in fluids: the smoke of a cigarette or over a fire exhibits a disordered behaviour characteristic of the motion of the air which transports it. The wind is subject to abrupt changes in direction and velocity, which may have dramatic consequences for the seafarer or the hang-glider. During air travel, one often hears the word turbulence generally associated with the fastening of seat-belts. Turbulence is also mentioned to describe the flow of a stream, and in a river it has important consequences concerning the sediment transport and the motion of the bed. The rapid flow of any fluid passing an obstacle or an airfoil creates turbulence in the boundary layers and develops a turbulent wake which will generally increase the drag exerted by the flow on the obstacle (and measured by the famous C_x coefficient): so turbulence has to be avoided in order to obtain better aerodynamic performance for cars or planes. The majority of atmospheric or oceanic currents cannot be predicted accurately and fall into the category of turbulent flows, even in the large planetary scales. Small-scale turbulence in the atmosphere can be an obstacle towards the accuracy of astronomic observations, and observatory locations have to be chosen in consequence. The atmospheres of planets such as Jupiter and Saturn, the solar atmosphere or the Earth's outer core are turbulent. Galaxies look strikingly like the eddies which are observed in turbulent flows such as the mixing layer between two flows of different velocity, and are, in a manner of speaking, the eddies of a turbulent universe. Turbulence is also produced in the Earth's outer magnetosphere, due to the development of instabilities caused by the interaction of the solar

wind with the magnetosphere. Numerous other examples of turbulent
flows arise in aeronautics, hydraulics, nuclear and chemical engineering,
oceanography, meteorology, astrophysics and internal geophysics.

It can be said that a turbulent flow is a flow which is disordered in
time and space. But this, of course, is not a precise mathematical definition. The flows one calls "turbulent" may possess fairly different dynamics, may be three-dimensional or sometimes quasi-two-dimensional,
may exhibit well organized structures or otherwise. A common property
which is required of them is that they should be able to mix transported quantities much more rapidly than if only molecular diffusion
processes were involved. It is this latter property which is certainly the
more important for people interested in turbulence because of its practical applications: the engineer, for instance, is mainly concerned with
the knowledge of turbulent heat diffusion coefficients, or the turbulent
drag (depending on turbulent momentum diffusion in the flow). The
following definition of turbulence can thus be tentatively proposed and
may contribute to avoiding the somewhat semantic discussions on this
matter:

- a) Firstly, a turbulent flow must be unpredictable, in the sense
that a small uncertainty as to its knowledge at a given initial time will
amplify so as to render impossible a precise deterministic prediction of
its evolution.

- b) Secondly, it has to satisfy the increased mixing property defined
above.

- c) Thirdly, it must involve a wide range of spatial wave lengths.
Such a definition allows in particular an application of the term "turbulent" to some two-dimensional flows. It also implies that certain non
dimensional parameters characteristic of the flow should be much greater
than one: indeed, let l be a characteristic length associated to the large
energetic eddies of turbulence, and v a characteristic fluctuating velocity;
a very rough analogy between the mixing processes due to turbulence
and the incoherent random walk allows one to define a turbulent diffusion
coefficient proportional to $l\,v$. As will be seen later on, l is also called
the integral scale. Thus, if ν and κ are respectively the molecular diffusion coefficients[1] of momentum (called below the kinematic molecular
viscosity) and heat (the molecular conductivity), the increased mixing
property for these two transported quantities implies that the two dimensionless parameters $R_l = lv/\nu$ and lv/κ should be much greater
than one. The first of these parameters is called the Reynolds number,
and the second one the Peclet number. Notice finally that the existence
of a large Reynolds number implies, from the phenomenology developed
in Chapter VI, that the ratio of the largest to the smallest scale may be

---

[1] These coefficients will be accurately defined in Chapter II.

of the order of $R_l^{3/4}$. In this respect, the property b) stressed above implies c).

A turbulent flow is by nature unstable: a small perturbation will generally, due to the nonlinearities of the equations of motion, amplify. The contrary occurs in a "laminar" flow, as can be seen on Figure I-1, where the streamlines, perturbed by the small obstacle, reform downstream. The Reynolds number of this flow, defined as

$$Re = [\text{fluid velocity}] \times [\text{size of the obstacle}]/\nu$$

is in this experiment equal to $2.26\ 10^{-2}$. This Reynolds number is different from the turbulent Reynolds number introduced above, but it will be shown in chapter III that they both characterize the relative importance of inertial forces over viscous forces in the flow. Here the viscous forces are preponderant and will damp any perturbation, preventing a turbulent wake from developing.

Figure I-1: Stokes flow of glycerin past a triangular obstacle (picture by S.Taneda, Kyushu University; from Lesieur (1982), courtesy S. Taneda and "La Recherche")

There is a lot of experimental or numerical evidence showing that turbulent flows are rotational, that is, their vorticity $\vec{\omega} = \vec{\nabla} \times \vec{u}$ is non zero, at least in certain regions of space. Therefore, it is interesting to ask oneself how turbulence does in fact arise in a flow which is irrotational upstream[2]. It is obviously due to the viscosity, since an immediate consequence of Kelvin's theorem, demonstrated in Chapter II, is that zero-vorticity

---

[2] for instance, a uniform flow

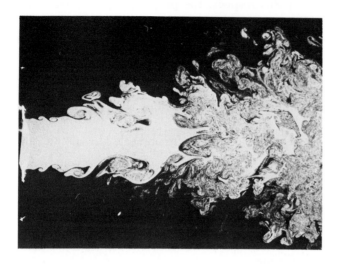

Figure I-2: turbulent jet (picture by J.L. Balint, M. Ayrault and J.P. Schon, Ecole Centrale de Lyon; from Lesieur (1982), courtesy J.P. Schon and "La Recherche")

Figure I-3: turbulence created in a wind tunnel behind a grid. Here turbulence fills the whole apparatus, and a localized source of smoke has been placed on the grid to visualize the development of turbulence (picture by J.L. Balint, M. Ayrault and J.P. Schon, Ecole Centrale de Lyon; from Lesieur (1982), courtesy "La Recherche")

is conserved following the motion in a perfect fluid[3]: the presence of

---

[3] The perfect fluid is an approximation of the flow where molecular

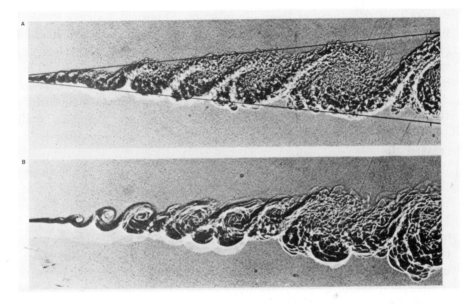

Figure I-4: turbulence in a mixing layer (Brown and Roshko, 1974). In I-4a, the Reynolds number (based on the velocity difference and the width of the layer at a given downstream position) is twice Figure I-4b's (courtesy A.Roshko and J. Fluid Mech.)

boundaries or obstacles imposes a zero-velocity condition which produces vorticity. Production of vorticity will then be increased, due to the vortex filaments stretching mechanism to be described later, to such a point that the flow will generally become turbulent in the rotational regions. In what is called grid turbulence for instance, which is produced in the laboratory by letting a flow go through a fixed grid, the rotational "vortex streets" behind the grid rods interact together and degenerate into turbulence. Notice that the same effect would be obtained by pulling a grid through a fluid initially at rest. In some situations, the vorticity is created in the interior of the flow itself through some external forcing or rotational initial conditions (as in the example of the temporal mixing layer presented later on).

---

viscous effects are ignored.

2 - Examples of turbulent flows

To illustrate the preceding considerations, it may be useful to display some flows which come under our definition of turbulence. Figure I-2 shows a turbulent air jet marked by incense smoke and visualized thanks to a technique of laser illumination. Figure I-3 shows a "grid turbulence" described above. Figure I-4, taken from Brown and Roshko (1974), shows a mixing layer between two flows of different velocities, which develop at their interface a Kelvin-Helmholtz type instability. This instability is eventually responsible for the generation of large quasi-two-dimensional structures, which are referred to as "coherent structures". Upon these structures are superposed three-dimensional turbulent small scales which seem to be more active when the Reynolds number is increased.

Plate 1-a (see the colour plates section in the middle of the book) shows a two-dimensional numerical calculation of the diffusion of a passive dye by the large structures of the mixing layer presented in Figure I-4, in a numerical resolution of the equations of the flow motion performed by Normand et al. (1988): the faster fluid is shown in blue, while the slower fluid is in red. The same coherent structures as in the experiment are present, which proves that the dynamics of these structures is essentially dominated by two-dimensional mechanisms. We will come back to this discussion in Chapter XIII. An advantage of the calculation over the experiment is to provide also the vorticity field[4], as shown on plate 1-b: comparison between the vorticity and passive dye contours shows that the coherent structures consist here in spiral concentrations of vorticity around which the convected scalar winds up. Finally, Plate 2 (see colour plates section) shows the vorticity field in a two-dimensional numerical simulation of a temporal mixing layer (that is, periodic in the mean flow direction), where the upper and lower currents are driven with velocities respectively $U\vec{x}$ and $-U\vec{x}$ (see Lesieur et al., 1988, and Comte, 1989). As already stressed, the resemblance of the pairing eddies to the spiral galaxies is striking.

The latter structures are called coherent because they can be found extremely far downstream, with approximately the same shape. But it is possible for them to become irregular and unpredictable, and thus constitute a quasi-two-dimensional turbulent field. Evidence for that is presented in Figure I-5, showing the vorticity contours in a two-dimensional calculation of a temporal mixing layer taken from Staquet et al. (1985): the evolution of the flow after 30 characteristic dynamic initial times is

---

[4] Indeed, experimental vorticity measurements are extremely difficult to perform, see however Balint et al. (1986).

presented for four independent initial small random perturbations superimposed upon the basic inflexional velocity shear: the structures display some important differences, since there are for instance four eddies in Figure I-5-d and only three eddies in Figure I-5-b. They therefore show some kind of unpredictability. In the rest of the book, we will define a *coherent structure* as possessing these two properties of shape preservation and unpredictability. This definition will apply also to other free-shear flows, or to wall flows.

The experimental and numerical visualizations shown in Figures I-4 and Plate 1 are a very good example of the revolution which has occurred in our understanding of turbulence over the last 15 years: we are now able to see the details of the vortices which form inside the flow. Before, we were like blind people, clumsily trying to develop global theories from measurements or calculations we could not really interpret. Now, the increasingly-fast development of visualization devices, especially in Computational-Fluid Dynamics, sheds a totally new light upon the dynamics of turbulence. It does not mean that former theories have become obsolete, but they have to be re-interpreted with the aid of this new knowledge.

Let us stress that the formation of the mixing layers is extremely common in aeronautics, when for instance a boundary layer detaches, and more generally in the separated flows. It is one of the simplest and richest prototypes of turbulent free-shear flows. An example of separation is given on Plate 3, which represents the vorticity field in the two-dimensional direct-numerical simulation of the flow above a backwards-facing step. The same coherent eddies as in Plate 1 are shed and undergo successive pairings. They may also recirculate behind the step, or be reflected by the boundaries, which increases the degree of turbulence and unpredictability. More details on this flow will be given in Chapter XIII.

In the mixing-layer experiment of Figure I-4, the turbulence in the small scales could be called fully developed turbulence, because it might have forgotten the mechanisms of generation of turbulence, i.e. the basic inflexional shear. On the contrary, the large structures depend crucially on the latter, and the terminology of "developed" cannot be used for them.

Similar large structures can be found in the turbulence generated in a rapidly rotating tank by an oscillating grid located at the bottom of the tank. Plate 4 shows a section of the tank perpendicular to the axis of rotation. Here, the effect of rotation is to induce two-dimensionality in the flow, and to create strongly-concentrated eddies with axes parallel to the axis of rotation (Hopfinger et al., 1982). These eddies could have some analogy with tornadoes in the atmosphere.

As already mentioned earlier, atmospheric and oceanic flows are

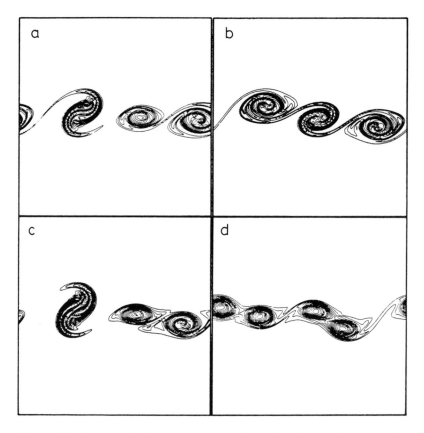

Figure I-5: Isovorticity lines, in a two-dimensional periodic mixing-layer calculation: with four initial velocity fields differing only slightly, a decorrelation
develops. This indicates that the coherent structures are unpredictable (courtesy C. Staquet, Institut de Mécanique de Grenoble)

highly unpredictable and fall into the category of turbulent flows. Their
dynamics in the large scales is strongly influenced by their shallowness
(the ratio of vertical scales to the horizontal extension of planetary scales
is of the order of 10^{-2} in the Earth's atmosphere), by the Earth's sphericity and rotation, by differential heating between the equator and the
poles, and by topography. The pressure highs or lows, that can be
observed every day on the meteorological maps in the medium latitudes,

correspond to respectively anticyclonic or cyclonic[5] large-scale vortices in the atmosphere. Figure I-6 shows for example an impressive vortex corresponding to a depression above the Atlantic. Plate 5 shows also the eddy field which can be seen from satellites in the Alboran sea, west of the Mediterranean. Similar eddies develop in the Gulf Stream, or in the Kuroshio extension, east of Japan. The simplified model of two-dimensional and geostrophic turbulence will be considered in chapter IX so as to study the particular dynamics associated with these flows.

On a planet such as Jupiter which, like the Earth, is rapidly rotating (this concept of rapid rotation can be defined with respect to the smallness of a dimensionless parameter, the Rossby number, which will be defined in Chapter III), the mean circulation is strikingly simple, since it consists of zonal jets going eastwards or westwards. This has not yet been convincingly explained, since some interpretations are based on geostrophic turbulence (see Williams, 1978, and Chapter IX), while others propose an interaction mechanism between thermal convection and rotation (Busse, 1983). Plate 6 shows the turbulence generated in the vicinity of the great red spot. The origin of the latter is also quite mysterious, and not well understood yet. Notice that the same kind of intense vortex has more recently[6] been discovered on Neptune (Neptune's great dark-blue spot) by NASA's Voyager 2 probe. Notice also that Jupiter's atmosphere, unlike the Earth, is extremely deep, and shallow-water theories have to be handled with care in this case.

3 - Fully developed turbulence

The word "developed" has already been employed for the small-scale three-dimensional turbulence which appears in the mixing-layer experiments. Fully-developed turbulence is a turbulence which is free to develop without imposed constraints. The possible constraints are boundaries, external forces, or viscosity: one can easily observe that the structures of a flow of scale comparable with the dimensions of the domain where the fluid evolves cannot deserve to be categorized as "developed". The same remark holds for the structures directly created by the external forcing, if any. So no real turbulent flow, even at a

---

[5] On a rotating sphere of solid-body rotation $\vec{\Omega}$, cyclonic eddies are defined in order to have a relative vorticity $\vec{\omega}$ such that $\vec{\omega}.\vec{\Omega} > 0$. Therefore, they rotate clockwise in the Northern hemisphere and anticlockwise in the Southern hemisphere.

[6] in 1989

Figure I-6: Image taken by METEOSAT II on 18th September 1983, and showing a strong depression above the Atlantic, centered 58^0 north and 18^0 east; (courtesy METEO-FRANCE)

high Reynolds number, can be "fully-developed" in the large energetic scales. At smaller scales, however, turbulence will be fully-developed if the viscosity does not play a direct role in the dynamics of these scales.[7] This will be true if the Reynolds number is high enough so that an "inertial-range" can develop[8]. In the preceding experimental examples of the jet and the mixing layer, one actually obtains fully-developed turbulence at scales smaller than the large energetic scales and larger than the dissipative scales. On the contrary, in the majority of grid-turbulence experiments, the Reynolds number is not high enough to enable an inertial-range to develop. The small three-dimensional tur-

---

[7] It will be seen in Chapter VI that the concept of local energy cascade in high Reynolds number three-dimensional turbulence implies that inertial forces transfer energy from large to small scales without any influence of viscosity, up to the so called "dissipative scales" where the kinetic energy is finally dissipated by viscous forces.

[8] See Chapter VI.

bulent scales of the Earth's atmosphere and oceans, Jupiter or Saturn are certainly fully-developed. But the planetary scales of these flows are not, because of constraints due to the rotation, thermal stratification[9] and finite size of planets. In this monograph, the term "developed" will mainly be used for three-dimensional flows, though it could be generalized to some high Reynolds number two-dimensional flows constrained to two-dimensionality by some external mechanism which does not affect the dynamics of the two-dimensional eddies once created.

Finally, we stress that it is possible, for theoretical purposes, to assume that turbulence is fully developed in the large scales also, when studying a freely-evolving statistically homogeneous turbulence: there is in this case no external force or boundary action.

4 - Fluid turbulence and "chaos"

The definition of turbulence we have given here is extremely broad, and there does not seem to be a clear distinction between "turbulence" and "chaos". Nevertheless, the word chaos is now mainly used in mechanics to describe a particular behaviour pertaining to dynamical systems with a limited number of degrees of freedom: some of these systems, under particular conditions, exhibit solutions which are chaotic in the sense that two points in the phase-space, initially very close, will separate exponentially. The characteristic rate of evolution of the exponential is called a Liapounov exponent, and must be positive in order to obtain a chaotic behaviour. In the case of dissipative systems, this behaviour is generally associated with the existence of strange attractors around which the trajectory of the point representing the system will wind up. One of the most famous examples of that is the Lorenz attractor, a three-mode dynamical system derived from the equations of thermal convection (Lorenz, 1963). Figure I-7, taken from Lanford (1977), shows for instance a numerical simulation of the Lorenz attractor for values of the parameters corresponding to the following dynamical

---

[9] Both effects of rotation and a stable thermal stratification are responsible for the development of an instability called the "baroclinic instability". This instability converts potential energy of the flow into horizontal kinetic energy, and is one of the kinetic energy sources of these flows (see Chapter IX).

system:

$$\frac{dx}{dt} = -10\ x + 10\ y$$

$$\frac{dy}{dt} = 28\ x - y - xz \qquad\qquad (I - 4 - 1)$$

$$\frac{dz}{dt} = -\frac{8}{3}\ z + xy$$

Chaos has now become an entire discipline in itself, covering domains that are sometimes extremely far from fluid dynamics. Its relations with the latter are up to now limited to some aspects of the transition to turbulence, in the thermal convection problem in particular, and it is not our intention to include this topic in the present monograph. The reader is referred to Bergé et al. (1984) for further details on this point of view. An attempt to apply these theories to the wall-region in the developed turbulent boundary layer has been undertaken by Lumley and colleagues (see Aubry et al., 1988), who use a Galerkin projection of the velocity field on a proper set of eigenmodes of the Reynolds tensor $< u'_i u'_j >$, where the u' refer to the fluctuations of the velocity with respect to the mean velocity $< u >$: once having been properly truncated in order to retain only longitudinal vortex structures[10], the system exhibits a chaotic behaviour resembling the occurrence of intermittent bursts in a turbulent boundary layer. The difficulty with this approach lies in the necessity of knowing in advance the eigenmodes, which requires the knowledge of the Reynolds tensor, either from experiments or numerical simulations: hence, the analysis is not predictive in the sense that it needs the problem to have already been investigated experimentally or numerically. It necessitates also the modelling of small-scale turbulence, as in the subgrid-scale modelling problem which will be discussed in Chapter XII.

Another chaotic approach to turbulence introduces the concept of chaotic advection (see Aref, 1983, 1985, for a review), where even a low number of point vortices in a two-dimensional flow can stir passive tracers in a chaotic manner. However, it is not clear, up to now, how these concepts can be applied to developed turbulence.

We would like to stress, however, that there is a-priori no contradiction between the "chaos philosophy" and the point of view which will be presented here: while dynamical systems limit their space dependance to a small number of degrees of freedom, and are only chaotic in time, a

---

[10] Longitudinal eddies have been observed experimentally close to the wall (Kline et al., 1967) in a turbulent boundary layer, and numerically (Moin and Kim, 1982, Kim et al., 1987) in a channel flow. We will come back to the origin of these structures in Chapter III.

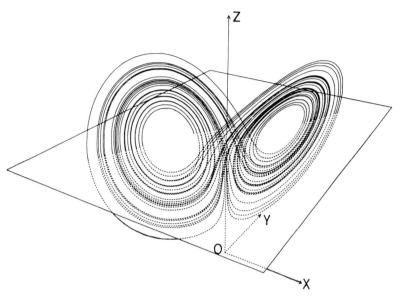

Figure I-7: computer plot of the Lorenz attractor, from Lanford (1977) (courtesy Springer-Verlag)

turbulent flow has generally a large number of spatial degrees of freedom and exhibits a chaotic behaviour in both time and space. As a matter of fact, fluid turbulence is sometimes referred to as a "spatio-temporal chaos" (see Favre et al., 1988, for a discussion on this point). Like chaotic dynamical systems, fluid turbulence also displays unpredictability, as already stressed: the degree of unpredictability can be measured locally in space for a given spatial wave length, and exchanges of predictability exist between the spatial scales (see also Chapter XI).

5 - "Deterministic" and statistical approaches

We will show in Chapter II how the fluid satisfies, for macroscopic scales large in front of the microscopic molecular scales, partial-differential equations called the Navier-Stokes equations. There is ample evidence that these equations describe properly turbulent flows, even in hypersonic situations up to a Mach number of the order of 15. The smallest macroscopic scale δl_d is smaller than the Kolmogorov dissipation scale l_d, characteristic of the dissipative scales already introduced, and much larger than the mean-free path of molecules. In fact, Navier-Stokes equations refer to quantities such as the velocity, pressure, temperature, density, which are spatially averaged on elementary control volumes of size δl_d. From a mathematical viewpoint, however, the spatial scales in these equations can be as low as necessary.

We begin with the postulate that fluid turbulence satisfies Newton's principle of determinism: if the initial positions and velocities were known, for a given time t_0 , at all scales[11], then there exists only one possible state for the flow at any time $t > t_0$. Mathematically, this is nothing more than an assumption of existence and uniqueness for the solutions of the Navier-Stokes equations: such a result exists in two dimensions, but holds only for finite times in three dimensions (Leray, 1934, see Temam, 1977, for a review). Physically, it is nevertheless to be expected that the presence of molecular viscosity in the Navier-Stokes equations will smooth the solutions sufficiently, in order to prevent the appearance of any singularity and the bifurcation[12] to another solution.

We conclude from these considerations that fluid turbulence is a deterministic phenomenon, although it evolves with time in a very complicated way, due to the non-linear interactions. It seems of course impossible to consider theoretically for arbitrary times the deterministic evolution of a given turbulent flow, starting with a given field of initial conditions. Nevertheless such an approach will be shown here as becoming promising, due to the significant recent development of scientific calculators: indeed, this last decade has seen extraordinary progress in the speed and capacity of computers, to such a degree that the numerical resolution of the Navier-Stokes equations is now at hand in some turbulent situations, for moderate Reynolds number flows. These direct-numerical simulations are an attempt towards solving the problem deterministically. At higher Reynolds numbers, however, the simulations generally deal only with the large scales of the flow, and contain errors due to the inaccuracy of the numerical schemes, to our current ignorance vis-a-vis the small scales, and to the lack of detail concerning the initial and boundary conditions. These errors are generally amplified by the non linearities of the equations, and after a period of time the predicted turbulent flow will differ significantly from the actual field. It follows that, even for a deterministic system, unpredictability and randomness will be introduced. These deterministic large-eddy simulations (L.E.S.) are nevertheless extremely useful, for they generally predict the shape (but not the phase – or position) of the large structures existing in the flow . They also often contain the statistical information needed by the

---

[11] with a given set of boundary conditions and a proper external forcing

[12] Here, we consider the evolution of a time-dependent system with fixed external parameters. The theory of "bifurcations" concerns the exchange from one to another stationary solution in a system when one of the external parameters is varied (see Iooss and Joseph, 1980, for a review).

engineer, which can be derived from a single realization without any recourse to ensemble averaging. This point of view will be developed in Chapter XII.

On the other hand, it is also very useful to employ stochastic tools and consider the various fluctuating quantities as random functions. For fully developed turbulence, these functions will be assumed to be statistically invariant under translations (homogeneity) and rotations (isotropy). This monograph will extensively study the dynamics of homogeneous isotropic turbulence, in particular the energy transfers between the various scales of motion. Emphasis will be given to the analytical statistical theories (also called stochastic models or two-point closures) developed in particular by Kraichnan and Orszag. More details on these methods can also be found in Leslie (1973) and Orszag (1977).

To conclude this section, and at the risk of becoming repetitive, we stress again that it might be erroneous to oppose the so-called deterministic and statistical points of view of turbulence: the deterministic approach can be, computationally speaking, extremely expensive, and a statistical theory or modelling may prove to be very useful. In Chapter III we will come back to a discussion about turbulence, order and chaos, once some basic results about the instability theory have been introduced.

6 - Why study isotropic turbulence?

One might argue that no real turbulent flow is isotropic or even homogeneous in the large scales. Isotropy and homogeneity can even be questionable in the small scales. But these assumptions will allow us to easily easily the analytical statistical theories mentioned above. Such theories are extremely powerful in the sense that they permit one to deal with strong non-linearities when departures from gaussianity are not too high. The point of view developed here is that these techniques describe satisfactorily the dynamics of the small three-dimensional scales of a turbulent flow, and also allow one to model their action on the anisotropic large scales. The latter will generally require the numerical large-eddy simulations mentioned above. So, the major problem of turbulence as far as the applications (that is the prediction, at least statistical, of the large scales) are concerned, is, as already emphasized, the question of how to model the action of the small scales (not explicitly simulated) onto the large scales. This point will be discussed in Chapter XII. In those particular cases when the turbulence is constrained to quasi-two-dimensionality, such as in large-scale atmospheric or oceanic situations,

the two-point closures may also be a very good tool to study the statistics
of the large scales and their degree of predictability: it is by using one of
these closures[13], that Lorenz (1969) could show how, in two-dimensional
isotropic turbulence, the initial uncertainty in the small scales was transferred to the large scales in a period of time which, converted into atmospheric parameters, gave a limit of 10 days to the predictability of
the Earth's atmosphere (see Chapter XI for details). Such information
about the spatial inverse cascade of error could never have been obtained
from the dynamical system approach followed earlier by Lorenz (1963).

7 - One-point closure modelling

There is another approach of turbulence, mainly developed in order
to model inhomogeneous flows in practical applications, known as "one-
point closure modelling": it starts with the exact Reynolds equations for
the mean motion $< \vec{u} >$, where the turbulent fluctuations are introduced through their second-order moments at the same point in space.
The latter need to be modelled in terms of the mean flow, and the simplest way for doing that is to use an eddy-viscosity assumption, where
the eddy-viscosity is calculated with the aid of Prandtl's mixing-length
concept (Prandtl, 1925, see Schlichting, 1968, for details). This allows
one to calculate the mean characteristics (velocity profiles, spreading
rates) of simple turbulent shear layers such as jets, wakes, mixing layers,
or boundary layers on flat plates. More refined modelling techniques
involve supplementary evolution equations allowing one to determine
either the eddy-viscosity, such as the $K - \epsilon$ model (see Launder and
Spalding, 1972), or the second-order moments themselves (see Lumley,
1978). These techniques, which will not be discussed here, can be extremely efficient for flows of engineering interest when numerous similar
calculations under various conditions have to be repeated, in order to
find a quick optimal solution to a problem such as the design of an airfoil or of a heat exchanger in a nuclear plant. But they do not give one
an understanding of the physical processes really involved. It is certain
that the development of direct-numerical simulations or large-eddy simulations will help to assess and improve the classical one-point closures, by
allowing a direct calculation of the various Reynolds stresses. Also, information or new concepts derived from two-point closure approaches may
be useful for one-point closure modelling. This is true in particular for
the decay laws of kinetic energy and passive-scalar variance, obtained in
Chapters VII and VIII with the aid of the stochastic models, and which
are useful in the implementation of the $K - \epsilon$ method.

---

[13] the Quasi-Normal approximation, see Chapter VII

8 - Outline of the following chapters

The present monograph is organized as follows: we give in Chapter II a review of the basic principles of Fluid Dynamics, focusing on the various approximations relating in particular to Geophysical-Fluid Dynamics situations, with emphasis given to the dynamics of vorticity and potential vorticity, and the role of rotation and stratification. Internal and external inertial or gravity waves will also be considered.

Chapter III will consider the problem of transition to turbulence: we will first summarize the main results of the linear hydrodynamic stability theory, relating to free or bounded shear flows. This will allow us to introduce a discussion on the various instabilities participating in the process of transition to turbulence in shear flows, and the associated vortex dynamics. The role of viscosity, rotation and stratification will be discussed, introducing the non-dimensional parameters known as the Reynolds, Rossby, Froude and Rayleigh numbers. We will also try to point out the analogies between the vortices which form during the transition stage, and the coherent structures which may be found when the turbulence has developed. We will finally push further the discussion about the concepts of coherence, order and chaos, and how they can apply to fluid turbulence.

Chapters IV and V will introduce the mathematical tools of the spectral statistical analysis of turbulence, which are needed in the remainder of the study. Chapter VI will present the phenomenological theories of turbulence, theories which are quite attractive, but need to be supported by quantitative models: these will be provided by the stochastic analysis of Chapter VII, and applied to turbulent diffusion problems in Chapter VIII. A brief presentation of the so-called Renormalization Group Techniques will also be given in Chapter VII.

Since two-dimensional turbulence corresponds to a lowest-order approximation of the dynamics of atmospheres and oceans on a rapidly-rotating planet, it will be extensively studied in Chapter IX, as well as the so-called "quasi-geostrophic" (or simply "geostrophic") approximation from which it is derived.

Chapter X will present the Statistical Mechanics of the truncated Euler equations for a perfect fluid, and discuss to what extent this study applies to real dissipative systems, both in three and two dimensions. Chapters XI and XII will consider the problem of turbulence from the points of view of statistical predictability theory and of large-eddy simulations, questions having important practical applications to the modelling of a wide class of flows, either for industrial or geophysical purposes.

The last chapter will have a slightly different philosophy, for it will consider more extensively two important classes of flows, namely the stably-stratified turbulence and the mixing layer, whose dynamics are not, up to now, completely understood, but which constitute challenging examples of application for the above theories. These flows relate also to some of the motivations of this monograph, namely the oceans-atmosphere dynamics and the concept of coherence in turbulence. The influence of compressibility on free-shear layers is also of prior importance for the design of hypersonic or highly supersonic planes.

This book does not claim to be exhaustive in all the subjects which will be considered. Its aim is to point out that the dynamics of a large class of turbulent flows can be understood with the aid of the analytical statistical theories of turbulence allied to numerical large-eddy simulations. It will also provide to students, scientists and engineers a basic knowledge of rotational fluid mechanics and vortex dynamics in transitional or turbulent situations, with applications to either aerospace engineering or meteorology and oceanography. Readers will also find advanced developments on the modelling of turbulence which may provide them with new tools and help to improve the dynamical understanding of turbulence wherever it occurs.

Usually, Fluid Mechanics textbooks simply ignore turbulence, or present it as an artefax allowing to increase the molecular viscosity in order to get theoretical predictions in good agreement with the experiments. We believe, on the contrary, that much important phenomena characteristic of the dynamics of slightly-viscous fluids are present in a turbulent flow: in this sense, we have tried to question fluid dynamics from the point of view of turbulence.

Chapter II

BASIC FLUID DYNAMICS

As already stressed in Chapter I, the validity of the Navier-Stokes equations in describing the phenomenon of turbulence in fluids is no longer a topic for serious debate. We will recall in this chapter the basic equations of fluid mechanics and the dynamics of vorticity, with the possible influence of a solid-body rotation $\vec{\Omega}$, due for instance to the rotation of the Earth (when one considers the motion of oceans or atmosphere. We will discuss the approximation of incompressibility for the velocity field, which will be used in most of this book: it will allow us to discard the acoustic waves, but will take into consideration various heated or stably density-stratified flows. The Boussinesq approximation, in particular, allows one to study the effects of buoyancy. The reader is referred to numerous textbooks (see e.g. Batchelor, 1967, Gill, 1982) for the complete derivation of these equations.

1 - Eulerian notation and Lagrangian derivatives

Let us consider an orthonormal reference frame which can be at rest (that is, Galilean) or rotating with a solid-body rotation $\vec{\Omega}$ with respect to a "fixed" frame. A "fluid particle", of size large in comparison to the molecular scales and small in comparison to the Kolmogorov dissipation scale introduced in Chapter I, and located in $\vec{x}$ at time t, will have a velocity $\vec{u}(\vec{x},t)$ with respect to the reference frame. The components of the velocity will be $u_i(\vec{x},t)$. Let $\rho(\vec{x},t)$ be the density of the fluid element passing by $\vec{x}$ at time t. This notation corresponds to the Eulerian formulation. Let $A(\vec{x},t)$ be any quantity associated with the motion of the fluid. When the fluid particle considered above moves, it produces a variation of A , and the derivative of A following the fluid motion in the reference frame will be denoted DA/Dt . The operator D/Dt is the Lagrangian derivative ("Lagrangian" means here "following the motion", and is not to be confused with Lagrange variational approaches in analytical mechanics). One can show very easily that

$$\frac{DA}{Dt} = \frac{\partial A}{\partial t} + \vec{u}.\vec{\nabla}A \qquad (II-1-1)$$

Now let δV be the volume of the small fluid particle. It can easily be shown, for instance by the change of coordinates $\vec{x} \rightarrow \vec{x}_0$, where $\vec{x}_0$ is the original position at some initial time t_0 of the fluid particle located in $\vec{x}$ at time $t \geq t_0$ (see Lamb, 1932), that the divergence of the velocity is given by

$$\vec{\nabla}.\vec{u} = \frac{1}{\delta V}\frac{D\,\delta V}{Dt} \qquad (II-1-2)$$

2 - The continuity equation

This equation is the mass conservation equation: let $\delta m = \rho\,\delta V$ be the mass of the fluid particle. It is conserved following the fluid motion, since average exchanges of mass with the surrounding fluid, which are due to molecular diffusion across the boundary $\partial(\delta V)$ of δV , will be zero for macroscopic time scales large in comparison to the molecular time scales. Hence, the logarithmic Lagrangian derivative of δm will also be zero, and one obtains

$$\frac{1}{\rho}\frac{D\rho}{Dt} + \frac{1}{\delta V}\frac{D\,\delta V}{Dt} = 0 \qquad (II-2-1)$$

or equivalently, because of (II-1-2)

$$\frac{1}{\rho}\frac{D\rho}{Dt} + \vec{\nabla}.\vec{u} = 0 \qquad (II-2-2)$$

which is the continuity equation. The particular case of incompressibility (conservation of volumes following the fluid motion) reduces to

$$\vec{\nabla}.\vec{u} = 0, \text{ or } \frac{D\rho}{Dt} = 0 \qquad (II-2-3)$$

Notice that, at this level, incompressibility does not imply *a-priori* that the density is uniform in space: a counter example is given in ocean dynamics, where the motion is approximately incompressible, but where there exists a thermal stratification responsible for spatial density variations.

3 - The conservation of momentum

The second law of motion is obtained by applying to the fluid particle the fundamental principle of Newtonian mechanics, namely

$$\delta m \, \frac{D\vec{u}}{Dt} = [\text{body forces }] + [\text{surface forces }] \qquad (\text{II} - 3 - 1)$$

The body forces applied to the fluid particle are gravity $\delta m \, \vec{g}$, the Coriolis force (if any) $-2 \, \delta m \, \vec{\Omega} \times \vec{u}$, and possible other external forces (like the Lorentz force in the case of an electrically-conducting flow). We recall that $\vec{\Omega}$ may be the Earth's (or another planet's) rotation, or the rotation of an experimental apparatus in the laboratory, and that we are working in a relative reference frame of solid body rotation $\vec{\Omega}$. The gravity $\vec{g}$ is irrotational, and includes both the Newtonian gravity and the centrifugal force implied by rotation. Thus, $\vec{g}$ may have spatial variations while being irrotational. The possible variation of $\vec{\Omega}$ with time has been neglected. Notice that such an assumption might be questionable in the context of Earth climatic studies, which may involve periods of time of several thousand years or more. The reader is referred to Munk and MacDonald (1975) for a detailed discussion of the variation of $\vec{\Omega}$.

As shown in Batchelor (1967), the fact that the surface forces applied to the fluid particle have to be proportional to δV in order to balance the two other terms of (II-3-1) implies the existence of a strain tensor σ_{ij} such that the force exerted on a small surface $d\Sigma$ oriented by a normal unit vector $\vec{n}$ is

$$df_i = \sigma_{ij} \, n_j \, d\Sigma \qquad (\text{II} - 3 - 2)$$

In a Newtonian fluid, the strain tensor is assumed to be linear with respect to the deformation tensor

$$e_{ij} = \frac{1}{2}\left(\frac{\partial u_i}{\partial x_j} + \frac{\partial u_j}{\partial x_i}\right) \qquad (\text{II} - 3 - 3)$$

and isotropic. Hence, it is found (see e.g. Batchelor, 1967) that

$$\sigma_{ij} = -p \, \delta_{ij} + \mu[(\frac{\partial u_i}{\partial x_j} + \frac{\partial u_j}{\partial x_i}) - \frac{2}{3}\vec{\nabla}.\vec{u} \, \delta_{ij}] \qquad (\text{II} - 3 - 4)$$

where δ_{ij} is the Kronecker tensor and μ the dynamic viscosity. Following Batchelor (1967), we have defined the pressure p with the aid of the trace of the strain tensor ($p = -(1/3)\sigma_{ii}$, with summation upon the i). The dynamic viscosity μ may vary with the physical properties of the fluid, as in gases at high temperature, or as in the Earth's outer mantle, where cellular thermal convective motions are responsible for the sea floor spreading and the existence of dorsal and subduction zones (see e.g. Allègre, 1983).

In fact, the notion of a viscous strain is in Newton's principles, recalled in Rouse and Ince (1957), which state that "*the resistance arising [...] in the parts of a fluid is [...] proportional to the velocity with which the parts of the fluid are separated from one another*". Such a simple idea leads in particular to the significant result that, in the case of a parallel flow along a material plane surface, the tangential force exerted per unit surface area by the flow on the surface and which tends to entrain it is proportional to $\mu \, \partial U / \partial n$, where $\vec{n}$ is the unit vector normal to the surface (directed towards the flow) and U the velocity component normal to $\vec{n}$. This result can be easily deduced from (II-3-4), to which it gives a firm physical basis. In this case, the normal force exerted on the surface is $-p \, \vec{n}$, in agreement with the physical role of the pressure. Notice also that, in the general case, the force normal to $d\Sigma$ is equal to $-[p + (2/3)\mu\vec{\nabla}.\vec{u}] \, \vec{n}$, and contains a small viscous contribution which is zero only when the fluid is nondivergent.

Therefore (II-3-1) becomes, after integration of the surface forces over the surface of the fluid particle

$$\frac{Du_i}{Dt} = (\vec{g} - 2\vec{\Omega} \times \vec{u})_i + \frac{1}{\rho} \frac{\partial \sigma_{ij}}{\partial x_j} \qquad (II-3-5)$$

or equivalently

$$\frac{Du_i}{Dt} = (\vec{g} - 2\vec{\Omega} \times \vec{u})_i - \frac{1}{\rho} \frac{\partial p}{\partial x_i} + \frac{1}{\rho} \frac{\partial}{\partial x_j} \mu [(\frac{\partial u_i}{\partial x_j} + \frac{\partial u_j}{\partial x_i}) - \frac{2}{3}\vec{\nabla}.\vec{u} \; \delta_{ij}]$$
$$(II-3-6)$$

which is the momentum equation for a compressible fluid. The flows we will consider in the present monograph will have a constant uniform dynamic viscosity μ, so that (II-3-6) yields in this case

$$\frac{D\vec{u}}{Dt} = \vec{g} - 2\vec{\Omega} \times \vec{u} - \frac{1}{\rho}\vec{\nabla}p + \nu[\nabla^2\vec{u} + \frac{1}{3}\vec{\nabla}(\vec{\nabla}.\vec{u})] \; , \qquad (II-3-7)$$

with

$$\frac{D\vec{u}}{Dt} = \frac{\partial \vec{u}}{\partial t} + \vec{u}.\vec{\nabla}\vec{u} \; , \qquad (II-3-8)$$

where $\nu = \mu/\rho$ is called the kinematic viscosity. It is this viscosity that has already been introduced in Chapter I to define the Reynolds number. The term proportional to ν in the r.h.s. of eq. (II-3-7) characterizes the diffusion of momentum due to molecular exchanges between the fluid particle and the fluid surrounding it. When the velocity is nondivergent, which will very often be the case in the present monograph, the dissipative term reduces to $\nu\nabla^2\vec{u}$. Notice that if other driving forces exist, they have to be added to the r.h.s. of (II-3-7).

The system of equations (II-2-2) and (II-3-6), supplemented by an energy equation (see next section) is called the Navier-Stokes equations. The momentum equation (II-3-6) without the viscous and Coriolis terms is called the Euler equation, and was derived by Leonhard Euler in the middle of the 18th century. History has been quite unfair to Euler, since the only role of Navier was, 70 years later, to model the missing viscous term of the Euler equation, using physical ideas contained in Newton's principles. As for Stokes, the latter had no role in the establishment of the equations of motion, but studied them in strongly viscous cases. For more details on these historic aspects of fluid dynamics, the reader is referred to the very interesting book by Rouse and Ince (1957).

Eq.(II-3-7) can be written in an alternative way, by introducing the geopotential Φ, such that

$$\vec{g} = -\vec{\nabla}\Phi \qquad\qquad (II-3-9)$$

and which contains both the effects of Newtonian gravity and of the centrifugal force. Noticing that

$$(\vec{\nabla}\times\vec{u})\times\vec{u} = \vec{u}.\vec{\nabla}\vec{u} - \vec{\nabla}\frac{\vec{u}^2}{2} \qquad\qquad (II-3-10)$$

(II-3-7) can be written as

$$\frac{\partial\vec{u}}{\partial t} + (\vec{\omega} + 2\vec{\Omega})\times\vec{u} = -\frac{1}{\rho}\vec{\nabla}p - \vec{\nabla}(\Phi + \frac{\vec{u}^2}{2}) + \nu[\nabla^2\vec{u} + \frac{1}{3}\vec{\nabla}(\vec{\nabla}.\vec{u})]$$
$$(II-3-11)$$

where $\vec{\omega}$ is the vorticity of the fluid (in the rotating frame):

$$\vec{\omega} = \vec{\nabla}\times\vec{u} \qquad\qquad (II-3-12)$$

One notices also in (II-3-11) the appearance of the absolute vorticity

$$\vec{\omega}_a = \vec{\omega} + 2\,\vec{\Omega} \qquad\qquad (II-3-13)$$

which is the vorticity of the fluid in an "absolute" reference frame, and will be seen to play an interesting role with respect to the relative flow $\vec{u}$ in the rotating reference frame.

The conservation of mass, leading to eq. (II-2-2), and the conservation of momentum (II-3-6) provide two equations for three unknown variables $\vec{u}, \rho, p$. The last equation will come from thermodynamic principles developed in the following section.

A last remark concerns Bernoulli's theorem, which has very important applications in fluid dynamics: assume that the flow is inviscid ($\nu = 0$), incompressible according to (II-2-3), and time-independent ($\partial/\partial t = 0$, which implies $D/Dt = \vec{u}.\vec{\nabla}$). Hence, taking the scalar product of eq. (II-3-11) by $\vec{u}$, one obtains:

$$\frac{D}{Dt}\left(p + \rho\Phi + \rho\frac{\vec{u}^2}{2}\right) = 0 \quad . \tag{II-3-14}$$

Therefore the quantity $(p + \rho\Phi + \rho u^2/2)$ is conserved following the motion. The consequences of the theorem are numerous: in aeronautics for instance, if a flow separates around an assymmetric airfoil and reforms behind, the velocity differences implied by the different paths followed by the fluid particles on each side of the airfoil will generate pressure differences between both sides, and hence lift (see Figure II-1-a). In meteorology, let us consider two air masses initially very close (same pressure and velocity), which are entrained into respectively a cyclonic and an anticyclonic perturbation: the fluid particle trapped into the depression will acquire a higher velocity than the one going into the pressure high. This is one of the reasons why winds tend to be strong in the troughs and weak in the highs (Figure II-I-b).

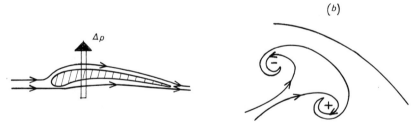

Figure II-1: applications of Bernoulli's theorem to a) lift exerted on an airfoil; b) winds generated in pressure highs and troughs.

4 - The thermodynamic equation

Let us first take the scalar product of eq. (II-3-5) by $\rho\vec{u}$, and integrate over δV. After some manipulations, and provided the geopotential Φ is time independent, one obtains:

$$\frac{D}{Dt}\int_{\delta V}\left(\frac{u^2}{2}+\Phi\right)\rho dV = \int_{\partial(\delta V)} u_i\sigma_{ij}n_j \, d\Sigma - \int_{\delta V}\sigma_{ij}\frac{\partial u_i}{\partial x_j} \, dV \tag{II-4-1}$$

with summation over the repeated indices. This is nothing more than the kinetic energy theorem for the fluid particle. Since the first term of the r.h.s., $P_e = \int_{\partial(\delta V)} u_i \sigma_{ij} n_j \, d\Sigma$, represents the rate of work done by the surface stresses, the second term $P_i = -\int_{\delta V} \sigma_{ij}(\partial u_i / \partial x_j) \, dV$ is the rate of work done by the forces internal to δV. Using eq. (II-3-4), this term turns out to be equal to

$$P_i = \int_{\delta V} [p \, \vec{\nabla}.\vec{u} - 2\mu(e_{ij}e_{ij} - \frac{1}{3}e_{ii}e_{jj})] \, dV \qquad (II-4-2)$$

where e_{ij} is defined in eq. (II-3-3). P_i is thus decomposed into an internal pressure forces contribution, and a negative viscous contribution, corresponding to an internal dissipation of kinetic energy due to viscous stresses inside δV.

Now, let e be the internal energy per unit mass of the fluid particle. The first law of thermodynamics yields

$$\frac{D}{Dt} \int_{\delta V} (\frac{u^2}{2} + \Phi + e) \, \rho dV = \delta m \, \dot{Q} - [\text{molecular diffusion heat loss }] + P_e$$

$$(II-4-3)$$

where $\dot{Q}$ is the rate of heat per mass unit supplied to the system, e.g. by chemical reaction. Let λ be the thermal diffusivity, such that the rate of heat transport across a small surface $d\Sigma$ oriented by the unit normal vector $\vec{n}$ is equal to, according to Fourier law:

$$-\lambda \frac{\partial T}{\partial n} \, d\Sigma = -\lambda \, (\vec{\nabla} T). \vec{n} \, d\Sigma \qquad (II-4-4)$$

where T is the temperature. The rate of heat loss of the fluid particle is then given by $-\vec{\nabla}.(\lambda \vec{\nabla} T) \, \delta V$. Using (II-4-1), one obtains

$$\frac{De}{Dt} = \dot{Q} + \frac{1}{\rho}\vec{\nabla}.(\lambda \vec{\nabla} T) - \frac{P_i}{\rho \, \delta V} \qquad (II-4-5)$$

or, using (II-4-2):

$$\frac{De}{Dt} = \dot{Q} + \frac{1}{\rho}\vec{\nabla}.(\lambda \vec{\nabla} T) - \frac{p}{\rho} \, \vec{\nabla}.\vec{u} + 2\nu(e_{ij}e_{ij} - \frac{1}{3}e_{ii}e_{jj}) \qquad (II-4-6)$$

The last term in the r.h.s. of (II-4-6) corresponds to the heating of the fluid particle coming from the internal molecular viscous dissipation of kinetic energy, as shown by eq. (II-4-2). This term is generally a very small contribution in the thermal balance of the fluid particle, except when strong compressibility effects exist. We will neglect the thermal forcing $\dot{Q}$ (which could nevertheless be introduced if necessary), and assume that λ is constant. Thus one gets

$$\frac{De}{Dt} = \frac{\lambda}{\rho}\nabla^2 T - \frac{p}{\rho}\,\vec{\nabla}.\vec{u} + 2\nu(e_{ij}e_{ij} - \frac{1}{3}e_{ii}e_{jj})\,. \qquad (II-4-7)$$

We also introduce the enthalpy of the fluid

$$h = e + \frac{p}{\rho}\,, \qquad (II-4-8)$$

allowing one to derive the following enthalpy equation:

$$\frac{Dh}{Dt} = \frac{1}{\rho}\frac{Dp}{Dt} + \frac{\lambda}{\rho}\nabla^2 T + 2\nu(e_{ij}e_{ij} - \frac{1}{3}e_{ii}e_{jj})\,. \qquad (II-4-9)$$

Notice also that, within the assumption of a time-independent perfect fluid[1], it may easily be shown from eqs. (II-3-11) and (II-4-9) that

$$\frac{D}{Dt}(h + \frac{1}{2}\vec{u}^2 + \Phi) = 0\,, \qquad (II-4-10)$$

which is a generalized Bernoulli theorem. Let us mention finally that, for a perfect barotropic fluid[2], the momentum equation reduces to:

$$\frac{D\vec{u}}{Dt} = -\vec{\nabla}(h + \Phi) - 2\vec{\Omega} \times \vec{u}\,. \qquad (II-4-11)$$

Now, we consider successively the case of a liquid and of an ideal gas: for a liquid, one has approximately

$$e = C_p\,T \qquad (II-4-12)$$

where C_p is the specific heat at constant pressure. C_p will be considered as constant. Thus, eq. (II-4-7) yields

$$\frac{DT}{Dt} = \kappa\nabla^2 T - \frac{p}{\rho C_p}\vec{\nabla}.\vec{u} + \frac{2\nu}{C_p}(e_{ij}e_{ij} - \frac{1}{3}e_{ii}e_{jj}) \qquad (II-4-13)$$

where

$$\kappa = \frac{\lambda}{\rho\,C_p} \qquad (II-4-14)$$

---

[1] We will define a perfect fluid as an approximation of the motion where molecular viscous and conductive effects are neglected, that is, $\nu = 0$ and $\lambda = 0$.

[2] defined such as p is a function of ρ only

is the molecular conductivity, already introduced in Chapter I to define the Peclet number. Finally, since liquids are very slightly compressible, the evaluation of the last two terms of the r.h.s. of (II-4-13) shows that they are negligible in comparison to the first. To a very good approximation, the thermodynamic equation needed for a liquid is (in the absence of thermal forcing)

$$\frac{DT}{Dt} = \kappa \nabla^2 T \qquad (II-4-15)$$

or equivalently, since in a liquid temperature and density are approximately linearly related

$$\frac{D\rho}{Dt} = \kappa \nabla^2 \rho \qquad (II-4-16)$$

Eq. (II-4-16) provides then the third equation which was needed to close the system of equations of motion of the liquid. Notice finally in this case that if the motion is adiabatic ($\kappa = 0$), eq. (II-4-16) reduces to $D\rho/Dt = 0$, which implies from the continuity equation (II-2-2) that the velocity is nondivergent.

For an ideal gas, things are somewhat more complicated because the second term on the r.h.s. of (II-4-7) can no longer be neglected, the gas being more compressible than a liquid. To a good approximation, one now has

$$e = C_v T \qquad (II-4-17)$$

where C_v is the specific heat at constant volume, which will also be assumed constant. Let C_p be the specific heat at constant pressure, related to C_v by

$$R = C_p - C_v \qquad (II-4-18)$$

with the state equation

$$\frac{p}{\rho} = R\,T \quad . \qquad (II-4-19)$$

We also introduce the coefficient

$$\gamma = \frac{C_p}{C_v} \quad . \qquad (II-4-20)$$

Therefore, the enthalpy is here equal to:

$$h = C_p T \quad , \qquad (II-4-21)$$

and satisfies:

$$\frac{Dh}{Dt} = \frac{1}{\rho}\frac{Dp}{Dt} + \kappa \nabla^2 h + 2\nu\left(e_{ij}e_{ij} - \frac{1}{3}e_{ii}e_{jj}\right) , \qquad (II-4-22)$$

where the conductivity κ is still defined by eq. (II-4-14). Eq. (II-4-22), together with the continuity equation and the momentum equation, closes the problem, since $h = (C_p/R)(p/\rho)$ is a function of p and ρ only. Notice that, for a barotropic ideal gas within the perfect flow approximation, it may be shown that p/ρ^γ is constant and uniform throughout the fluid; the enthalpy is in this case $\sim \rho^{\gamma-1}$, and the motion is totally described by the momentum equation (II-4-11) and the continuity equation, which is then written

$$\frac{1}{h}\frac{Dh}{Dt} + (\gamma - 1)\,\vec{\nabla}.\vec{u} = 0 \quad . \tag{II-4-22'}$$

Returning to the general case, one can also write the thermodynamic equation in the form:

$$\frac{D\Theta}{Dt} = \frac{\Theta}{T}\,[\kappa\nabla^2 T + 2\frac{\nu}{C_p}(e_{ij}e_{ij} - \frac{1}{3}e_{ii}e_{jj})] \quad , \tag{II-4-23}$$

where Θ is the potential temperature:

$$\Theta = T(\frac{p_0}{p})^{1-\gamma^{-1}} \tag{II-4-24}$$

which is the temperature of the gas if it was brought adiabatically to a reference level of pressure p_0. Θ is conserved following the motion[3] for a perfect fluid, and plays for the gas the role of the temperature in an adiabatic liquid. Except for hypersonic gases, the heating due to the viscous-dissipation term in the r.h.s. of eq. (II-4-23) may be neglected, and the equation reduces to

$$\frac{T}{\Theta}\frac{D\Theta}{Dt} = \kappa\nabla^2 T \quad . \tag{II-4-25}$$

5 - The incompressibility assumption

Let ρ_0 be a basic density in the fluid, ρ' the characteristic value of the density fluctuation about ρ_0, and U and L a characteristic velocity and length. In the continuity equation (II-2-2), the term $(1/\rho)(D\rho/Dt)$ is of the order of $(U/L)(\rho'/\rho_0)$, and $\vec{\nabla}.\vec{u}$ of the order of U/L. Hence, when $\rho'/\rho_0 \ll 1$, and to the lowest order with respect to this small parameter, the continuity equation reduces to

$$\vec{\nabla}.\vec{u} = 0 \tag{II-5-1} \quad .$$

---

[3] which is equivalent to the conservation of p/ρ^γ following the motion

This is the case in particular for a barotropic perfect fluid at low Mach number: indeed, expanding eqs (II-4-11) and (II-4-22) to the first order about a basic state defined by h_0, ρ_0 and p_0, and neglecting gravity, rotation and molecular diffusivity, it is found for the fluctuations h', ρ' and $p' = c^2 \rho'$ (where $c = (dp/d\rho)^{1/2}$ is the velocity of sound)

$$\frac{D\vec{u}}{Dt} = -\vec{\nabla} h' \quad ; \quad \frac{Dh'}{Dt} = \frac{c^2}{\rho_0} \frac{D\rho'}{Dt} \quad ,$$

which shows that $h' \sim U^2$, and $\rho'/\rho_0 \sim p'/p_0 \sim M^2$, where $M = U/c$ is the Mach number. Therefore, in a fluid which is not strongly heated[4], the incompressibility assumption (II-5-1) is acceptable up to $\rho'/\rho_0 <\approx 0.1$, that is up to a Mach number of the order of 0.3. This condition is in practice always satisfied for liquids. It fixes for gases the limit under which a gas and a liquid obey the same equations of motion. An advantage of the incompressibility assumption is the suppression of all the sound waves, which greatly reduces the computational times in numerical simulations. Such a simplification does not prevent variation of the density through the thermodynamic equation: density variations will in turn affect the velocity through the momentum equation. In that sense density is not a passive scalar.

Within the incompressibility approximation, the above equations of motion will reduce to:

$$\frac{D\vec{u}}{Dt} = -\frac{1}{\rho}\vec{\nabla} p + \vec{g} - 2\vec{\Omega} \times \vec{u} + \nu\nabla^2 \vec{u} \qquad (II-5-2)$$

$$\vec{\nabla}.\vec{u} = 0 \qquad (II-5-3)$$

$$\frac{D\rho}{Dt} = \kappa\nabla^2\rho \text{ for a liquid} \qquad (II-5-4)$$

$$\frac{T}{\Theta}\frac{D\Theta}{Dt} = \kappa\nabla^2 T \text{ for a gas} \qquad (II-5-5)$$

These equations are extremely general, and describe very well the motion of neutral (not electrically-conducting) fluids at low Mach numbers in the majority of laboratory and environmental situations, even with heat exchange, provided the temperature fluctuations are not too strong (the incompressibility assumption is certainly not valid for combustion problems). We stress again that the incompressibility assumption (II-5-3) is not in contradiction with the possibility of density or temperature variations offered by equations (II-5-4) or (II-5-5). When the density is uniform and equal to ρ_0, the gravity term in eq. (II-5-2) can be included in a modified pressure term P, using the geopotential Φ defined by eq. (II-3-9)

$$P = p + \rho_0\Phi \quad , \qquad (II-5-6)$$

---

[4] so that gravity may be neglected

and the equations of motion become

$$\frac{D\vec{u}}{Dt} = -\frac{1}{\rho_0}\vec{\nabla}P - 2\vec{\Omega} \times \vec{u} + \nu\nabla^2\vec{u}$$

$$\vec{\nabla}.\vec{u} = 0 \qquad\qquad\qquad (II - 5 - 7)$$

where the pressure is no longer proportional to the strain tensor trace, and now includes gravitational and possible centrifugal effects: it mainly acts to maintain the incompressibility of the velocity field.

A less drastic step of simplification than (II-5-7) resides in a simplification of the Navier-Stokes equations, still taking into account the effects of density stratification, and known as the Boussinesq approximation: this approximation will be given later on. But beforehand it is of interest to study the dynamics of vorticity within the general framework of the Navier-Stokes equations.

6 - The dynamics of vorticity

The results given in this section can be found in a more extensive form in Batchelor (1967) and Pedlosky (1979). We consider eq. (II-3-11), which is valid even for compressible fluids. By taking the curl of (II-3-11), and neglecting the possible spatial variations[5] of ν, one obtains:

$$\frac{\partial\vec{\omega}}{\partial t} + \vec{\nabla} \times (\vec{\omega}_a \times \vec{u}) = \frac{1}{\rho^2}\vec{\nabla}\rho \times \vec{\nabla}p + \nu\nabla^2\vec{\omega} \qquad (II - 6 - 1)$$

or equivalently, since $\partial\vec{\omega}_a/\partial t = \partial\vec{\omega}/\partial t$:

$$\frac{\partial\vec{\omega}_a}{\partial t} + \vec{\nabla} \times (\vec{\omega}_a \times \vec{u}) = \frac{1}{\rho^2}\vec{\nabla}\rho \times \vec{\nabla}p + \nu\nabla^2\vec{\omega} \qquad (II - 6 - 2)$$

$$\frac{D\vec{\omega}_a}{Dt} = \vec{\omega}_a.\vec{\nabla}\vec{u} - (\vec{\nabla}.\vec{u})\,\vec{\omega}_a + \frac{1}{\rho^2}\vec{\nabla}\rho \times \vec{\nabla}p + \nu\nabla^2\vec{\omega} \qquad (II - 6 - 3)$$

The term $(1/\rho^2)\vec{\nabla}\rho \times \vec{\nabla}p$ is called the *baroclinic* vector. It is zero when the fluid is barotropic: in this case, the isopycnal surfaces $[\rho = \text{constant}]$ are also isobaric surfaces $[p = \text{constant}]$.

---

[5] This assumption is not very constraining at this point, and allows in particular the recovery of the exact compressible Euler equation for a perfect fluid.

It can be checked that if $\vec{A}$ is a non-divergent vector field, and if Σ is a surface closed by a contour C, with differential surface area $d\Sigma$ oriented by a unit vector $\vec{n}$, and moving with the fluid, then

$$\frac{D}{Dt} \iint_\Sigma \vec{A}.\vec{n}\, d\Sigma = \iint_\Sigma [\frac{\partial \vec{A}}{\partial t} + \vec{\nabla} \times (\vec{A} \times \vec{u})].\vec{n}\, d\Sigma \qquad (\text{II} - 6 - 4)$$

If we calculate the flux of the l.h.s. of (II-6-2) across Σ and apply (II-6-4), we obtain

$$\frac{D}{Dt} \iint_\Sigma \vec{\omega}_a.\vec{n}\, d\Sigma = \iint_\Sigma (\frac{1}{\rho^2}\vec{\nabla}\rho \times \vec{\nabla}p + \nu\nabla^2\vec{\omega}).\vec{n}\, d\Sigma \qquad (\text{II} - 6 - 5)$$

Let us consider first the case with no rotation ($\vec{\omega}_a = \vec{\omega}$). Eq. (II-6-5) shows that the vorticity flux across Σ, which is the circulation of the velocity around the close contour C (or equivalently the intensity of the vortex tube[6] on which C is drawn), is a constant of motion in the absence of viscosity and if the fluid is barotropic. This result is the well known Kelvin theorem, and implies that, with the same restrictive conditions (perfect barotropic fluid), vortex tubes and filaments are *material* and move with the fluid particles they contain. The consequence is that if the vortex tube is stretched (and hence if its cross section decreases) its mean vorticity across the section will increase. If one then considers a thin vortex tube embedded in turbulence in a real (i.e. slightly viscous) barotropic flow, it will be both stretched by turbulence, as would be a material line of dye injected into the fluid, and diffused by molecular viscosity due to the second term of the r.h.s. of (II-6-5). This point of view leads one finally to consider the turbulence as a collection of thin vortex tubes stretched by the corresponding velocity field (which will be referred to as the "induced velocity field"). This vortex-tube stretching might lead to the formation of regions of space characterized by a high vorticity (and therefore, as will be seen later, by a high dissipation of kinetic energy), surrounded by nearly irrotational fluid with low vorticity. Such a state of the fluid, i.e. highly dissipative structures embedded into an irrotational flow, corresponds to what is called internal intermittency. Conjectures have been made as to the topology of these structures: Corrsin (1962) proposed they would form elongated sheets, while Tennekes (1968) proposed a tube form. In fact, it may be that these structures should be distributed on sets of fractal dimensions, following the ideas of Mandelbrot (1975, 1977). This phenomenology is totally invalid for two-dimensional flows, where there is no stretching

---

[6] A vortex line is, at a given time, a line whose points have a vorticity tangential to it. A vortex tube is composed of generating lines which are vortex lines.

of vortex tubes since the vorticity is conserved following the motion. Let us conclude these considerations by mentioning that some numerical methods based on vortex dynamics have been developed in two (Zabusky and Deem, 1971) and three dimensions (Léonard, 1985, Coüet, 1985).

Equation (II-6-3), multiplied by ρ^{-1} , can be written differently, using the continuity equation (II-2-2):

$$\frac{D}{Dt}(\frac{\vec{\omega}_a}{\rho}) = \frac{\vec{\omega}_a}{\rho}.\vec{\nabla}\vec{u} + \frac{1}{\rho^3}\ \vec{\nabla}\rho \times \vec{\nabla}p + \frac{\nu}{\rho}\nabla^2\vec{\omega} \qquad (II-6-6)$$

Hence, for a perfect barotropic fluid, $\vec{\omega}_a/\rho$ satisfies the equation of evolution of a small vector $\delta\vec{M} = M\vec{M}'$ when M and M' follow the fluid motion: indeed

$$\frac{D}{Dt}M\vec{M}' = \vec{u}(M') - \vec{u}(M) = M\vec{M}'.\vec{\nabla}\vec{u} \qquad (II-6-7)$$

characterizes the "passive vector" equation. This is a further argument for the stretching of vorticity by turbulence, in the same way as a pair of Lagrangian tracers are dispersed. For Magneto-Hydrodynamic turbulence, it can be shown that the magnetic field satisfies an equation similar to (II-6-7). Notice finally that eqs (II-4-11) and (II-6-7) provide a straightforward derivation of the Kelvin theorem (without rotation). Indeed, we have:

$$\frac{D}{Dt}\int_C \vec{u}.\vec{\delta l} = \int_C \vec{u}.\frac{D\vec{\delta l}}{Dt} + \int_C \frac{D\vec{u}}{Dt}.\vec{\delta l} \ , \qquad (II-6-8)$$

and the two terms of the r.h.s. are zero.

The effects of stratification and compressibility will be mainly examined in Chapters III, IX and XIII. Some of them are nevertheless contained in the next section.

7 - The generalized Kelvin theorem

Let us now consider the case where $\vec{\Omega}$ can be non-zero. For a perfect barotropic fluid, eq. (II-6-5) allows an immediate generalization of the Kelvin theorem in a rotating frame: in this case the flux of the absolute vorticity $\vec{\omega}_a$ across Σ is conserved following the fluid motion $\vec{u}$. It follows therefore that *absolute* vortex tubes are material tubes with respect to the *relative* velocity field. It means also that the circulation of the *absolute* velocity on a close contour is conserved following the relative motion. The implications of this result for the dynamics of a turbulence submitted to a solid body rotation are not totally understood. However, in the case of a rapid rotation characterized by $|\vec{\omega}|/2\Omega << 1$

(this parameter will later on be associated with the Rossby number), it is straightforward that the absolute vortex filaments are very close to a set of straight lines parallel to Ω. It follows that material lines (marked with a dye for instance) following initially absolute vortex lines (and hence quasi-two-dimensional), will evolve with time keeping their quasi-two-dimensional character. On the contrary, the same material lines drawn in an initially identical non-rotating flow will lose very quickly their two-dimensionality if the initial relative flow is turbulent. A more precise argument associating strong rotation to quasi-two-dimensionality will be given in Chapter III with the Proudman-Taylor theorem. We will see also in this chapter the possible destabilizing effects of a moderate rotation on turbulence.

On the other hand, another consequence of equation (II-6-5), sometimes referred to as Ertel's theorem (see Pedlosky, 1979), carries some important effects of rotation and stratification.

This theorem states that if $\sigma(\vec{x},t)$ is a scalar quantity conserved following the relative fluid motion $\vec{u}$, if the fluid is perfect and if one of the two following conditions is fulfilled:

i) the fluid is barotropic

ii) σ is a function of p and ρ only,

then the "potential vorticity" associated with σ, defined as

$$\Pi_\sigma = \frac{\vec{\omega}_a.\vec{\nabla}\sigma}{\rho} \qquad (II-7-1)$$

is conserved following the relative motion $\vec{u}$. The derivation is straightforward: we consider an iso-σ-surface S_{σ_0}, corresponding to $\sigma = \sigma_0$. This surface is material, since σ is conserved following the motion. It may however move and deform with time. One considers a small closed contour δC drawn on S_{σ_0} and enclosing an area $d\Sigma$ oriented by $\vec{n}$. Eq. (II-6-5) implies that

$$\frac{D}{Dt}\left(\vec{\omega}_a.\vec{n}\, d\Sigma\right) = 0 \qquad (II-7-2)$$

since the mixed product $(\vec{\nabla}p, \vec{\nabla}\rho, \vec{n})$, is zero in both cases i) and ii): indeed, the condition $\sigma(p,\rho)$ implies that $\vec{\nabla}\sigma = \partial\sigma/\partial p\,\vec{\nabla}p + \partial\sigma/\partial\rho\,\vec{\nabla}\rho$, and hence $\vec{n}$, parallel to $\vec{\nabla}\sigma$, is normal to the baroclinic vector. Therefore, $\vec{\omega}_a.\vec{n}\, d\Sigma$ is conserved on S_{σ_0}, following $\vec{u}$. Let us consider another iso-σ surface $S_{\sigma_0-d\sigma}$, where $d\sigma$ is a small fixed increment of σ_0. Let dn be the distance between the two iso-surfaces in the neighbourhood of $d\Sigma$. Let us consider at a given time a small fluid cylinder of section $d\Sigma$, contained between the two iso-surfaces (see Figure II-2): since the latter are material, the conservation of mass for the cylinder is

$$\frac{D}{Dt}\left(\rho\, d\Sigma\, dn\right) = 0 \quad , \quad d\Sigma \sim \frac{1}{\rho\, dn} \quad , \qquad (II-7-3)$$

and hence

$$\frac{D}{Dt}\left(\frac{\vec{\omega}_a \cdot \vec{n}}{\rho \, dn}\right) = 0 \qquad\qquad (\text{II} - 7 - 4)$$

Multiplying this equation by $d\sigma$ (which is fixed), gives the desired result

$$\frac{D}{Dt}\left(\frac{\vec{\omega}_a \cdot \vec{\nabla}\sigma}{\rho}\right) = 0 \quad . \qquad\qquad (\text{II} - 7 - 5)$$

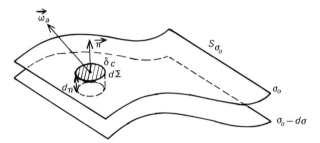

Figure II-2: sketch of iso-σ surfaces used in the derivation of Ertel's theorem.

The consequences of this theorem are numerous for geophysical flows: they imply in particular, for large-scale horizontal barotropic atmospheric and oceanic motions, the conservation of $\vec{\omega}_a \cdot \vec{n}$, where $\vec{n}$ is the unit vector defining the local vertical on the Earth: indeed, one may take σ as the potential temperature for the atmosphere, and as the density for the ocean, and assume approximately that $\vec{\nabla}\sigma // \vec{n}$. Let f be the projection of the *planetary* vorticity $2\vec{\Omega}$ on $\vec{n}$, and equal to

$$f = 2\Omega \sin\varphi \quad , \qquad\qquad (\text{II} - 7 - 6)$$

where φ is the local latitude which may vary with the motion. The potential vorticity is proportional to $f + \vec{\omega} \cdot \vec{n}$, which is conserved following the horizontal fluid motion. Ertel's theorem thus implies vorticity is exchanged between the relative motion and the horizontal planetary vorticity. These exchanges are in particular responsible for the propagation of waves called Rossby waves in the atmosphere and the oceans: indeed, the effective Coriolis force acting on the horizontal motion is equal to $-f\vec{n} \times \vec{u}$, and varies with the latitude, resulting in a restoring force causing a meridional oscillation of the fluid and the propagation of waves. How these waves interact with horizontal turbulent large scale motions will be discussed in Chapter IX. Notice finally that $f/2$ is the frequency of rotation of the Foucault pendulum oscillation plane.

As mentioned by Holton (1979), the conservation of $f + \omega$ (where $\omega = \vec{\omega}.\vec{n}$) implies that uniform westerly zonal flows must remain zonal, otherwise they would violate the potential vorticity conservation. On the other hand, easterly flows may deviate either to the North or to the South while conserving their potential vorticity. Therefore, westerly zonal flows tend to be more stable than the easterly ones.

In a viscous fluid, we still define the potential vorticity with eq. (II-7-1). Its conservation following the fluid motion is no more exact, since the vorticity, as well as quantities chosen for σ, diffuse molecularly. The exact potential vorticity evolution equation including molecular-diffusion effects can be derived in the following way: we suppose that σ is still a function of p and ρ only (or that the fluid is barotropic), and diffuses as

$$\frac{D\sigma}{Dt} = \kappa_\sigma \, \nabla^2 \sigma \quad , \qquad (II-7-7)$$

where κ_σ is a molecular diffusion coefficient relative to σ. As an exercise, the reader can show that, for any scalar σ,

$$\frac{D}{Dt} \vec{\nabla}\sigma = \vec{\nabla}\frac{D\sigma}{Dt} - (\vec{\nabla}.\vec{u})\vec{\nabla}\sigma \quad . \qquad (II-7-8)$$

Using eq. (II-6-6), the potential vorticity can be shown to satisfy

$$\frac{D}{Dt} \frac{\vec{\omega}_a.\vec{\nabla}\sigma}{\rho} = \frac{\nu}{\rho}\nabla^2\vec{\omega}.\vec{\nabla}\sigma + \frac{\vec{\omega}_a}{\rho}.\vec{\nabla}\frac{D\sigma}{Dt} \quad , \qquad (II-7-9)$$

or, using (II-7-7),

$$\frac{D}{Dt} \frac{\vec{\omega}_a.\vec{\nabla}\sigma}{\rho} = \frac{\nu}{\rho}\nabla^2\vec{\omega}.\vec{\nabla}\sigma + \frac{\kappa_\sigma}{\rho}\vec{\omega}_a.\vec{\nabla}(\nabla^2\sigma) \quad . \qquad (II-7-10)$$

This result is valid even in the absence of rotation. However, molecular diffusion effects could be of significance in three-dimensional developed turbulence, and lead to a production of potential vorticity, as noticed by Staquet and Riley (1989). These diffusion effects may be much less important however in quasi-two-dimensional situations, for example geostrophic turbulence, resulting from a strong rotation and stratification. Notice finally that for an ideal gas, the potential temperature Θ does not diffuse exactly according to eq. (II-7-7), but to (II-4-23).

8 - The Boussinesq approximation

These equations are an abridgement of the Navier-Stokes equations which enables one to take into account variations of density when the pressure and the density are not too far from a hydrostatic state. We start with eqs. (II-5-2) and (II-5-3) and define $\bar{p}(\vec{x})$ and $\bar{\rho}(\vec{x})$ as a basic hydrostatic pressure and density distribution obtained by setting $\vec{u} = \vec{0}$ in the equation of motion and thus satisfying

$$-\frac{1}{\bar{\rho}}\vec{\nabla}\bar{p} + \vec{g} = \vec{0} \qquad (II-8-1)$$

Therefore, the gravity $\vec{g}$ is normal to the isobaric surfaces of the basic state. Notice also that, by taking the curl of (II-8-1), $\vec{\nabla}\bar{p} \times \vec{\nabla}\bar{\rho} = \vec{0}$ since $\vec{g}$ is irrotational. Thus, the isobaric and isopycnal surfaces of the basic state coincide[7] and are normal to $\vec{g}$; the vectors $\vec{\nabla}\bar{p}$, $\vec{\nabla}\bar{\rho}$ and $\vec{g}$ are parallel.

The essence of the Boussinesq approximation is to assume that the actual pressures, densities and potential temperatures [8] are close to the basic profiles:

$$p(\vec{x}, t) = \bar{p}(\vec{x}) + p'(\vec{x}, t)$$
$$\rho(\vec{x}, t) = \bar{\rho}(\vec{x}) + \rho'(\vec{x}, t) \qquad (II-8-2)$$
$$\Theta(\vec{x}, t) = \bar{\Theta}(\vec{x}) + \Theta'(\vec{x}, t)$$

with $|p'| << \bar{p}, |\rho'| << \bar{\rho}$ and $|\Theta'| << \bar{\Theta}$. Let us define an "average level" in the layer, of density ρ_0, temperature T_0 and pressure p_0. For the gas, the reference pressure which allows the definition of potential temperature in (II-4-17) will be taken equal to the same p_0, in such a way that the potential temperature of this average level will be $\Theta_0 = T_0$.

Now, we perform a first order expansion of eq. (II-5-2) with respect to the small parameters $p'/\bar{p}$ and $\rho'/\bar{\rho}$, and obtain the momentum equation

$$\frac{D\vec{u}}{Dt} = -\frac{1}{\bar{\rho}}\vec{\nabla}p' + \frac{\rho'}{\bar{\rho}}\vec{g} - 2\vec{\Omega} \times \vec{u} + \nu\nabla^2\vec{u} \qquad (II-8-3)$$

to which one will associate the zero-divergence condition. Assuming that the thickness D of the layer is small with respect to the total depth of the fluid H (so that $\bar{\rho}$ is close to ρ_0), this equation can be replaced by:

$$\frac{D\vec{u}}{Dt} = -\frac{1}{\rho_0}\vec{\nabla}p' + \frac{\rho'}{\rho_0}\vec{g} - 2\vec{\Omega} \times \vec{u} + \nu\nabla^2\vec{u} \quad , \qquad (II-8-4)$$

---

[7] in other words, the hydrostatic basic state is barotropic
[8] for a gas

where we recall that ρ' is the density fluctuation with respect to $\bar{\rho}$ (and not ρ_0). Notice that an equation equivalent to (II-8-4) is:

$$\frac{D\vec{u}}{Dt} = -\frac{1}{\rho_0}\vec{\nabla}p + \frac{\rho}{\rho_0}\vec{g} - 2\vec{\Omega} \times \vec{u} + \nu\nabla^2\vec{u} \quad , \qquad (II-8-4')$$

which may be useful for direct-numerical simulations where time variations of $\bar{\rho}$ are envisaged.

The momentum equation (II-8-4) is valid both for a liquid and an ideal gas: for a liquid, the thermodynamic equation will be (II-5-4), which can be written as:

$$\frac{D\rho'}{Dt} + \vec{u}.\vec{\nabla}\bar{\rho} = \kappa\nabla^2\rho' \qquad (II-8-5)$$

(we have neglected the contribution coming from $\bar{\rho}$ in the Laplacian, which is usually very small, if not zero).

For an ideal gas, the thermodynamic equation is given by (II-5-5). Here, the relative dynamic pressure fluctuations (due to velocity differences) are, from section 5, of the order of M^2, while the relative pressure fluctuations due to gravity are of the order of $(\rho'/\rho_0)(D/H)$. Hence, at low Mach number, the hydrostatic pressure dominates, and $p'/p_0 << \rho'/\rho_0$ since $D/H << 1$. It can easily be checked in this case that

$$\frac{T'}{\bar{T}} = \frac{\Theta'}{\bar{\Theta}} = -\frac{\rho'}{\bar{\rho}} \qquad (II-8-6)$$

which implies $\Theta' = T'$ (to the lowest order), since $\Theta_0 = T_0$. Hence, the thermodynamic equation (II-4-16) gives, if one neglects the $\nabla^2\bar{T}$ term in the right-hand side, to the lowest order:

$$\frac{D\Theta'}{Dt} + \vec{u}.\vec{\nabla}\bar{\Theta} = \frac{\Theta}{T}(\kappa\nabla^2 T') \approx \kappa\nabla^2\Theta' \qquad (II-8-7)$$

It turns out that the Boussinesq approximation yields the same equations for the liquid and the gas: indeed, setting $\tilde{p} = p'/\rho_0$, these equations are:

$$\frac{D\vec{u}}{Dt} = -\vec{\nabla}\tilde{p} + \tilde{\rho}\,\vec{g} - 2\vec{\Omega} \times \vec{u} + \nu\nabla^2\vec{u}$$

$$\vec{\nabla}.\vec{u} = 0$$

$$\frac{D\tilde{\rho}}{Dt} + \vec{u}.\vec{\nabla}\bar{\rho}_* = \kappa\nabla^2\tilde{\rho} \qquad (II-8-8)$$

$$\tilde{\rho} = \frac{\rho'}{\rho_0}, \ \bar{\rho}_* = \frac{\bar{\rho}}{\rho_0} \quad \text{for a liquid}$$

$$\tilde{\rho} = -\frac{\Theta'}{\Theta_0}, \ \bar{\rho}_* = -\frac{\bar{\Theta}}{\Theta_0} \quad \text{for a gas}$$

Here the density is not a passive scalar, since it influences the velocity through the buoyancy force.

When the buoyancy force is neglected in the Boussinesq approximation, and the mean profile $\bar{\rho}_*$ assumed to be uniform, one obtains the Navier-Stokes equations with *constant* density in the momentum equation, and with fluctuations of density (or temperature, or potential temperature) satisfying the diffusion equation (II-4-16): density behaves as a passive scalar, and is uncoupled from the velocity which now satisfies eqs. (II-5-7). The latter equations describe very well non-heated (or slightly-heated) flows. They are independent of gravity (including centrifugal effects), and the temperature is only a passive quantity which marks the flow, as would a dye. An appreciable part of this monograph will be devoted to such flows.

One may wonder about the vorticity dynamics within the Boussinesq approximation (II-8-8): neglecting the molecular diffusion, the analogue of the vorticity equation (II-6-2) is now

$$\frac{\partial \vec{\omega}_a}{\partial t} + \vec{\nabla} \times (\vec{\omega}_a \times \vec{u}) = \vec{\nabla}\tilde{\rho} \times \vec{g} \qquad (II-8-9)$$

to which one has to associate the density[9] transport equation

$$\frac{D\rho_*}{Dt} = 0 \qquad (II-8-10)$$

with $\rho_* = \bar{\rho}_* + \tilde{\rho}$. It is then easy to check that Ertel's theorem derived above is no longer valid for any conserved quantity σ, except if σ is taken equal to ρ_*: indeed, in this latter case, the flux of the r.h.s. of (II-8-9) across a small surface $\delta\Sigma$ drawn on an isopycnal surface (surface of constant total density ρ_*) is zero since, because of the remark following eq. (II-8-1), $\vec{\nabla}\bar{\rho}_* \times \vec{g} = \vec{0}$, and the vector

$$\vec{\nabla}\rho_* \times \vec{g} = \vec{\nabla}\tilde{\rho} \times \vec{g} \qquad (II-8-11)$$

is tangential to the isopycnal surface. Thus Ertel's theorem is valid within the Boussinesq equations only with a potential vorticity

$$(\vec{\omega} + 2\vec{\Omega}).\vec{\nabla}\rho_* \quad ,$$

where we recall that $\rho_* = \rho/\rho_0$ for the liquid, and $\rho_* = -\Theta/\Theta_0$ for the ideal gas.

---

[9] For a gas, "density" means here the opposite of the potential temperature.

9 - Internal inertial-gravity waves

Another interesting application of the Boussinesq approximation is the possibility of obtaining inertial-gravity waves after a linearization about a rest state. These waves are internal in the sense that they propagate within the fluid. Surface waves presenting analogies with the internal waves can be found on the free surface of a rotating fluid. For the internal waves, one obtains to the lowest order in $\vec{u}, \vec{\omega}$ and $\tilde{\rho}$, neglecting molecular diffusion:

$$\frac{\partial \vec{\omega}}{\partial t} + \vec{\nabla} \times (2\vec{\Omega} \times \vec{u}) = \vec{\nabla}\tilde{\rho} \times \vec{g} \qquad (\text{II} - 9 - 1)$$

$$\frac{\partial \tilde{\rho}}{\partial t} + w\frac{d\bar{\rho}_*}{dz} = 0 \qquad (\text{II} - 9 - 2)$$

We also assume that $\vec{\Omega}$ is parallel to $\vec{g}$, which defines the vertical coordinate z :

$$\vec{\Omega} = \Omega\, \vec{z} \qquad (\text{II} - 9 - 3)$$

This latter hypothesis does not apply *a-priori* to a flow on a rotating sphere already considered above (and where gravity is normal to the sphere), nor to a rotating flow in a laboratory experiment when centrifugal effects are significant and produce a non negligible spatial variation of the effective gravity field. We will show later on that the following analysis can nevertheless be generalized to geophysical flows.

With the assumption (II-9-3), and using the zero velocity divergence condition, (II-9-1) reduces to

$$\frac{\partial \vec{\omega}}{\partial t} = 2\Omega \frac{\partial \vec{u}}{\partial z} + \vec{\nabla}\tilde{\rho} \times \vec{g} \qquad (\text{II} - 9 - 4)$$

Notice that (II-9-4) shows in particular that, when Ω is zero, the vertical component of $\partial\vec{\omega}/\partial t$ is zero. The gravity waves, which will be shown below to propagate, are vertically irrotational: indeed the time-independent vertical component of the vorticity corresponds to a permanent motion which has no direct influence on the propagation of the wave and can be eliminated. This property that the gravity waves induce irrotational horizontal motions has to be related to the wave-vortex decomposition of the flow, which will be derived in Chapter IV.

Taking now the curl of (II-9-4), and noticing that, since $\vec{u}$ is non-divergent,

$$\vec{\nabla} \times \vec{\omega} = -\nabla^2 \vec{u} \qquad (\text{II} - 9 - 5)$$

One finally obtains

$$\frac{\partial \nabla^2 \vec{u}}{\partial t} + 2\Omega \frac{\partial \vec{\omega}}{\partial z} = g\frac{\partial \vec{\nabla}\tilde{\rho}}{\partial z} + (\nabla^2 \tilde{\rho})\, \vec{g} \qquad (\text{II} - 9 - 6)$$

Differentiating eq. (II-9-6) with respect to time, projecting it onto the z axis, and making use of (II-9-4), one finds for the vertical velocity component w (see Gill, 1982):

$$\frac{\partial^2}{\partial t^2}\nabla^2 w + 4\Omega^2\frac{\partial^2 w}{\partial z^2} = -g\frac{\partial}{\partial t}\left(\frac{\partial^2\tilde{\rho}}{\partial x^2} + \frac{\partial^2\tilde{\rho}}{\partial y^2}\right) \qquad (II-9-7)$$

Then, using eq (II-9-2), one finds finally

$$\frac{\partial^2\nabla^2 w}{\partial t^2} + 4\Omega^2\frac{\partial^2 w}{\partial z^2} + N^2\left(\frac{\partial^2 w}{\partial x^2} + \frac{\partial^2 w}{\partial y^2}\right) = 0 \qquad (II-9-8)$$

with

$$N^2 = -g\frac{d\bar{\rho}_*}{dz} \quad . \qquad (II-9-9)$$

Thus, for a liquid, $N^2 = -(g/\rho_0)(d\bar{\rho}/dz)$, while for an ideal gas $N^2 = (g/\Theta_0)(d\bar{\Theta}/dz)$.

Let us first consider the case of the non-rotating fluid: If $N^2 < 0$, the solutions of (II-9-8) amplify exponentially, and the system is unstable (unstable stratification). This corresponds to the onset of thermal convection in the absence of conductivity (that is the infinite Rayleigh number case, as will be seen in the next chapter). Notice that a neutral atmosphere such that $N^2 = 0$ will be in adiabatic equilibrium ($d\bar{\Theta}/dz = 0$), which may easily be shown to correspond to $d\bar{T}/dz = -g/C_p$: this represents the dry adiabatic lapse rate of the atmosphere (see e.g. Holton, 1979). If N^2 is positive (stable stratification), N is called the Brunt-Vaisala frequency[10], and the system admits gravity waves. When N is independent of z , the dispersion relation of these waves is, from (II-9-8)

$$\varpi^2 = N^2\frac{k_1^2 + k_2^2}{k_1^2 + k_2^2 + k_3^2} \qquad (II-9-10)$$

where ϖ is the frequency and k_1, k_2, k_3 the components of the wave-vector. When Ω is non-zero, waves are inertial-gravity waves, and the dispersion relation becomes

$$\varpi^2 = \frac{N^2(k_1^2 + k_2^2) + 4\Omega^2 k_3^2}{k_1^2 + k_2^2 + k_3^2} \qquad (II-9-11)$$

In the case of a uniform density fluid ($N = 0$), the waves are purely inertial. The reader is referred to Lighthill (1978), Holton (1979) or Gill (1982) for further details on these waves, which play a particularly important role in the dynamics of the meso-scale and middle atmosphere.

---

[10] or the *buoyancy frequency*

An example of these waves corresponds to "lee waves", which are a particular internal gravity waves field in the lee of a mountain and stationary with respect to it. These lee waves are usually marked by regularly spaced stationary clouds, and may be of great help to the flight of gliders. More generally, one of the main problems arising from the internal waves is their interaction with small-scale turbulence, and possibly with large-scale quasi-two-dimensional turbulence. This question will be examined in Chapter XIII.

An important characteristic length in regard to the ocean or planetary atmospheres dynamics is the internal Rossby radius of deformation, which can be defined in the following way: let

$$D = 2\pi k_3^{-1} \quad , \quad L = 2\pi(k_1^2 + k_2^2)^{-1/2} \quad ,$$

be respectively a characteristic vertical and horizontal length scale of the motion. Thus, the relative importance of the "inertial" to the "gravity" contribution in the r.h.s. of (II-9-11) is equal to $(L/r_I)^2$ where the internal Rossby radius of deformation r_I is given by

$$r_I = \frac{ND}{f} \qquad (II-9-12)$$

f being here equal to 2Ω.

Eq. (II-9-12) can be generalized to the Earth's atmosphere or ocean dynamics in the medium latitudes, f now being given by (II-7-6): indeed, let u, v, w, be the components of the velocity in a local frame of reference with axes directed respectively along a parallel, a meridian and the local vertical defined with the gravity field, the latter taking into account centrifugal effects[11]. The Boussinesq approximation (II-8-8) projected in this frame yields, if one neglects some sphericity corrections

$$\frac{Du}{Dt} = -\frac{\partial \tilde{p}}{\partial x} + fv - 2\Omega w \cos\varphi + \nu\nabla^2 u$$

$$\frac{Dv}{Dt} = -\frac{\partial \tilde{p}}{\partial y} - fu + \nu\nabla^2 v \qquad (II-9-13)$$

$$\frac{Dw}{Dt} = -\frac{\partial \tilde{p}}{\partial z} - \tilde{\rho}g + 2\Omega u \cos\varphi + \nu\nabla^2 w$$

---

[11] Because of the centrifugal force due to its rotation, the Earth has evolved towards an ellipsoidal shape such that the gravity field (gravitation + centrifugal effects) is, in the mean, normal to the surface. In fact, the geoid presents deviations from this shape due to the non-homogeneous distribution of masses, caused in particular by the motions in the Earth's mantle.

We assume that r_I given by (II-9-12) and (II-7-6) is of the order of magnitude of the characteristic scale under which the rotation effects are negligible compared with the stratification effects, and over which the stratification becomes insignificant compared with the rotation. So for scales smaller than r_I, it is not a severe approximation to neglect in (II-9-13) the vertical Coriolis force component as well as the $\Omega w \cos\varphi$ term in the Du/Dt equation, since the Coriolis force is anyhow negligible. From the values of r_I given below, it appears that the motions at scales larger than r_I are quasi-two-dimensional, and of vertical amplitude D small compared with the horizontal amplitude L. Therefore, the vertical velocity w is at most of the order of $(D/L)U$, U being a characteristic horizontal velocity, and $2\Omega w \cos\varphi$ is negligible compared with fv in the Du/Dt equation. Finally in the Dw/Dt equation, Ωu is negligible compared with $g\tilde{\rho}$ if $\Omega << N/F, F = U/ND$ being a non-dimensional parameter (the Froude number) which will be discussed in the next chapter, and is smaller than or of the order of one. Since $\Omega << N$ in the geophysical dynamics conditions considered here, the vertical component of the Coriolis force turns out to be negligible compared with the buoyancy force: this permits us to replace (II-9-13) by

$$\frac{D\vec{u}}{Dt} = -\vec{\nabla}\tilde{p} + \tilde{\rho}\vec{g} - f\vec{z} \times \vec{u} + \nu\nabla^2\vec{u} \qquad (II-9-14)$$

where the actual solid-body vorticity $2\vec{\Omega}$ has been replaced by its projection on the local vertical on the sphere. The analysis of the inertial-gravity waves arising from eq. (II-9-14) is the same as that done previously in this section, provided the parameter 2Ω be replaced by $f = 2\Omega \sin\varphi$. This justifies a-posteriori the local value of the internal radius of deformation given by (II-9-12) and (II-7-6). Notice finally that the approximation which has led us to replace the Coriolis force by $-f\vec{z} \times \vec{u}$ in (II-9-14) could also be done usefully on the Navier-Stokes equations themselves[12]

As already stressed, the length r_I characterizes the horizontal scales of motion over which internal waves are mainly inertial, and under which they are gravity dominated. It will be seen in Chapter IX that this length is also characteristic of the baroclinic instability due to the combined effects of rotation and stratification. It is difficult to determine r_I precisely, mainly because the Brunt-Vaisala frequency N may vary appreciably with stratification profiles. Average values of r_I are of 50 km in the oceans and 1000 km in the Earth's atmosphere.

Eq (II-9-14) allows one to consider the motion of the atmosphere or the ocean at a given latitude φ as a rotating fluid of rotation $(f/2)\vec{z}$,

---

[12] However, this analysis is valid only for a shallow layer. For a fluid of arbitrary depth, the complete Coriolis force $-2\vec{\Omega} \times \vec{u}$ must be considered.

the equation being valid both for small three-dimensional scales and larger quasi-two-dimensional scales. This approximation will be seen in Chapter IX to be compatible, in the case of rapid rotation, with the geostrophic approximation, and is certainly a very good candidate in the atmosphere or the oceans for studying the interaction between the large quasi-two-dimensional geostrophic currents and the smaller three-dimensional motions.

It seems from these considerations that the Navier-Stokes equations within the Boussinesq approximation are quite satisfactory to study rotating stably-stratified flows, even in situations related to geophysical fluid dynamics. The Boussinesq approximation is simpler mathematically. Its incompressibility character (suppressing the acoustic waves) greatly simplifies the numerical simulations done using pseudo-spectral methods (see Chapter IV), and also allows us to apply the stochastic models of turbulence (see Chapter VII) in the stably-stratified case. The Boussinesq approximation is, on the contrary, certainly not valid in situations with strong heat release such as in combustion or reacting flows, and more generally, with sharp density gradients.

10 - Barré de Saint-Venant equations

10.1 Derivation of the equations

These equations are also known as the "shallow-water equations". We consider the Navier-Stokes equations locally on a sphere for a fluid of constant uniform density ρ_0, with the same approximation as in the preceding section for the Coriolis force, that is

$$\frac{D\vec{u}}{Dt} = -\frac{1}{\rho_0}\vec{\nabla}p + \vec{g} - f\vec{z} \times \vec{u} + \nu\nabla^2\vec{u} \qquad (\text{II} - 10 - 1)$$

$$\vec{\nabla}.\vec{u} = 0 \qquad (\text{II} - 10 - 2)$$

$\vec{g}$ and $\vec{z}$ being parallel. The fluid is assumed to have a free surface of mean elevation H and to lie above a topography of height $\tau(x,y)$. $h(x,y,t)$ is the depth of the fluid layer[13] and $\eta(x,y,t)$ the elevation of the free surface with respect to H (see Figure II-3). Hence we have

$$h(x,y,t) + \tau(x,y) = H + \eta(x,y,t) \qquad (\text{II} - 10 - 3)$$

---

[13] This depth h is not to be confused with the enthalpy considered above, although analogies exist, as stressed below.

The pressure at the free surface is uniform and equal to p_0. The assumption of shallowness actually means that one assumes the pressure is hydrostatically distributed along the vertical, that is,

$$p(x, y, z, t) = p_0 + \rho_0 \, g \, (h + \tau - z) \quad , \qquad (\text{II} - 10 - 4)$$

and that the horizontal velocity field $\vec{u}_H = (u, v, 0)$ depends only on the horizontal space variables x and y and on the time. The vertical velocity $w(x, y, z, t)$ still depends on the vertical coordinate, in order to allow vertical variations of the free surface. With these assumptions, and integrating the continuity equation (II-10-2) along the vertical, one obtains Barré de Saint-Venant equations

$$\frac{D_H u}{Dt} = -g \frac{\partial \eta}{\partial x} + fv + \nu \nabla_H^2 u \qquad (\text{II} - 10 - 5)$$

$$\frac{D_H v}{Dt} = -g \frac{\partial \eta}{\partial y} - fu + \nu \nabla_H^2 v \qquad (\text{II} - 10 - 6)$$

$$\frac{D_H h}{Dt} = -h \, \vec{\nabla}_H . \vec{u} \qquad (\text{II} - 10 - 7)$$

where

$$\frac{D_H}{Dt} = \frac{\partial}{\partial t} + u \frac{\partial}{\partial x} + v \frac{\partial}{\partial y} \qquad (\text{II} - 10 - 8)$$

is the derivative following the horizontal motion, and ∇_H^2 and $(\vec{\nabla}_H.)$ stand respectively for the horizontal Laplacian and divergence operators.

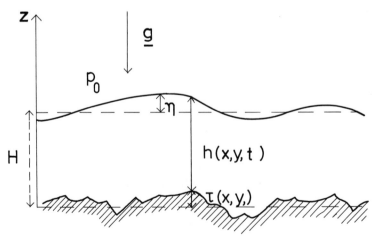

Figure II-3: schematic vertical cross-section of the shallow layer: $\tau(x, y)$ is the height of the topography, $h(x, y, t)$ the depth of the fluid, H the average height of the free surface, and $\eta(x, y, t)$ the elevation of the free surface.

10.2 The potential vorticity

Now, let us write in this case an analogous potential vorticity conservation result: let

$$\vec{\omega} = \omega \, \vec{z} = \vec{\nabla} \times \vec{u}_H \qquad (II-10-9)$$

be the vorticity of the horizontal velocity, with

$$\omega = \frac{\partial v}{\partial x} - \frac{\partial u}{\partial y} \qquad (II-10-10)$$

Paralleling eq. (II-3-11), Barré de Saint-Venant equations may be written in the form:

$$\frac{\partial \vec{u}_H}{\partial t} + (f+\omega) \, \vec{z} \times \vec{u}_H \;=\; -g\vec{\nabla}_H \eta - \vec{\nabla}_H \frac{u_H^2}{2} + \nu \nabla_H^2 \vec{u}_H \qquad (II-10-11)$$

Taking the three-dimensional curl of (II-10-11), one obtains, after projection onto the z axis:

$$\frac{\partial \omega}{\partial t} + u \frac{\partial}{\partial x}(\omega + f) + v \frac{\partial}{\partial y}(\omega + f) + (\omega + f)\, \vec{\nabla}_H . \vec{u}_H = \nu \nabla_H^2 \vec{u}_H \quad ,$$
$$(II-10-12)$$

that is,

$$\frac{D_H}{Dt}(\omega + f) + (\omega + f)\, \vec{\nabla}_H . \vec{u}_H = \nu \nabla_H^2 \vec{u}_H \quad . \qquad (II-10-13)$$

Finally, making use of the continuity equation (II-10-7), we obtain

$$\frac{D_H}{Dt} \frac{\omega + f}{h} = \frac{\nu}{h} \nabla_H^2 \omega \quad . \qquad (II-10-14)$$

When the viscous effects are neglected, we find that

$$\frac{\omega + f}{h} = \frac{(\vec{\omega} + 2\vec{\Omega}).\vec{z}}{h}$$

is conserved following the horizontal motion $\vec{u}_H$. In fact, this is a result which could also have been obtained directly from Ertel's theorem, assuming a shallow fluid layer, and taking $\sigma = p = p_0$ on the free surface: assuming quasi-two-dimensionality, eq. (II-7-2) leads to

$$\frac{D_H}{Dt} \left[(\vec{\omega} + 2\vec{\Omega}).\vec{z} \, d\Sigma \right] = 0 \quad , \qquad (II-10-15)$$

and the mass conservation written for a fluid cylinder of depth h and section $d\Sigma$ yields

$$\frac{D_H}{Dt}(h\,d\Sigma) = 0 \quad , \qquad\qquad (\text{II} - 10 - 16)$$

which proves the result and justifies calling $(\omega + f)/h$ the potential vorticity. Such a result allows the recovery of the relative vorticity changes due to the variation of f (differential rotation) and of h (by stretching or compression of the relative vortex tubes). In this case, a "positive" topography (mountain) will tend to create anticyclonic eddies which may lock on it (see Chapter IX).

10.3 Inertial-gravity waves

A surface-wave analysis, analogous to what has been done previously for the case of internal waves, can easily be performed on the Barré de Saint-Venant equations. One finds the following dispersion relation

$$\varpi^2 = f^2 + (k_1^2 + k_2^2)\,gH \quad , \qquad\qquad (\text{II} - 10 - 17)$$

where k_1 and k_2 are the components of the horizontal wave vector $\vec{k}$. This allows definition of a horizontal scale r_E, called the external Rossby radius of deformation:

$$r_E = \frac{1}{f}\sqrt{gH} \quad . \qquad\qquad (\text{II} - 10 - 18)$$

For motions of horizontal length scale L smaller than r_E, surface-gravity waves are predominant, and (II-10-17) yields

$$\varpi = k\,\sqrt{gH} \quad , \qquad\qquad (\text{II} - 10 - 19)$$

an expression originally due to Lagrange, and which corresponds to non-dispersive waves propagating at the constant phase velocity $\sqrt{gH}$. For L larger than r_E, surface-inertial waves of frequency f dominate. Let us take for instance the Earth's atmosphere, and consider it as a homogeneous shallow layer with a free surface: taking $H \approx 10\ km$ and $f = 10^{-4}\ rd\ s^{-1}$ (which corresponds to medium latitudes), the external Rossby radius of deformation is of the order of $3000\ km$, and the velocity of external gravity waves is $\approx 300\ m/s$. This is of the order of the sound velocity in the air, and much higher than the typical values $\approx 30\ m/s$ for the synoptic perturbations.

10.3.1 Analogy with two-dimensional compressible gas

We recall at this point the barotropic equations of motion (II-4-11) for a barotropic perfect fluid in the case of an ideal gas. If one considers a two-dimensional flow, the momentum equation for u and v reduces to inviscid Barré de Saint-Venant momentum equations (without topography), provided the enthalpy of the gas be replaced by $gh(x, y, t)$ in the shallow-water equations. Since, as already stressed, the enthalpy of the gas is proportional to $\rho^{\gamma-1}$, the gas continuity equation (II-4-22') reduces to (II-10-7) only if $\gamma = 2$. Therefore, Barré-de Saint-Venant equations may be considered as a special case of two-dimensional compressible turbulence, but with a not very realistic value of γ (equal to 1.4 in the air). However, the analogy (called the shallow-water analogy) is interesting, since the gravity-waves velocity $\sqrt{gH}$ may be interpreted as the sound velocity for the associated barotropic gas (when rotation is neglected). For atmospheric synoptic motions of length scale smaller than the external Rossby radius of deformation, the equivalent Mach number is of the order of 0.1, and the equivalent gas motion may be considered as incompressible. Thus, Barré de Saint-Venant equations reduce in this case to two-dimensional Navier-Stokes equations in a layer of constant depth, and gravity waves have no influence on the horizontal motion.

11 - Gravity waves in a fluid of arbitrary depth

The calculation of surface-gravity waves can be made for a fluid of arbitrary depth H (see e.g. Lighthill, 1978). We ignore rotation, consider a perfect flow and assume a uniform density ρ_0 and a vertical constant gravity g. If p_0 is the pressure at the free surface, we define a modified pressure

$$\tilde{P} = p + \rho_0 gz - p_0 \quad , \qquad (\text{II} - 11 - 1)$$

choosing the rest state of the free surface as $z = 0$. $\tilde{P}$ is the pressure fluctuation with respect to the hydrostatic rest state. Eq. (II-5-7), after linearization with respect to the velocity fluctuations, becomes

$$\frac{\partial \vec{u}}{\partial t} = -\frac{1}{\rho_0} \vec{\nabla} \tilde{P} \quad , \qquad (\text{II} - 11 - 2)$$

and the linearized vorticity equation is

$$\frac{\partial \vec{\omega}}{\partial t} = 0 \quad . \qquad (\text{II} - 11 - 3)$$

In fact, we will assume that initially the flow is irrotational, and hence it will remain irrotational for any time. The velocity is such that $\vec{u} = \vec{\nabla}\Phi$, and is nondivergent. Therefore the potential $\Phi(x, y, z, t)$ satisfies a Laplace equation

$$\nabla^2 \Phi = 0 \quad , \qquad (\text{II} - 11 - 4)$$

to which the equation of motion reduces. From (II-11-2), Φ may be chosen such that

$$\tilde{P} = -\rho_0 \frac{\partial \Phi}{\partial t} \quad . \tag{II-11-5}$$

The boundary condition at the free surface for the linearized problem is

$$\frac{\partial \eta}{\partial t} = \frac{\partial \Phi}{\partial z}(x, y, 0, t) \quad , \tag{II-11-6}$$

where $\eta(x, y, t)$ (defined as in Barré de Saint-Venant equations), is the elevation of the free suface with respect to its rest level. Furthermore, we assume that the flow is hydrostatic close to the free surface, which yields

$$\tilde{P}(x, y, 0, t) = \rho_0 \ g \ \eta(x, y, t) \quad . \tag{II-11-7}$$

Hence, using (II-11-5), it is found that

$$g \ \eta(x, y, t) = -\frac{\partial \Phi}{\partial t}(x, y, 0, t) \quad ,$$

which when differentiated with respect to t yields, with the aid of (II-11-6):

$$\frac{\partial^2 \Phi}{\partial t^2}(x, y, 0, t) + g\frac{\partial \Phi}{\partial z}(x, y, 0, t) = 0 \quad . \tag{II-11-8}$$

Eq. (II-11-4) can then be solved with the boundary condition (II-11-8) using a normal-mode approach: one looks for solutions of the type

$$\Phi(x, y, z, t) = A(z) \ \exp i(\vec{k}.\vec{x} - \varpi t) \quad , \tag{II-11-9}$$

where $\vec{k}$ is here a horizontal wave vector which corresponds to horizontally-propagating waves. One obtains:

$$A''(z) - k^2 A(z) = 0 \quad , \tag{II-11-10}$$

$$-\varpi^2 \ A(0) + g \ A'(0) = 0 \quad . \tag{II-11-11}$$

For a fluid of depth H with a zero vertical velocity $\partial \Phi/\partial z$ at the bottom $z = -H$, the solution of eq. (II-11-10) is

$$A(z) = A_0 \cosh[k(z + H)] \quad , \tag{II-11-12}$$

and hence it is found, from (II-11-11):

$$\varpi^2 = gk \ \tanh(kH) \quad . \tag{II-11-13}$$

The phase velocity $c = \varpi/k$ is given by

$$c = \sqrt{gH} \ [\frac{\tanh(kH)}{kH}]^{1/2} \quad . \tag{II-11-14}$$

In the limit $kH \to 0$ (long-wave approximation), which corresponds to horizontal wave lengths large in comparison to the depth of the layer, one recovers the shallow-layer result $\sqrt{gH}$. On the contrary, and in the limit $kH \to \infty$ (fluid of infinite depth),

$$c = \sqrt{\frac{g}{k}} \quad . \tag{II-11-15}$$

Chapter III

TRANSITION TO TURBULENCE

As already stressed in Chapter I, turbulence can only develop in rotational flows: it is due to the existence of shear in a basic flow that small perturbations will develop, through various instabilities, and eventually degenerate into turbulence. Some of these instabilities, at least during the initial stage of their development, may be understood within the framework linear-instability theory, the main results of which will be recalled in this chapter. The non-linear instability studies may prove to be useful in the future in understanding transition to turbulence, but, to date, they are still in progress and have not led to any unified theory of transition. On the contrary, an extremely useful tool to understand the transition, and assess the various theories claiming to describe it, is the direct-numerical simulations of the Navier-Stokes equations: numerous examples of these calculations will be given here.

Among the shear-basic flows which will be considered, we will make a distinction between the free-shear flows, such as mixing layers, jets or wakes on the one hand, and wall-bounded flows such as boundary layers, pipe flows or channel flows on the other hand. In free-shear flows, primary instabilities leading to the formation of coherent vortices are *inviscid*, in the sense that they are not affected by molecular viscosity, if it is small enough. In wall-bounded flows, on the contrary, the linear instabilities depend critically upon the viscosity, and vanish when the latter is zero.

In some situations, transition will be characterized by a critical value of a non-dimensional parameter depending on the motions or forces imposed upon the flow. That is why certain sections of the chapter will be devoted to these various non-dimensional parameters, which will be associated to a particular physical situation.

Finally, it has to be emphasized that the concept of *transition to turbulence* is not very well defined: generally, what experimentalists call transition corresponds to the development of small-scale three-dimensional turbulence within the fluid. Actually, we will see that the whole process of transition to turbulence may involve several successive stages: in the plane mixing layer behind a splitter plate for instance, these stages observed experimentally are: a) the growth of two-dimensional coherent structures; b) the merging of these structures together (pairing); c) a catastrophic breakdown into three-dimensional turbulence. But, as already discussed in Chapter I and further justified in Chapter XIII, the two-dimensional coherent structures, which appear before the transition to three-dimensional turbulence, may themselves be considered as a field of two-dimensional turbulence.

1 - The Reynolds number

The most famous (but certainly not the most characteristic) experiment on the transition to turbulence is the Reynolds experiment of a flow in a circular pipe (circular Poiseuille flow). Let U be an average velocity of the flow across the tube section, D the diameter of the tube, and ν the molecular viscosity. Reynolds (1883) introduced the non-dimensional parameter

$$R = \frac{U\,D}{\nu} \qquad\qquad (III-1-1)$$

and showed experimentally that there was a critical value of R above which the flow inside the tube became turbulent. This was done by varying independently the velocity U, the diameter D of the pipe, or considering fluids of various viscosities. The critical value R_c found by Reynolds was of the order of 2000. For $R < R_c$, the flow remained regular ("laminar"), and for $R > R_c$ it became turbulent. These observations are corroborated by the measurements of the pressure drop coefficient $\Delta p/\rho\, U^2$ in the pipe flow, whose dependence upon R showed a drastic change of behaviour, from the R^{-1} dependence predicted theoretically with the Poiseuille parabolic velocity profile, to a much more gentle decrease above the critical Reynolds number. Though, as we shall see, the notion of a critical Reynolds number is extremely ambiguous, this parameter can nevertheless be shown to characterize the relative importance of non-linear interactions developing in the fluid: indeed, let us consider a fluid particle of velocity u going across the pipe: the necessary time for this to occur is the "inertial" time

$$T_{in} = \frac{D}{u} \qquad\qquad (III-1-2)$$

provided the fluid particle is not prevented from moving by viscous effects: the latter will act on the distance D in a time of the order of

$$T_\nu = \frac{D^2}{\nu} \qquad\qquad (III-1-3)$$

as can be seen by considering the simple diffusion equation

$$\frac{\partial u}{\partial t} = \nu \, \nabla^2 u \qquad (III - 1 - 4)$$

Thus a transverse perturbation u of the flow U in the tube will be able to cross it only if the ratio of the inertial frequency over the viscous frequency is greater than 1. This condition corresponds to:

$$\frac{T_\nu}{T_{in}} = \frac{u\,D}{\nu} > 1 \qquad (III - 1 - 5)$$

If one assumes that the order of magnitude of u is small with respect to U, it can be concluded that the velocity fluctuations will be able to develop in the flow only if the Reynolds number UD/ν based on the basic flow is much greater than 1.

From this oversimplified phenomenological analysis, we retain the idea that the Reynolds number characterizes the relative importance of non-linear effects over viscous effects in the Navier-Stokes equation. In other words, it represents the relative importance of inertial forces over viscous forces.

The same experiment repeated in a plane Poiseuille flow (plane channel) shows, as in Reynolds experiment, a transition to turbulence at a critical Reynolds[1] number of the order of 2000. For a plane Couette flow, the critical Reynolds number[2] is of the order of 1000. We will discuss below in detail the case of the plane boundary layer.

It is, nevertheless, difficult to give a firm basis to this notion of a critical Reynolds number: In fact, experiments show that the transitional Reynolds number to turbulence depends on the intensity of turbulence existing in the incoming flow (residual turbulence). Furthermore, no stability analysis[3] has, up to now, predicted satisfactorily the critical Reynolds number for the transition to turbulence in the above-quoted flows.

In spite of the limitations of this approach, it is of interest to go back to the hydrodynamic instability theory, in order to understand some basic mechanisms of the transition to turbulence in free-shear or wall-bounded flows. This theory also permits us to understand the mechanisms of generation of coherent structures in some free-shear flows such as mixing layers, wakes or jets.

---

[1] based on the width of the channel and the average velocity across its section.

[2] based on the entrainment velocity of the moving plane and the distance between the two planes.

[3] linear or nonlinear

To conclude this section, let us stress that, in order to accurately
define the Reynolds number, one has to choose a characteristic velocity
and length. According to these choices, the various Reynolds numbers
shown on Table 1 can be constructed. The Taylor microscale λ is useful
when studying isotropic turbulence.

Velocity:	Mean flow	turbulent	turbulent
Length:	Apparatus	integral scale	Taylor microscale
Reynolds Number:	"external"	turbulent	R_λ

Table 1: Definition of the Reynolds numbers based on various lengths
and velocities.

Note finally that the Reynolds number can evolve following the motion of the fluid particle if the characteristic scale and velocity depend
upon space or time: in the spatially-growing mixing layer or boundary
layer, for instance, the Reynolds number based on the thickness of the
layer increases with this scale in the downstream direction.

2 - Linear-instability theory

We recall here some important results of linear-hydrodynamic instability theory which are of interest in understanding the transition to
turbulence. For a more complete presentation of the theory, the reader
is referred to Drazin and Reid (1981). In a fluid of uniform density ρ_0 ,
we will restrict our attention to the stability of a parallel flow with components $\bar{u}(y), 0, 0$, and upon which is superposed a small perturbation.
This perturbation is assumed to be two-dimensional in the (x, y) plane.
We consider a stream function $\tilde{\psi}(x, y, t)$ of the form

$$\tilde{\psi} = \epsilon \, \Phi(y) \, \exp i \, \alpha(x - ct) \quad , \qquad (III - 2 - 1)$$

corresponding to a perturbed velocity field with components $\tilde{u} = \partial\tilde{\psi}/\partial y$
and $\tilde{v} = -\partial\tilde{\psi}/\partial x$, i.e.,

$$\tilde{u} = \epsilon \, \frac{d\Phi}{dy} \, \exp i \, \alpha(x - ct), \, \tilde{v} = -\epsilon \, i\alpha\Phi(y) \, \exp i \, \alpha(x - ct) \, , \, (III - 2 - 2)$$

where $\epsilon << 1$ is a small non-dimensional parameter. α is real, and is the
spatial longitudinal wave number of the perturbation: this is a temporal

analysis, as opposed to a spatial analysis where α is complex. In this temporal study, c is complex, of real and imaginary parts c_r and c_i. αc_r is the phase-speed of the perturbation, while αc_i is its temporal-growth rate. Notice that, within the linear-instability analysis, it is not necessary to consider the growth of three-dimensional perturbations, which are always less amplified than two-dimensional perturbations (Squire's theorem, see Drazin and Reid, 1981). This is, however, no longer true in rotating or compressible flows: for instance, the temporal-compressible mixing layer admits, at a Mach number $M_c = U/c > 0.6$ (U is half the velocity difference), three-dimensional oblique waves which are more unstable than the two-dimensional waves (see e.g. Sandham and Reynolds, 1989).

2.1 The Orr-Sommerfeld equation

We assume the reference frame is not rotating, and that the total velocity field (basic flow + pertubation) is two-dimensional (no z-dependence). Its vorticity is

$$\vec{\omega} = \omega \ \vec{z} = (-\frac{d\bar{u}}{dy} + \tilde{\omega}) \ \vec{z} \ ; \ \tilde{\omega} = -\nabla_H^2 \tilde{\psi} \qquad (III-2-3)$$

The vorticity equation in this two-dimensional case reduces, from eq. (II-6-3), to

$$\frac{D_H}{Dt}\omega = \nu\nabla^2\omega \quad , \qquad (III-2-4)$$

since the flow is incompressible ($\vec{\nabla}.\vec{u} = 0$), and $(\vec{z}.\vec{\nabla})\vec{u} = \vec{0}$. Notice that, when studying the stability of a two-dimensional geophysical flow on a rotating sphere, one can make use of Barré de Saint-Venant equations with constant depth, which yields

$$\frac{D_H}{Dt} (\omega + f) = \nu\nabla^2\omega \quad . \qquad (III-2-5)$$

This equation will be used later on in order to obtain the so-called Kuo equation, allowing one to study the influence of differential rotation upon the stability of parallel flows.

Let us come back to eq. (III-2-4), where we perform a normal-mode analysis: it is expanded to the first order with respect to ϵ. We assume that the basic flow is a solution of the Navier-Stokes equation, and thence its vorticity satisfies also eq. (III-2-4). Therefore, the following is readily obtained:

$$\frac{\partial\tilde{\omega}}{\partial t} + \bar{u}(y)\frac{\partial\tilde{\omega}}{\partial x} + \tilde{v}\frac{d\bar{\omega}}{dy} = \nu\nabla_H^2\tilde{\omega} \quad . \qquad (III-2-6)$$

Making use of eqs. (III-2-1) to (III-2-3), this equation becomes (assuming that $\alpha \neq 0$):

$$[\bar{u}(y) - c] \left(\frac{d^2\Phi}{dy^2} - \alpha^2\Phi\right) - \frac{d^2\bar{u}}{dy^2} \Phi = -\frac{i\nu}{\alpha} \left(\frac{d^2}{dy^2} - \alpha^2\right)^2 \Phi \quad , \quad (III - 2 - 7)$$

which is the traditional form of the Orr-Sommerfeld equation. The boundary conditions at the two boundaries $y = y_A$, $y = y_B$ (which may also be moved to infinity) are of the no-slip type, and hence, from eq. (III-2-2): $\Phi = 0$; $d\Phi/dy = 0$. Before studying the stability of viscous flows using eq. (III-2-7), we will look at what can be said from the point of view of a perfect flow.

2.2 The Rayleigh equation

More specifically, we assume that viscous effects have a negligible influence in the evolution of the perturbed flow described by ψ: hence, the above Orr-Sommerfeld equation simplifies to the Rayleigh equation:

$$[\bar{u}(y) - c] \left(\frac{d^2\Phi}{dy^2} - \alpha^2\Phi\right) - \frac{d^2\bar{u}}{dy^2} \Phi = 0 \quad , \quad\quad\quad (III - 2 - 8)$$

with free-slip boundary counditions $\Phi = 0$ at $y = y_A$, $y = y_B$. Notice that we will not exclude using the Rayleigh equation for a viscous basic flow satisfying the Navier-Stokes equation; in some cases, it will prove to be useful, as will be seen below.

The problem to be solved is of the eigen-value type: for a given $\bar{u}$, we have to determine the values of α (real) and c (complex) for which eq. (III-2-8) admits complex eigensolutions Φ. A first and important remark concerning the possible solutions is that, if $\Phi_{\alpha,c}$ is a solution corresponding to the couple (α, c), $\Phi_{\alpha,c^*} = \Phi^*_{\alpha,c}$ is also a solution (here $*$ denotes the complex conjugate). Since without loss of generality we may restrict ourselves to positive values of α, it turns out that, for any damped solution (such that $\alpha c_i < 0$), one may associate an amplified solution such that $\alpha(-c_i) > 0$. In other words, the research of unstable modes is reduced to the research of non-neutral modes ($c_i \neq 0$). This is of course not valid for the Orr-Sommerfeld equation. A second remark concerns the critical points y (see e.g. Maslowe, 1981), such that $\bar{u}(y) = c$. These points can exist only in the case of a neutral mode, since $\bar{u}(y)$ is evidently real. Therefore, eigen solutions corresponding to amplified modes[4] do satisfy eq. (III-2-8) divided by $[\bar{u}(y) - c]$:

$$\Phi'' = [\frac{\bar{u}''}{\bar{u} - c} + \alpha^2] \Phi \quad , \quad\quad\quad (III - 2 - 9)$$

---

[4] if they exist

where the suffix " ' " stands for the operator d/dy . In order to look for necessary conditions of instability, we assume hence that $c_i \neq 0$, and multiply eq. (III-2-9) by Φ^* . Integrating from y_A to y_B , we obtain, after an integration by parts and making use of the boundary conditions $\Phi = 0$:

$$-\int_{y_A}^{y_B} (|\Phi'|^2 + \alpha^2 |\Phi|^2)\, dy = \int_{y_A}^{y_B} |\Phi|^2\, \frac{\bar{u}''(\bar{u} - c^*)}{|\bar{u} - c|^2}\, dy \quad . \quad (\mathrm{III} - 2 - 10)$$

The imaginary part of the r.h.s. must be zero, which yields

$$\int_{y_A}^{y_B} |\Phi|^2\, \frac{\bar{u}''}{|\bar{u} - c|^2}\, dy = 0 \quad , \qquad (\mathrm{III} - 2 - 11)$$

since $c_i \neq 0$ and $|\Phi|$ cannot be zero everywhere. This result implies that $\bar{u}''$ must change sign at least once on the $[y_A, y_B]$ interval, or, equivalently, that $\bar{u}$ admits at least one inflection point[5]: this necessary criterion of instability, known as the *Rayleigh inflection-point inviscid-instability criterion,* shows that basic velocity profiles such as the Blasius boundary-layer velocity profile on a flat plane, or the parabolic plane Poiseuille flow in a channel, are unconditionally stable from an inviscid point of view. In fact, it will be seen below that they are subject to viscous instabilities.

This criterion can be improved in the following manner: one still considers an unstable solution ($c_i \neq 0$), and looks at the real part of eq. (III-2-10), which is written:

$$-\int_{y_A}^{y_B} (|\Phi'|^2 + \alpha^2 |\Phi|^2)\, dy = \int_{y_A}^{y_B} |\Phi|^2\, \frac{\bar{u}''(\bar{u} - c_r)}{|\bar{u} - c|^2}\, dy \quad . \quad (\mathrm{III} - 2 - 12)$$

We notice that the constant c_r arising in the r.h.s. may be replaced by any other constant, due to eq. (III-2-11). This is true in particular if this constant is set equal to $\bar{u}(y_s)$, where y_s is any of the inflection points of $\bar{u}$. The result is that

$$\int_{y_A}^{y_B} |\Phi|^2\, \frac{\bar{u}''[\bar{u} - \bar{u}(y_s)]}{|\bar{u} - c_r|^2}\, dy < 0 \quad , \qquad (\mathrm{III} - 2 - 13)$$

and $\bar{u}''[\bar{u} - \bar{u}(y_s)]$ must be negative somewhere, in order for the flow to be unstable. This is the *Fjortoft criterion,* which implies that $|d\bar{u}/dy|$, the absolute value of the basic vorticity, must have a local maximum

---

[5] The case $\bar{u}'' = 0$ everywhere, corresponding to a plane Couette flow, leads from eq. (III-2-10) to $\Phi = 0$ on $[y_A, y_B]$. Therefore, there is in this case no amplified mode within the Rayleigh analysis.

at y_s (see Drazin and Reid, 1981, for details). From the Rayleigh and Fjortoft criteria, it is to be expected that a temporal mixing layer

$$\bar{u}(y) = U \tanh \frac{y}{\delta_0} \quad , \qquad (III - 2 - 14)$$

or a plane jet, or a wake, will be unstable from an inviscid point of view. Notice however that the two above criteria give only *necessary conditions* for the instability. These conditions are not always *sufficient*: for instance, the basic velocity profile $\sin y$ is stable.

2.2.1 Kuo equation

Let us consider eq. (III-2-5), describing the two-dimensional horizontal motion of a flow on a rotating sphere. The Coriolis parameter f is given by (II-7-6). Let us consider a reference parallel of latitude φ_0, serving as an origin for the meridional coordinate y. The β-plane approximation consists in expanding f about φ_0 as

$$f = f_0 + \beta y \quad , \qquad (III - 2 - 15)$$

$\beta = df/dy$ being assumed to be constant. If we perform a linear-stability analysis about a parallel flow $\bar{u}(y)$, the term $\tilde{v} \, d\bar{\omega}/dy$ in eq. (III-2-6) has to be replaced by $\tilde{v}(\beta + d\bar{\omega}/dy)$, and the analogous Rayleigh equation for this problem, the Kuo equation, is (see Howard and Drazin, 1964)

$$[\bar{u}(y) - c] \left(\frac{d^2\Phi}{dy^2} - \alpha^2\Phi \right) + \left(\beta - \frac{d^2\bar{u}}{dy^2} \right) \Phi = 0 \quad . \qquad (III - 2 - 16)$$

It is easy to check that eq. (II-2-10) is now transformed into

$$- \int_{y_A}^{y_B} \left(|\Phi'|^2 + \alpha^2|\Phi|^2 \right) dy = \int_{y_A}^{y_B} |\Phi|^2 \, \frac{(\bar{u}'' - \beta)(\bar{u} - c^*)}{|\bar{u} - c|^2} \, dy \quad . \qquad (III - 2 - 17)$$

Therefore, the necessary criterion of instability is now that $(\bar{u}'' - \beta)$ must change sign. For a hyperbolic-tangent velocity profile of the form (III-2-14), this is not possible if

$$\beta > \frac{4}{3\sqrt{3}} \frac{U}{\delta_0^2} = 0.77 \frac{U}{\delta_0^2} \quad , \qquad (III - 2 - 18)$$

which is a critical value above which the instability cannot exist. We will discuss later on in this chapter and in Chapter IX the physical signification of this result.

3 - Transition in shear flows

3.1 Free-shear flows

3.1.1 Mixing layers

We start by considering the temporal mixing layer[6]. Its stability diagram, obtained from a numerical solution of the Orr-Sommerfeld equation, is shown on Figure III-1 (taken from Betchov and Szewczyk, 1963).

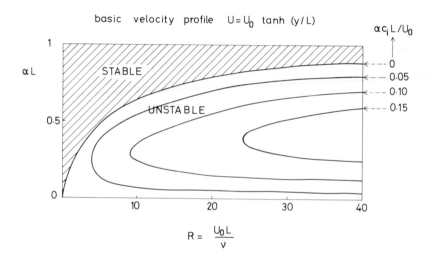

Figure III-1: linear-stability diagram of the mixing-layer instability in the (α, R) domain (non-stratified case). The different curves correspond to various rates of amplification αc_i. Under the neutral curve ($c_i = 0$), the flow is unstable (from Betchov and Szewczyk, 1963, courtesy of The Physics of Fluids)

Here, U_0 and L correspond to U and δ_0 in eq. (III-2-14). For a given Reynolds number, there exists a range of unstable longitudinal modes α. It may be shown that their amplification rate αc_i is maximum for a wave number α_a, called the most-amplified wave number. It is this mode which is expected to be selected if the initial perturbation, superposed upon the basic hyperbolic-tangent velocity profile, contains an equal amount of energy in all the unstable modes: indeed, the perturbations grow exponentially with time, and the one with the highest amplification rate will grow to finite amplitude before the other unstable modes will

---

[6] that is, periodic in the x direction

have time to develop. This has been verified for instance in the case of a white-noise perturbation by Lesieur et al. (1988).

When the Reynolds number exceeds values of the order of $30 \sim 40$, the amplification rates are no longer affected by viscosity, and the instability becomes inviscid, in the sense that it is well described by the Rayleigh equation. The most amplified wave number, calculated by Michalke (1964), is given by

$$\alpha_a = 0.44 \; \delta_0^{-1} \quad , \qquad\qquad (III-3-1)$$

corresponding to a spatial wave length

$$\lambda_a = 14 \; \delta_0 \quad . \qquad\qquad (III-3-2)$$

The mechanism of instability may be described in the following manner (see Batchelor, 1967, and Drazin and Reid, 1981): let us consider a longitudinal stripe of rotational fluid (approximately of width $2\delta_0$) in the x, y plane, separating the two irrotational regions corresponding to the uniform flows of respective velocity U and $-U$ (see Figure III-2). Suppose that this rotational zone is perturbed and undulates about the line $y = 0$ with a longitudinal wave length λ_a. Since $d^2\bar{u}/dy^2 = 0$ at the inflection point, and using the inviscid equation (III-2-6), the total vorticity is convected by the basic flow $\bar{u}(y)$. Hence, the crests of disturbance (in the region $y > 0$) and the troughs of disturbance ($y < 0$) will travel in opposite directions. This effect, taken together with the vorticity-induction phenomena, will contribute to the formation of spiral vortices, initially of wave length λ_a. This instability will be called here the Kelvin-Helmholtz instability, the resulting vortices being Kelvin-Helmholtz vortices. The roll up of the vortices may also be understood using the so-called displaced fluid particle arguments: a fluid particle located initially at $y = 0$ (where the vorticity is maximum) and moved towards the regions $y > 0$ or $y < 0$ will keep its vorticity[7], and therefore will be surrounded by fluid of weaker vorticity, which will induce the roll up by vorticity induction.

Afterwards, the two-dimensional evolution of the layer can be investigated by means of direct-numerical simulations: Plates 7 and 8, taken from Comte (1989) present the time evolution of a temporal mixing layer in a square domain of size equal to 8 fundamental wave lengths: the vorticity (left) shows how the fundamental eddies form, then undergo successive pairings. A passive scalar (right) plays the role of a numerical dye marking the two streams, and shows how the coherent structures mix

---

[7] We recall that, in two dimensions, vorticity is transported following the velocity field.

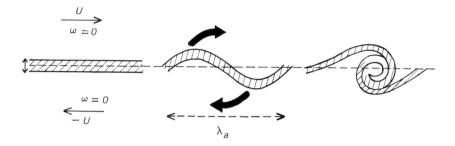

Figure III-2: schematic illustration showing the formation of spiralling Kelvin-Helmholtz vortices in a mixing layer

the two flows. The first pairing occurring in the row of primary vortices may be understood in terms of a subharmonic instability of wave length $2\lambda_a$: indeed, a perturbation at this wave length will push vortices lifted into the upper region ($y > 0$) against vortices brought into the region $y < 0$, allowing the vorticity induction to act. Notice also that the features of the passive-scalar field in this calculation are reminiscent of the experimental visualizations of a spatially-growing mixing layer between two reacting flows carried out by Koochesfahani and Dimotakis (1986).

Temporal mixing-layer experiments are difficult to realize experimentally, although they may be much more relevant to geophysical situations. However, there is a less known aspect of Reynolds' 1883 historical paper, where he remarked that the pipe flow was difficult to destabilize: he then performed another experiment involving two fluids of different densities ρ_1 and ρ_2 in a slightly inclined tube (with the lighter fluid on the top), and stressed that such a flow could be destabilized much more easily. This is due to the fact that gravity accelerates one fluid with respect to the other, creating a mixing layer, which we have seen to be linearly unstable whatever the value of the Reynolds number[8]. A schematic picture of this second Reynolds experiment in a tilted pipe is shown on Figure III-3. This experiment has been redone by Thorpe (1968), who shows the formation of a row of Kelvin-Helmholtz vortices very much resembling the passive scalar distribution of the calculation shown in Plate 7, when the primary vortices form.

We have already presented in Plate 1 a two-dimensional numerical

---

[8] Notice, however, that such a flow will become stable again if the density difference is too high, as will be discussed later.

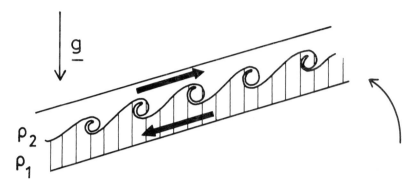

Figure III-3: formation of Kelvin-Helmholtz billows and transition to turbulence in Reynolds' stratified tilted pipe.

simulation of a spatially-growing mixing layer, performed by Normand et al. (1988). This calculation is forced by a small random perturbation at the inflow, superposed upon the incoming hyperbolic-tangent velocity profile. This is done in order to model the turbulent fluctuations brought by the boundary layers developing along the splitter plate in the experiment. In this calculation, Kelvin-Helmholtz instability develops following the mean flow: as in the temporal case, Kelvin-Helmholtz vortices form by roll up, pair, and are finally carried away downstream, out of the computational domain. Analogous calculations, where the inflow is forced at the fundamental with a random phase, were made by Lowery and Reynolds (1986).

   Although, as already stressed in Chapter I, the calculation seems to represent quite well the large-scale coherent-structure dynamics of the experiments, it cannot of course simulate the transition to small-scale fully-developed turbulence. Experiments (see e.g. the review by Ho and Huerre, 1984) do show that this transition occurs after the first pairing. Thin longitudinal vortex tubes of opposite sign interconnecting the Kelvin-Helmholtz billows, and discovered experimentally by the Caltech group (see e.g. Konrad, 1976, Bernal, 1981, Breidenthal, 1981, Jimenez, 1983, and Bernal and Roshko, 1986) seem to play a role in this transition. The same vortex filaments were found in a three-dimensional direct-numerical simulation performed by Metcalfe et al. (1987).

   The origin of these filaments is still subject to controversy: In Breidenthal's (1981) experiment, this author notes a *wiggle disturbance* in the outer edge of the billows, from which the longitudinal streaks originate, due to "the global strain field of the flow" (sic). This is shown on Figure III-4, taken from this reference. In this experiment, the spanwise wave length of the oscillation is 1.1 λ, where λ is the longitudinal wave length of the local Kelvin-Helmholtz billows.

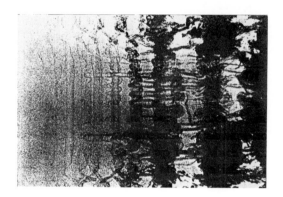

Figure III-4: plan view of Breidenthal's (1981) mixing-layer experiment, showing the development of a sinuous spanwise instability of the Kelvin-Helmholtz billows. Simultaneously, longitudinal vortices form downstream (courtesy of Cambridge University Press).

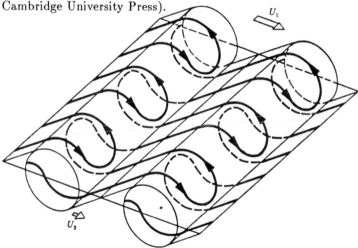

Figure III-5: topology of streamwise vortex lines in the mixing layer, as proposed by Bernal and Roshko (1986) from their experiments (courtesy of Cambridge University Press).

The topology of these streaks in the mixing layer downstream, when small-scale turbulence has developed, has been experimentally reconstructed by Bernal and Roshko (1986), who propose a hairpin-shaped vortex filament, "which loops back and forth between adjacent primary vortices" (sic, see Figure III-5) with an average spanwise wave length of 0.67λ . This spanwise spacing scales downstream with λ. A cross section view of these longitudinal vortices in Bernal and Roshko's (1986) experiment is shown on Figure III-6.

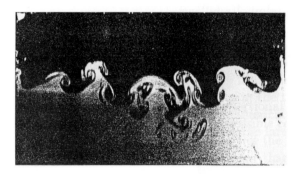

Figure III-6: cross-section of the longitudinal vortices in the experiments reported in Figure III-5; the flow direction is normal to the figure (courtesy of
Cambridge University Press).

It is tempting to interpret these longitudinal filaments as a consequence of the *translative instability*, discovered by Pierrehumbert and
Widnall (1982), on the basis of a three-dimensional linear-stability analysis performed on a row of two-dimensional Stuart vortices: this instability consists of a global sinuous spanwise oscillation of the primary
billows, and is triggered by vorticity perturbations in the stagnation region between two billows. The most-amplified spanwise wave length λ_s
of the instability is found to be

$$\lambda_s \approx \frac{2}{3}\lambda \quad , \qquad\qquad (III - 3 - 3)$$

where λ is the longitudinal wave length of the two-dimensional vortices. This accords well with Bernal and Roshko's (1986) experiments,
although they refer to the developed mixing layer, while the translative
instability is based on a linear theory describing the early stages of three-
dimensionality (where we recall that Breidenthal (1981) found a spanwise
wave length of 1.1λ). Furthermore, experiments show clearly that the
hairpin vortices are much thinner than the primary billows. It is therefore difficult to consider the hairpin vortices as the eventual evolution of
a sinuous oscillation of the billows.

There is another explanation of the streaks, which may be found
in Corcos and Lin (1984), Lasheras et al. (1986), and Lasheras and
Choi (1988). It can be explained as follows: residual three-dimensional
turbulence existing in the mixing layer will contain longitudinal vorticity of both signs. If such fluctuations exist in the stagnation region
between two billows, the corresponding vortex filaments will be stretched
between the two billows, leading to formation of longitudinal vortices
along the braids. The sign of these vortices will thus depend on the

sign of the initial vorticity fluctuations. It is clear in particular that, if the vorticity in the stagnation region is distributed along a spanwise sinuous filament, it will be stretched into hairpin-shaped vortices of same topology as described by Bernal and Roshko (1986). There is here no preferred spanwise wave length, but it could be imposed by the above-described translative instability of the billows, as proposed by Sandham and Reynolds (1989).

We will show in Chapter XIII some numerical evidence that, in a mixing layer forced initially by a three-dimensional random perturbation of small amplitude, strong spanwise decorrelations develop during the roll up and the subsequent pairings. This leads to a Λ-shaped structure of the Kelvin-Helmholtz billows, and to the production of thinner longitudinal hairpin vortices shed from the former. On the contrary, a quasi-two-dimensional perturbation affects the billows only slightly, and generates strained longitudinal vortices as in the experiments (Comte and Lesieur, 1990). More details will be given in Chapter XIII.

3.1.2 Mixing layer with differential rotation

Let us mention finally the study of the two-dimensional temporal mixing-layer submitted to differential rotation (β-effect), carried out by Maslowe et al. (1989), using direct-numerical simulations: these calculations show that no rollup occurs for $\beta > 0.2$, although, from eq. (III-2-18), the linear instability may develop[9]. For $0.05 < \beta < 0.2$, the roll up occurs, but the pairing is suppressed. As will be shown in Chapter IX, differential rotation is responsible for the propagation of Rossby waves, which limit the meridional extent of motions: the inhibition of the pairing is thus one of the aspects of this limitation. This range might correspond to realistic values of β for the Earth's atmosphere[10], and also for Jupiter. This might explain why cyclonic depressions above the northern Atlantic rarely pair (André, 1988), and could question the theories interpreting Jupiter's great red spot as the result of pairings of smaller-size eddies created in the shear between two neighbouring jets (Somméria et al., 1988).

3.1.3 Plane jets and wakes

It is well known from the Prandtl viscous boundary-layer equations (see e.g. Schlichting, 1968) that a laminar plane jet issuing from point source will have, at a certain distance downstream x, a longitudinal velocity given by

$$u = \frac{U}{\cosh^2 y/\delta_0} \quad , \qquad (III-3-4)$$

---

[9] β is here nondimensionalized with U/δ_0^2

[10] δ_0 and U being properly chosen with respect to the coherent eddies

where $U(x)$ is the velocity at the centre of the jet, and $\delta_0(x)$ a characteristic width. The same expression holds approximately for a turbulent plane jet, as can be demonstrated using Prandtl's mixing-length theory (1925), and is well verified experimentally. This plane-jet velocity profile is sometimes referred to as the Bickley jet.

The linear stability of the velocity profile given by eq. (III-3-4) has been investigated by Drazin and Howard (1966) on the basis of the Rayleigh equation. It was found that two classes of unstable solutions exist, corresponding respectively to the *even modes* (called also *sinuous modes*), where the amplitude $\Phi(y)$ in eq. (III-2-1) is even, and the *odd modes* (called *varicose modes*), where it is odd.

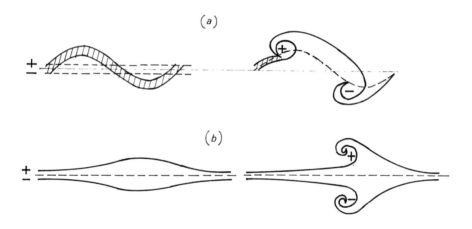

Figure III-7: growth in a Bickley jet of a) the sinuous mode, leading to the formation of a Karman street; b) the varicose mode

The schematic evolution of the jet in both cases is shown on Fig. III-7: in the sinuous mode, vorticity of alternatively positive and negative signs is transported into the upper and lower irrotational regions, tending, by vorticity-induction effects, to form a Karman-like vortex street. On the contrary, the varicose mode will form a pattern symmetric with respect to the x-axis. The stability analysis shows that the most-amplified sinuous mode grows three times faster than the varicose one.

In the viscous case, and if the Reynolds number is high enough, it may be shown that the instability is of the inviscid type (see Drazin and Reid, 1981). Therefore, it is natural that the two-dimensional direct-numerical simulations of the Bickley jet, submitted initially to a random white-noise perturbation of small amplitude, show the formation of only the sinuous mode, at the most-amplified wave length predicted by the linear-stability theory, that is

$$\lambda_a = 6.54 \; \delta_0 \quad . \qquad\qquad (III - 3 - 5)$$

This is shown on Plate 9, taken from Comte et al. (1987). It represents the vorticity during respectively the initial stage, the growth of the sinuous instability, and the formation of the Karman street. The red colour corresponds to positive vorticity, and blue colour to negative vorticity. This calculation also shows the tendency for eddies of same sign to pair, thus increasing the wave length of the coherent structures.

A plane wake is characterized by a gaussian deficit velocity profile, both in the viscous laminar and the turbulent cases. Its linear-stability analysis in the temporal case has been carried out by Sato and Kuriki (1961). The results are qualitatively similar to the Bickley jet case, with dominant sinuous modes resulting in a Karman street. This is to be expected, since, within the temporal approximation, a wake and a jet differ only in the form of respectively the basic velocity and the basic deficit velocity[11]. Plate 10 shows such a street obtained in the three-dimensional direct-numerical simulation of a periodic wake, starting from a gaussian basic velocity profile destabilized by a random perturbation of small amplitude.

How do these temporal calculations apply to spatially-growing situations: for a wake of external velocity U_e, and whose deficit velocity is small compared to U_e, a Galilean transformation of velocity U_e allows one to associate a temporal problem to the spatial evolution. On the contrary, there is no downstream transport velocity in a jet, whose velocity on the axis decays downstream. In fact, two-dimensional spatially-growing calculations of both a Bickley jet and a Gaussian wake, presented in Fig III-8, confirm the experimental fact that the jet is more unstable than the wake, but still displays pairing interactions between eddies of same sign.

In this calculation, the resulting *two-dimensional turbulent jet* grows proportionally to x, while the wake expands only like $\sqrt{x}$, as predicted by both theory (see Chapter VI-1) and experiments. As a comparison, Plate 11, taken from Werlé (1974), presents the experimental Karman street obtained behind a splitter plate. A last remark to be made is that the formation of a Karman street behind an obstacle seems to be independent of the shape of the obstacle: for instance, a cylinder, a splitter plate or a wedge give rise to the same type of wake. This is easily understandable if one interprets the formation of the Karman street as the result of the development of the sinuous instability upon the doubly-inflectional basic velocity profile created downstream of the obstacle: the

---

[11] Indeed, the temporal approximation allows both a Galilean transformation and the reversal of the velocity.

Figure III-8: direct-numerical simulations of a two-dimensional spatially growing a) Bickley jet; b) Gaussian wake. Both are submitted to a white-noise
upstream forcing of small amplitude; (courtesy P. Alexandre, I.M.G.)

perturbations which trigger this instability come from the shear layers
existing in the close vicinity of the obstacle. Behind a cylinder for instance, experiments show immediately downstream a pair of symmetric
(with respect to the x-axis) small vortices (see e.g. Kourta et al., 1987),
before the appearance further downstream of the Karman street.

It is finally of interest, from a two-dimensional point of view, to
consider also the axisymmetric jets or wakes. The Orr-Sommerfeld and
Rayleigh stability analysis can be generalized in this case. In the two-
dimensional case, the unstable modes will give rise to vortex rings, which
are the axisymmetric analogues of Kelvin-Helmholtz vortices. Numerous
experimental examples are given (see e.g. Crow and Champagne, 1971;
Hussain and Zaman, 1981; Favre-Marinet and Binder, 1989; Van Dyke,
1982, pp 34 and 60). Possibilities of pairing between vortex rings of same
sign exist, as shown by Yamada and Matsui (1978, quoted by Van Dyke,
1982, p 46): in the same way as Kelvin-Helmholtz vortices rotate about
each other before merging, vortex rings *leapfrog* during the pairing. Such
a pairing might also be visible in the round jet, shown in Fig. I-2 before
its breakdown into three-dimensional turbulence. As quoted by Winant
and Browand (1974), *Laufer, Kaplan and Chu (1973) have made indirect
observations of vortex-ring pairing in an axisymmetric jet at Reynolds
numbers of order 160 000 (...) It is, in fact, proposed that pairing of the*

vortex rings is the primary mechanism responsible for the production of jet noise. Thus, the transition to turbulence in a round jet is analogous to the transition in a plane-mixing layer, that is, roll up, pairing and sudden breakdown.

Let us mention finally that there are numerous three-dimensional instabilities associated with the transition of wakes or jets into developed turbulence. For the plane wake for instance, experiments (see Breidenthal, 1981; Lasheras and Meiburg, 1990) and calculations (see Chen et al., 1990; Lasheras and Meiburg, 1990, and Gonze et al., 1990) show the formation of longitudinal streaks, which resemble those encountered in the mixing layer. These streaks are visible on Plate 12, taken from Gonze et al. (1990), which represents the time-evolved scalar field in the same calculation as in Plate 10. At higher Reynolds number, the further evolution of the wake is more complicated, and it becomes highly three-dimensional.

3.2 Wall flows

3.2.1 The boundary layer

Let us start with the Blasius velocity profile, corresponding to a laminar boundary layer over a semi-infinite flat plate. We define the displacement thickness as

$$\delta_1(x) = \frac{1}{U} \int_0^{+\infty} [U - \bar{u}(x, y)] \, dy \quad , \qquad (III - 3 - 6)$$

which is such that

$$\delta_1(x) = 1.73 \, (\frac{\nu x}{U})^{1/2} \qquad (III - 3 - 7)$$

(see e.g. Schlichting, 1968), x being the distance downstream of the leading edge, and U the free-stream velocity. In the temporal case, the Orr-Sommerfeld equation exhibits an unstable region in the [perturbation wave number-Reynolds number] domain, and a critical Reynolds number below which the flow is always linearly stable (see e.g. Drazin and Reid, 1981). The stability diagram of such a flow is shown schematically on Figure III-9. With the Reynolds number defined as $U\delta_1/\nu$, its critical value is equal to 520 (see Schlichting, 1968). In the downstream direction, the laminar boundary-layer thickness and the Reynolds number will grow like $\sqrt{x}$: when

$$\frac{Ux}{\nu} \geq (\frac{520}{1.73})^2 = 90347 \quad ,$$

the flow will become unstable to small perturbations, and unstable waves will begin to grow.

These waves are called Tollmien-Schlichting waves[12], from the work

---

[12] hereafter called T.S. waves

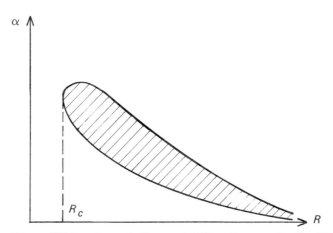

Figure III-9: schematic linear-stability diagram for the Blasius boundary-layer velocity profile, in the [α (longitudinal wave number of the perturbation) - R (Reynolds number of the basic flow)] domain. Inside the dashed area, the perturbations will amplify exponentially. When R goes to infinity, the unstable region will collapse onto the $\alpha = 0$ axis

of Tollmien and Schlichting[13], who solved the Orr-Sommerfeld equation for the boundary layer. The waves give rise to periodic oscillations in the boundary layer. Their existence was subsequently established experimentally by Schubauer and Skramstad (1948). The experimental critical Reynolds number for the growth of the waves is in good agreement with the theoretical value given above.

In practice, the transition to turbulence above a semi-infinite flat plate proceeds as follows[14] (see Klebanoff et al., 1962, Hinze, 1975, and Herbert, 1984): the layer is first laminar, up to $Ux/\nu \approx 10^5$; then T.S. waves begin growing. Afterwards, experiments (see Schubauer and Klebanoff, 1956, and Klebanoff et al., 1962) show the development of a spanwise oscillation of the wave, interpreted as the evidence of a hairpin-shaped vortex structure in the boundary layer. The peaks are characterized by high longitudinal velocity fluctuations, while the valleys correspond to low fluctuations. One observes the formation of longitudinal streaks of respectively slow and fast fluid in the peaks and valleys. This is easily understandable if one notes that longitudinal vortices of opposite vorticity on each side of a peak will pump slow fluid from the wall. The hairpin vortex is strained downstream, and the longitudinal velocity profile becomes inflexional in the peak. All these longitudinal and span-

---

[13] see Schlichting, 1968, for a review

[14] If the T.S. waves (defined above) are forced upstream. In Klebanoff et al. (1962) the boundary layer is forced initially with a vibrating ribbon at a fixed spanwise wave length.

wise inflexional instabilities contribute to the roll up and breakdown of T.S. waves into three-dimensional turbulence (Blackwelder, 1979).

In a naturally developing boundary layer, transition is intermittent in the sense that the above-described process occurs in turbulent spots which are spatially localized (Schubauer and Klebanoff, 1956). For $Ux/\nu > 10^6$, the boundary layer is turbulent everywhere. Up to now, no theory has been able to predict this last Reynolds number characterizing the transition to fully-developed turbulence in a boundary-layer.

3.2.2 Poiseuille flow

For the plane Poiseuille flow, the situation is even more frustrating: the Orr-Sommerfeld equation leads to a stability diagram which resembles the boundary-layer one, with a critical Reynolds number of $R_C = 5772$. On the other hand, experiments show a transition to fully-developed turbulence at a Reynolds number of 2000, as already mentioned above. They show also waves, analogous to the T.S. waves, which grow at lower Reynolds numbers. The instabilities developing below R_C are called subcritical instabilities. Clearly, they need finite-amplitude perturbations to develop. A two-dimensional nonlinear instability theory has been developed by Orszag and Patera (1983) for this case, predicting a lower critical Reynolds number, but still not enough to agree with the experiments. Let us mention finally the two-dimensional direct-numerical simulations of the periodic channel flow done by Jimenez (1987), which show the formation of a street of eddies of opposite sign[15]. These two-dimensional calculations give valuable information on the growth of two-dimensional nonlinear instabilities. It is however difficult in the case of wall-flows to apply their results to transitional processes, which seem to be highly three-dimensional.

For a circular Poiseuille flow, as well as for the plane Couette flow, the linear-stability analysis leads to stability, whatever the Reynolds number and the wavenumber of the perturbation. Again, the circular pipe flow will become turbulent from the development of finite-amplitude perturbations, which come from the residual incoming turbulence.

3.3 Transition, coherent structures and Kolmogorov spectra

It is not the aim of the present book to discuss at length all the stability problems involved in the transition to turbulence. How in practice will a real flow of high Reynolds number degenerate into turbulence? Here we will try to summarize the above information in order to give an explanation which certainly oversimplifies the problem, but enables one to understand how the transition occurs: it has already been stressed that turbulence is due to diffusion and stretching of vorticity, which

---

[15] each sign corresponding to one of the boundaries

could be created in the flow by various means, such as a boundary for instance. In fact, the transition to turbulence greatly depends on the manner in which this vorticity is created: if the basic velocity profile is inflexional, the inflection-point instability (also called barotropic instability), which is "inviscid" (i.e. nearly unaffected by viscosity if small enough) and linear, will give rise to the formation of large quasi-two-dimensional structures provided the small perturbations present in the flow contain energy in the unstable wave-numbers range (which is nearly always the case). Examples of such structures have already been given for the Kelvin-Helmholtz waves of the mixing layer. Other examples are given by jets or wakes. These large structures can be, as already pointed out for the mixing layer, unpredictable and we will not refuse to assign the denomination of turbulence to them. However, classical analyses of such flows prefer to reserve this appellation to small-scale three-dimensional turbulence. Sometimes the large structures can amalgamate and lead to the formation of larger structures. This is due to the vorticity conservation constraint characteristic of two-dimensional turbulence (see Chapter IX). Simultaneously, these structures degenerate into smaller and smaller structures, through some successive instabilities which are not very clearly understood, but which seem to agree very well in the small three-dimensional scales with the statistical phenomenological predictions of Kolmogorov (1941) (see Chapter VI). The reason for such good experimental correspondance with the theory certainly remains one of the great mysteries of modern fluid dynamics. These small scales "cascade" down to a dissipative scale where they are damped and die under the action of molecular viscosity. They are, as already emphasized in Chapter I, what can be called "fully-developed turbulence". But it would be erroneous to think that the large transitional structures disappear at a sufficiently high Reynolds number once small-scale turbulence has developed: modern methods of investigation and visualization of turbulent flows have contributed to a radical change in this former classical point of view, and it is, for instance, now widely recognized that they survive in the case of the mixing layer (Brown and Roshko, 1974) and of the jet (Crow and Champagne, 1971). This could also be true for the vortex streets in the wake of a cylinder for instance, or in a jet. This might be due to the linear-instability mechanisms, still developing on the inflexional mean shear flow. These large structures at high Reynolds number display some intermittent characteristics, due both to their intrinsic unpredictability already mentioned before, and to the sudden occurrence of three-dimensional turbulent bursts which destroy them. Further developments on these ideas will be given in Chapter XIII.

From what precedes, it seems necessary to discuss the validity of the concept of a critical Reynolds number applied to the transition to

where $\vec{u}_{2D}^{(0)}$ is the projection of $\vec{u}^{(0)}(x, y, t)$ onto the plane perpendicular to $\vec{\Omega}$, and v_3 the vertical component of $\vec{v}^{(1)}$. If we further assume that $\partial v_3/\partial z$ is zero somewhere along the vertical, it turns out that it is uniformly zero, and, from (III-5-4), the vertical vorticity of the basic flow satisfies a two-dimensional Navier-Stokes equation in the (x, y) plane. So, for such solutions, the rotation has no more effect on the motion and the "horizontal"(i.e. in planes perpendicular to the rotation axis) motions may give rise to inertial effects much larger than the molecular viscous effects, with a horizontal Reynolds number far greater than one. In that sense, one can speak of two-dimensional turbulence.

Notice that the above analysis applies to a flow of arbitrary depth and uniform density, and (III-5-3) is always valid (at high rotation) whatever the boundary conditions: if in particular some body is displaced within the flow, the whole column of fluid above and below will move with the body, leading to the famous phenomenon called the Taylor column and discovered by Taylor (1922).

When applied to a shallow layer of stratified flow on a rotating sphere (the Earth for instance)[24], it is the local planetary vorticity f defined by (II-7-6) and the horizontal characteristic length L which are generally employed to construct the local Rossby number U/fL . Thus planetary-scale atmospheric or oceanic motions in medium or high latitudes are characterized by a local Rossby number small with respect to one, and the quasi-geostrophic analysis (see Chapter IX) will show that rotation reinforces the tendency to two-dimensionality (on the sphere), due to the fact that these flows are in shallow layers. The combined effects of strong rotation and stratification in a fluid of arbitrary depth will be briefly looked at in Chapter IX-5.

5.1 Quasi-two-dimensional flow submitted to rotation

5.1.1 Linear analysis

We come back to the study of section 3.3.1, assuming a solid-body rotation $\vec{\Omega} = \Omega\vec{z}$. We define now the local Rossby number as

$$R_o = |\omega_{2D}|/2|\Omega| \quad . \qquad (III-5-5)$$

If the local Rossby number is of the order of one ($|\omega_{2D}| \sim |\Omega|$), the same quasi-two-dimensional expansion may be performed, leading to:

$$\frac{D_{2D}}{Dt}\underset{\sim}{\omega} = \underset{\sim}{\omega}.\vec{\nabla}\vec{u}_{2D}-(\omega_{2D}+2\Omega)\,\vec{z}\times\underset{\sim}{\omega}+(\omega_{2D}+2\Omega)\,\vec{\nabla}_H w_1 \quad . \quad (III-5-6)$$

---

[24] *shallow*, with respect to the Earth's radius, as already stressed in Chapter I

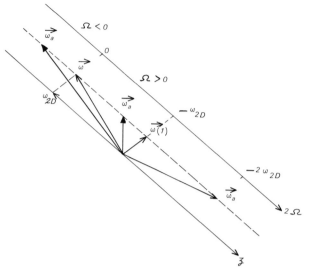

Figure III-12: influence of a solid-body rotation on the longitudinal straining of absolute vortex filaments by an initially quasi-two-dimensional shear flow

Consider a relative vortex filament of spanwise vorticity ω_{2D} which would be destabilized in the non-rotating case. If

$$\omega_{2D} + 2\Omega = 0 \quad , \qquad (III-5-7)$$

which corresponds to anticyclonic rotation[25] and a Rossby number of 1, the longitudinal straining will be enhanced with respect to the non-rotating case. Otherwise, it is difficult to draw definitive conclusions from this equation, which is not very general since it does not apply to low Rossby numbers. Therefore, it is preferable to come back to the original Kelvin theorem in a rotating frame.

5.1.2 The straining of absolute vorticity

We recall that Kelvin's theorem applies to the absolute vorticity, here equal to $(\omega_{2D} + 2\Omega)\vec{z} + \vec{\omega}^{(1)}$. Absolute vortex filaments are (neglecting viscosity) material, with respect to the total flow. We will assume that the straining of the absolute vortex filaments is mainly due to the basic two-dimensional flow. It is only by a longitudinal straining of the initial absolute vortex filament that longitudinal relative vorticity (corresponding to three-dimensionalization) may be produced. It is therefore essential to look at the initial distribution of the absolute vorticity. We assume $\omega_{2D} < 0$. The cases $\Omega < 0$ and $\Omega > 0$ correspond respectively to cyclonic and anticyclonic rotation. The following situations may be envisaged (see Figure III-12):

---

[25] of sign opposite to ω_{2D}

a) cyclonic rotation ($\Omega < 0$): the absolute vortex filament corresponding to the relative vorticity distribution $\omega_{2D}\vec{z}+\vec{\omega}^{(1)}$ will be closer to the z axis than the corresponding relative vortex filament. Therefore, the straining (if any) of the absolute vortex will be delayed, and a cyclonic rotation will have a stabilizing effect (compared with the non-rotating case).

b) anticyclonic rotation ($\Omega > 0$): when Ω increases by positive values starting from 0, the initial absolute vortex filament is more perturbed three-dimensionally than the relative one. Hence, rotation is destabilizing with respect to the non-rotating case. An intense three-dimensionalization of the layer occurs when $\omega_{2D} + 2\Omega \approx 0$, that is when the local Rossby number is of the order of one: indeed, the initial absolute vortex filament corresponds approximately to the vorticity distribution associated with $\vec{\omega}^{(1)}$, and is highly three-dimensional if $\vec{\omega}^{(1)}$ corresponds to a three-dimensionally turbulent perturbation. Hence, it is going to be immediately stretched by the basic flow, and intense longitudinal vorticity will be produced everywhere throughout the layer, as a highly distorted filament of dye would be mixed within the shear layer. Then, it is feasible that the longitudinal vorticity thus created by mixing of the absolute vorticity will strongly affect the basic flow, which will very quickly lose its two-dimensionality. This explosive mechanism is certainly extremely efficient to disrupt the coherent vortices.

When $\omega_{2D} + 2\Omega > |\vec{\omega}^{(1)}|$, the initial absolute vortex filament will see its orientation reversed with respect to the relative one. For

$$\omega_{2D} + 2\Omega > -\omega_{2D} \quad , \qquad (III-5-8)$$

that is $R_o < 0.5$, the absolute vortex is again less three-dimensional than its relative counterpart, and the anticyclonic rotation becomes stabilizing again. In fact, these criteria concern local Rossby numbers based on the vorticity of the vortex filament considered. This vortex filament results generally from the development of an instability of the basic flow, and one has to express the *local Rossby numbers* in terms of *global Rossby numbers*. In a mixing layer developing from a hyperbolic-tangent velocity profile of width δ_i for instance, the initially spanwise vortex filaments located in the stagnation region and strained longitudinally by the flow have a vorticity much smaller than the maximum spanwise vorticity $2U/\delta_i$. Hence the critical Rossby numbers based on this vorticity, $2U/2\Omega\delta_i$, should be higher than the values proposed above: in three-dimensional direct-numerical simulations of rotating mixing layers carried out by Yanase et al. (1990), it was confirmed that a cyclonic rotation is stabilizing with respect to the non-rotating case, while an anticyclonic rotation was destabilizing for $U/\Omega\delta_i >\approx 2$, and stabilizing otherwise. In this calculation, explosive growth of three-dimensionality is obtained for $U/\Omega\delta_i \approx 4$. This is qualitatively in agreement with the above phenomenology. More details on these calculations will be given in Chapter XIII.

These predictions are also in qualitative agreement with laboratory experiments on shear flows submitted to a constant rotation, which show stabilizing or destabilizing effects of the rotation according to the cyclonic or anticyclonic character of the eddies considered: it was shown for instance by Chabert d'Hières et al. (1988) that the wake of a cylinder is two-dimensionalized at low Rossby numbers, whereas cyclonic eddies are reinforced and anticyclonic eddies distroyed for Rossby numbers (based on the diameter of the cylinder) of the order of 1. Boundary-layer experiments (see the review by Tritton and Davies, 1985) do show in this case the same results (two-dimensionalization or disruption) according to the cyclonic or anticyclonic character of the shear layer. This may explain why, in the ocean, cyclonic eddies shed behind capes or islands are more commonly observed than anticyclonic ones. In the earth's atmosphere, the large-scale synoptic perturbations correspond to Rossby numbers of the order of 0.3, and hence rotation is stabilizing whatever their sign. However, at smaller scales, cyclonic perturbations are expected to be more stable than the anticyclonic ones.

6 - The Froude Number

There is an analogy between the case of a rotating fluid and a stably-stratified fluid. The characteristic frequency of the internal-gravity waves due to stratification is the Brunt-Vaisala frequency. The Froude number measures the relative importance of inertial effects (frequency U/D) over stratification effects (frequency N):

$$F = \frac{U}{ND} \qquad\qquad (III - 6 - 1)$$

thus for a low Froude number, stratification effects become preponderant. Their action is to reorganize the vertical velocities into internal gravity waves. But, as in the case of rotating fluids, stratification has no effect on quasi-two-dimensional motions which develop horizontally. One may then expect important horizontal non-linear transfers, at high Reynolds numbers. It might be possible that two-dimensional turbulent solutions exist for that problem . Actually, expansions with respect to F when $F \to 0$ have shown the possibility of such solutions (see Riley et al., 1981). However, the direct-numerical simulations of Métais and Herring (1989), and Herring and Métais (1989), show that the flow becomes quasi-two-dimensional in the sense that it develops horizontal layers more or less decorrelated vertically: the existence of these layers is responsible for important vertical shears between the horizontal velocities, causing a strong dissipation which might inhibit the inverse energy cascade within the horizontal layers.

To illustrate this point, let us consider the experiment where a grid is pulled with a velocity U through a channel initially containing a stably-stratified fluid with a constant Brunt-Vaisala frequency, or equivalently when the stratified fluid flows through a fixed grid with the velocity U (Stillinger et al., 1983). The phenomenological theory of what happens first has been given by Riley et al. (1981). The grid produces in its neighbourhood a three-dimensional turbulence of characteristic velocity and scale U_1 and D_1, and one can assume that initially this turbulence is negligibly affected by stratification. This corresponds to an initial Froude number $F_1 = U_1/ND_1$ that is large with respect to 1. Further from the grid, the turbulent velocity will decay with time t, due to viscous dissipation. The characteristic scale D will increase. It will be shown in Chapter VI that the ratio U/D decays like t^{-1} for isotropic three-dimensional turbulence. Therefore the instantaneous Froude number U/ND will be proportional to $(Nt)^{-1}$ and will reach values of order unity in a time of the order of N^{-1}. Stratification effects will then appear, with the propagation of internal gravity waves. To these waves will be superposed a quasi-two-dimensional turbulence.

Let us remark finally that, if the Brunt-Vaisala frequency is constant, the Froude number is equal to $R_i^{-1/2}$, where R_i, the Richardson number, is equal to

$$R_i = -g(\Delta\bar{\rho}_*) \frac{H}{U^2} \qquad (III-6-2)$$

In (III-6-2), $\Delta\bar{\rho}_*$ is, as in eq.(III-4-2), the variation of $\bar{\rho}/\rho_0$ (for a liquid) or $-\bar{\Theta}/\Theta_0$ (for a gas) on the height H. A negative value of the Richardson number will correspond to thermal convection, provided the Rayleigh number is high enough. For positive high values (with respect to one) of the Richardson number, the stratification is preponderant, since the Froude number is low. For positive small values of the Richardson number, inertial forces dominate. The transition from weakly to highly stratified situations can be illustrated in the problem of the stably-stratified mixing layer between two flows of different velocities U_1 and U_2 and different densities ρ_1 and ρ_2 : it may be shown, using the linear-stability analysis, that there is a critical value of the Richardson number, equal to 1/4, above which the Kelvin -Helmholtz instability cannot develop (see Drazin and Reid, 1981, for a complete discussion on this point).

7 - Turbulence, order and chaos

This discussion, in fact, belongs more to Chapter I than the present chapter, but since it requires some of the instability results which have just been presented, we have preferred to postpone it to the present chapter. This section will contain some historical and philosophical developments about turbulence, which are not really needed for an understanding of the rest of the book, but which could, nevertheless, be of some interest to the reader.

Actually, the concepts of "order", "disorder" and "chaos" are ill-defined when applied to fluid turbulence: in statistical thermodynamics for instance, "disorder" can be associated to the entropy of the system, and it is generally believed that the second principle of thermodynamics (that is the tendency for an isolated system to increase its entropy) implies a maximization of the disorder, and hence an evolution of the system from order to disorder. Valid or not, this last statement at any rate has proven useless for fluid turbulence, where no analogous entropy function has been defined[26]. As for the word "coherence", it is, as already stressed, generally used for structures having some kind of spatial organization, such as the mixing-layer large eddies, the boundary layer "hairpin vortices", or the "dissipative structures" of the internal intermittency envisaged in II-6 for instance. Some people are puzzled by the existence of such structures in turbulent flows, and tend either to reject their existence or to consider them separately from the rest of the flow, denying then the appellation of turbulence to them. Actually, it seems more reasonable to consider these structures, when they exist, as part of the turbulence itself: we have already seen that such structures are generally unpredictable, though they have a spatial coherence. It is erroneous to associate the concept of unpredictability to a spatial disorganization where no well-defined spatial structures would appear: in actual fact, when looking at a particular realization of a turbulent flow, one sometimes sees mainly a set of "coherent" structures, which nevertheless are unpredictable in phase (that is position in space), but which may conserve their geometrical shape for times much longer than the characteristic time of loss of predictability. Even in such situations a statistical analysis of turbulence using the statistical tools presented below can be performed, as will be seen in the following chapters.

With that in mind, it is not difficult to understand the points of view

---

[26] It has to be stressed however that in the case of the truncated Euler equations (Euler equations where only a finite number of modes has been retained), statistical thermodynamics apply, and the entropy of the system can be defined (see Chapter X and also Carnevale, 1982).

which associate "turbulence" to "order", if we interpret this latter word as meaning the existence of spatially organized "coherent" structures. This interpretation of turbulence was for instance contained in the Latin poet Lucretius' ideas, which were very aptly commented upon by Serres (1977): Lucretius interpreted the universe as a "turbulent order" which had emerged from an initial "Brownian-like chaos" through the development of what he called the "declination"[27], and which is exactly the infinitesimal perturbation in the instability theory. This initial "chaos" was assimilated to what we call a "laminar state", so that the usual scheme

"laminar yields turbulence"

was thus transformed into the provocative statement

$$\text{"order (i.e. turbulence) emerges from chaos"} \qquad (III - 7 - 1)$$

which could also be used to explain the appearance of life, the formation of the universe, and even the evolution of human societies.

Actually, there is not such a great gap between Lucretius' statement (III-7-1) and the general ideas on transition to turbulence which have been presented in section 1: Lucretius' philosophy contains both the idea of the development of a perturbation due to an instability, and perhaps also the idea of intermittency, where an initial random state distributed homogeneously in space would evolve towards spatially organized structures like the dissipative structures of turbulence. The appearance of dissipative structures as the result of the development of an instability was also emphasized by Prigogine (1980).

In the reality of fluid dynamics, it seems nevertheless difficult to accept blindly statements like (III-7-1) in order to explain turbulence. But perhaps this concept of order emerging from chaos can be adapted in the following way: let us start with the example of the mixing layer, where we superpose upon an inflectional velocity profile a small white-noise random perturbation, which possesses energy on all wave lengths and in particular in the unstable modes (see Plate 7): these latter modes will then grow, and the most unstable mode (that is with the highest amplification rate) will appear the first, corresponding to the coherent structures which are initially observed on Plate 7. In that sense, one can say that an "ordered structure" (the coherent eddy) has emerged from the chaos represented by the random perturbation. But the entire process was completely dependent on the existence of the linearly-unstable basic inflectional velocity profile, which in particular imposes the vorticity sign of all the large eddies which will successively appear. The same kind of analysis can be made for wakes or jets. There is also

---

[27] From the latin "clinamen".

some experimental and theoretical evidence (see Chapter XIII) that the same behaviour would be observed if, instead of the infinitesimal random perturbation, a finite amplitude three-dimensional turbulence (the "chaos") was superposed upon the basic inflectional velocity profile. In fact, (III-7-1) could be more correctly restated as

$$coherent\ structures\ emerge\ from\ chaos,$$
$$under\ the\ action\ of\ an\ external\ constraint \quad (III-7-2)$$

In the preceding examples, the external constraint was the instability of the inflectional basic velocity profile. In stratified turbulence, the constraint is the buoyancy which creates convective structures in the unstable case, and could tend to create two-dimensional turbulence in the stable case. In the experiment of rotating turbulence of Hopfinger et al. (1982) presented in Plate 4, the coherent structures are the high-vorticity eddies whose axes are parallel to the axis of rotation, the "chaos" consists of the three-dimensional turbulence created at the bottom of the tank, and the external constraint might be an instability similar to the thermal-convective instability in a rotating fluid heated from below, as proposed by Mory and Capéran (1987).

Chapter IV

THE FOURIER SPACE

When the turbulence is homogeneous, i.e. statistically invariant under translations, it is extremely useful to work in the Fourier space. In this chapter various Fourier representations of a statistically homogeneous turbulent flow will be presented, as well as the Navier-Stokes equations or the Boussinesq approximation projected in that space.

1 - Fourier representation of a flow

1.1 Flow "within a box":

The simplest mathematical way of introducing the Fourier representation of a homogeneous turbulent flow is to consider a fictitious ideal flow – the flow within a box –, defined in the following way: given a particular turbulent flow (which may not even be homogeneous), we consider in the physical space a cubic box of size L, chosen in such a way that it contains all the spatial features of the flow one wants to study. It is also assumed that the boundary conditions on the sides of the box are cyclic: this of course may pose some problems and will make this ideal flow differ from the actual flow. Once the cyclic flow within the box is constructed, we fill the whole space with an infinite number of identical boxes, so that one obtains a periodic flow of period L in the three directions of space. Thus this "flow within a box" is a periodic flow filling the whole space and whose features for scales smaller than L are close to the features of the real flow. Let $\vec{u}(\vec{x}, t)$ be the periodic velocity field. Since it is periodic of period L, it can be expanded as an infinite series

$$\vec{u}(\vec{x}, t) = (\frac{2\pi}{L})^3 \sum_{n_1, n_2, n_3 = -\infty}^{+\infty}$$

$$[\exp i\frac{2\pi}{L}(n_1 x_1 + n_2 x_2 + n_3 x_3)] \ \hat{\underline{u}}_B(n_1, n_2, n_3, t) \ , (IV - 1 - 1)$$

where $n_1 n_2 n_3$ are positive or negative integers. The coefficient in front of the r.h.s. of (IV-1-1) has been chosen for reasons of normalization. Introducing the wave-vector $\vec{k}$ of components

$$\vec{k} = [\frac{2\pi}{L}n_1, \frac{2\pi}{L}n_2, \frac{2\pi}{L}n_3] \quad ,$$

(IV-1-1) can be written as

$$\vec{u}(\vec{x},t) = (\delta k)^3 \sum_{n_1,n_2,n_3=-\infty}^{+\infty} [\exp i\vec{k}.\vec{x}] \ \hat{u}_B(\vec{k},t) \quad , \qquad (IV-1-2)$$

with $\delta k = 2\pi/L$.

$\hat{u}_B(\vec{k},t)$, the Fourier transform of the periodic velocity $\vec{u}(\vec{x},t)$, is only defined for wave-vectors whose components are multiples of the elementary wave-number δk. The next section will give an expression that allows one to determine $\hat{u}_B(\vec{k},t)$ in terms of $\vec{u}(\vec{x},t)$.

1.2 Integral Fourier representation

Let us now consider a flow $\vec{u}(\vec{x},t)$ defined in the whole physical space R^3 and not necessarily periodic. The integral Fourier transform of $\vec{u}(\vec{x},t)$ is defined as

$$\hat{\underline{u}}(\vec{k},t) = (\frac{1}{2\pi})^3 \int [\exp -i\vec{k}.\vec{x}] \ \vec{u}(\vec{x},t)d\vec{x} \quad , \qquad (IV-1-3)$$

with $d\vec{x} = dx_1 dx_2 dx_3 = d^3 x$. Generally, in homogeneous turbulence, $\vec{u}(\vec{x},t)$ does not decrease rapidly to infinity, and $\hat{\underline{u}}(\vec{k},t)$ has to be defined by referral to the theory of distributions[1] (see Schwartz, 1967). We use the inverse Fourier transform relation

$$\vec{u}(\vec{x},t) = \int [\exp i\vec{k}.\vec{x}] \ \hat{\underline{u}}(\vec{k},t)d\vec{k} \qquad (IV-1-4)$$

which gives in particular for the three-dimensional Dirac function $\delta(\vec{k})$, such that $\int d\vec{k} = 1$:

$$\delta(\vec{k}) = (\frac{1}{2\pi})^3 \int \exp -i\vec{k}.\vec{x} \ d\vec{x} \qquad (IV-1-5)$$

So a flow within a box possesses two different Fourier transforms, the integral one and the discrete one, and it may be interesting to determine a relation between these Fourier transforms. The following calculation is

---

[1] called also generalized functions

not essential for the understanding of the present chapter, but has been given here as an exercise allowing the reader to become accustomed to these notions: using (IV-1-2) and (IV-1-4), we obtain:

$$\underline{\hat{u}}(\vec{k},t) = (\frac{1}{2\pi})^3 \int \exp{-i\vec{k}.\vec{x}} \; (\frac{2\pi}{L})^3 \sum_{\vec{k}'} \exp{i\vec{k}'.\vec{x}} \; \underline{\hat{u}}_B(\vec{k}',t)d\vec{x} \; .$$

$$(IV-1-6)$$

In (IV-1-6), $\vec{k}$ is not necessarily of components multiple of δk ; $\vec{k}'$, on the contrary, must satisfy this condition. Then eq. (IV-1-6) is written, using (IV-1-5)

$$\underline{\hat{u}}(\vec{k},t) = (\frac{2\pi}{L})^3 \sum_{\vec{k}'} \delta(\vec{k}-\vec{k}') \; \underline{\hat{u}}_B(\vec{k}',t) \quad , \qquad (IV-1-7)$$

and $\underline{\hat{u}}(\vec{k},t)$ is a three-dimensional Dirac comb of "intensities"

$$(2\pi/L)^3 \; \underline{\hat{u}}_B(\vec{k}',t) \quad .$$

Let us consider now, in the Fourier space, an average of $\underline{\hat{u}}(\vec{k},t)$ on a cubic box B_i of center $\vec{k}_i$ (where $\vec{k}_i$ is one of the discrete wave-vectors for which $\underline{\hat{u}}_B$ is defined) and of sides $2\pi/L$:

$$\underline{\bar{\hat{u}}}(\vec{k}_i) = \frac{1}{\mathrm{Vol}(B_i)} \int_{B_i} \underline{\hat{u}}(\vec{k})d\vec{k} \quad . \qquad (IV-1-8)$$

We have

$$\underline{\bar{\hat{u}}}(\vec{k}_i) = \frac{1}{\mathrm{Vol}\,(B_i)} \int_{B_i} (\frac{2\pi}{L})^3 \sum_{\vec{k}'} \delta(\vec{k}-\vec{k}') \; \underline{\hat{u}}_B(\vec{k}') \; d\vec{k} \quad , \qquad (IV-1-9)$$

where $\mathrm{Vol}(B_i) = (2\pi/L)^3$ is the volume of box B_i. In (IV-1-9), $\vec{k}'$ must belong to B_i, otherwise $\delta(\vec{k}-\vec{k}')$ is always zero. Thus, using the identity

$$\int \delta(\vec{k}-\vec{k}_i) \; d\vec{k} = 1 \quad ,$$

we finally obtain

$$\underline{\bar{\hat{u}}}(\vec{k}_i,t) = \underline{\hat{u}}_B(\vec{k}_i,t) \quad . \qquad (IV-1-10)$$

Eq. (IV-1-10) shows that the discrete Fourier transform of a flow within a box is the integral Fourier transform of the flow averaged on the cubic box B_i, that is:

$$\underline{\hat{u}}_B(\vec{k}_i,t) = \frac{1}{\mathrm{Vol}(B_i)} \int_{B_i} (\frac{1}{2\pi})^3 d\vec{k} \int [\exp{-i\vec{k}.\vec{x}}] \; \vec{u}(\vec{x},t)d\vec{x} \quad .$$

$$(IV-1-11)$$

Therefore, Eq. (IV-1-11) enables us to calculate $\hat{\vec{u}}_B(\vec{k},t)$ in terms of $\vec{u}(\vec{x},t)$. In this book, we will mainly use the integral Fourier representation of the flow, but all the derivations could be given using the discrete representation. The latter is used in numerical spectral or pseudo-spectral methods. Notice also that, in two dimensions, the factors $(2\pi/L)^3$ and $(1/2\pi)^3$ have to be replaced by $(2\pi/L)^2$ and $(1/2\pi)^2$.

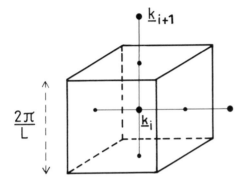

Figure IV-1: the discrete Fourier transform of a "flow within a box of size L" at the mode $\vec{k}_i$ is the average of the integral Fourier transform of the flow on an elementary cube B_i of side $2\pi/L$ surrounding $\vec{k}_i$.

2 - Navier-Stokes equations in Fourier space

This section will consider the case of a fluid of constant and uniform mean density ρ_0, without buoyancy or rotation Ω, which satisfies, from Chapter II:

$$\frac{\partial \vec{u}}{\partial t} + \vec{u}.\vec{\nabla}\vec{u} = -\frac{1}{\rho_0}\vec{\nabla}p + \nu\nabla^2\vec{u}$$

$$\frac{\partial \rho'}{\partial t} + \vec{u}.\vec{\nabla}\rho' = \kappa\nabla^2\rho' \qquad (IV-2-1)$$

$$\vec{\nabla}.\vec{u} = 0$$

ρ' is the temperature (resp. density, resp. potential temperature) fluctuation. We recall first that, for any function $f(x_1, x_2, x_3, t)$, the Fourier transform of $\partial f/\partial x_i$ is $ik_i\hat{f}(k_1, k_2, k_3, t)$, where $\hat{f}$ is the Fourier transform of f, and k_i the i-component of the wave vector $\vec{k}$. Thus, calling

"F.T" the Fourier transform operator, one has:

$$f(\vec{x},t) \quad \text{F.T} \quad \hat{f}(\vec{k},t)$$

$$\frac{\partial f}{\partial x_i} \quad \text{F.T} \quad ik_i \hat{f}(\vec{k},t)$$

$$\vec{\nabla} f \quad \text{F.T} \quad i\vec{k}\,\hat{f}(\vec{k},t)$$

$$\nabla^2 f = \frac{\partial^2 f}{\partial x_1^2} + \frac{\partial^2 f}{\partial x_2^2} + \frac{\partial^2 f}{\partial x_3^2} \quad \text{F.T} \quad -(k_1^2 + k_2^2 + k_3^2)\hat{f} = -k^2 \hat{f}$$

$$\vec{\nabla}.\vec{u} \quad \text{F.T} \quad i\,\vec{k}.\underline{\hat{u}}(\vec{k},t)$$

$$\vec{\nabla} \times \vec{u} \quad \text{F.T} \quad i\,\vec{k} \times \underline{\hat{u}}(\vec{k},t)$$

$$f(\vec{x},t)g(\vec{x},t) \quad \text{F.T} \quad [\hat{f} * \hat{g}](\vec{k},t)$$

$$(IV-2-2)$$

where $*$ is the convolution product $\int_{\vec{p}+\vec{q}=\vec{k}} \hat{f}(\vec{p},t)\hat{g}(\vec{q},t)d\vec{p}$. The incompressibility condition $\vec{\nabla}.\vec{u} = 0$ implies

$$\vec{k}.\underline{\hat{u}}(\vec{k},t) = 0 \quad , \qquad (IV-2-3)$$

and the velocity $\underline{\hat{u}}(\vec{k},t)$ is in a plane Π perpendicular to $\vec{k}$.

Now let us write Navier-Stokes equations in Fourier space: since $\underline{\hat{u}}(\vec{k},t)$ is in the plane perpendicular to $\vec{k}$, $\partial\underline{\hat{u}}(\vec{k},t)/\partial t$ and $\nu k^2 \underline{\hat{u}}$ also belong to that plane. On the contrary the pressure gradient $i\hat{p}\,\vec{k}$ is parallel to $\vec{k}$. The consequence is that the Fourier transform of

$$\vec{u}.\vec{\nabla}\vec{u} + (1/\rho_0)\vec{\nabla}p$$

is the projection on Π of the Fourier transform of $\vec{u}.\vec{\nabla}\vec{u}$. Let us introduce the tensor

$$P_{ij}(\vec{k}) = \delta_{ij} - \frac{k_i k_j}{k^2} \qquad (IV-2-4)$$

which allows a vector $\vec{a}$ to be projected on a plane perpendicular to $\vec{k}$:

$$a_j P_{ij}(\vec{k}) = i-\text{component of the projection of } \vec{a} \text{ upon } \Pi \quad (IV-2-5)$$

with the Einstein convention of summation upon repeated indices. Then, noticing that, due to incompressibility

$$\text{F.T}[u_j\frac{\partial u_i}{\partial x_j}] = \text{F.T}[\frac{\partial(u_i u_j)}{\partial x_j}] = ik_j \int_{\vec{p}+\vec{q}=\vec{k}} \hat{u}_i(\vec{p},t)\hat{u}_j(\vec{q},t)d\vec{p} \,,$$

$$(IV-2-6)$$

the i-component of $\vec{u}.\vec{\nabla}\vec{u} + (1/\rho_0)\vec{\nabla}p$ in Fourier space is equal to

$$ik_m P_{ij}(\vec{k}) \int_{\vec{p}+\vec{q}=\vec{k}} \hat{u}_j(\vec{p},t)\hat{u}_m(\vec{q},t)d\vec{p} \quad .$$

Finally the Navier-Stokes equation in Fourier space is written

$$(\frac{\partial}{\partial t} + \nu k^2)\hat{u}_i(\vec{k},t) = -ik_m P_{ij}(\vec{k}) \int_{\vec{p}+\vec{q}=\vec{k}} \hat{u}_j(\vec{p},t)\hat{u}_m(\vec{q},t)d\vec{p} \quad .$$

$$(IV-2-7)$$

The pressure has thus been eliminated. The evolution equation for the density fluctuation $\hat{\rho}'(\vec{k},t)$ is straightforward:

$$(\frac{\partial}{\partial t} + \kappa k^2)\hat{\rho}'(\vec{k},t) = -ik_j \int_{\vec{p}+\vec{q}=\vec{k}} \hat{u}_j(\vec{p},t)\hat{\rho}'(\vec{q},t)d\vec{p} \quad . \qquad (IV-2-8)$$

One can already notice that the non linear interactions involve triad interactions between wave vectors such that $\vec{k} = \vec{p} + \vec{q}$. A structure of wave length $2\pi/k$ will also often be associated with a wave number k.

In the pseudo-spectral methods used for direct-numerical simulations of turbulence (Orszag and Patterson, 1972), the Navier-Stokes equation for an incompressible non-rotating fluid in Fourier space is, from eq. (III-3-11), written as:

$$\frac{\partial}{dt}\hat{\vec{u}}(\vec{k},t) = \Pi(\vec{k}) \circ F[F^{-1}(\hat{\underline{u}}) \times F^{-1}(\hat{\underline{\omega}})] - \nu\, k^2\hat{\underline{u}}(\vec{k},t) \quad , \quad (IV-2-9)$$

where $\Pi(\vec{k})$ is the projector on the plane perpendicular to $\vec{k}$. F stands for a Fast Fourier transform operator. This procedure is much faster than a direct evaluation in Fourier space of the generalized convolution product in the r.h.s. of (IV-2-8). With these notations, the passive scalar equation writes

$$\frac{\partial}{dt}\hat{\rho}'(\vec{k},t) = -i\vec{k}.F[F^{-1}(\hat{\rho}')F^{-1}(\hat{\underline{u}})] - \kappa\, k^2\hat{\rho}'(\vec{k},t) \quad . \qquad (IV-2-10)$$

3 - Boussinesq approximation in the Fourier space

One can also write in Fourier space the Boussinesq approximation in a rotating frame of rotation $\vec{\Omega}$. This will be useful when studying stably-stratified turbulence in Chapter XIII. We recall the Navier-Stokes equations within the Boussinesq approximation in the physical space (assuming a constant mean density in the pressure gradient term):

$$\frac{\partial\vec{u}}{\partial t} + \vec{u}.\vec{\nabla}\vec{u} = -\vec{\nabla}\tilde{p} - \rho'_* g\vec{\beta} - 2\vec{\Omega} \times \vec{u} + \nu\nabla^2\vec{u}$$

$$\frac{\partial\rho'_*}{\partial t} + \vec{u}.\vec{\nabla}\rho'_* - \frac{N^2}{g}w = \kappa\nabla^2\rho'_* \qquad (IV-3-1)$$

$$\vec{\nabla}.\vec{u} = 0$$

where, according to the discussion of Chapter II, the vertical axis of coordinate $\vec{z}$ has been taken parallel to $\vec{g}$ and $\vec{\Omega}$, $\vec{\beta}$ being a vertical unit vector such that $\vec{g} = -g\vec{\beta}$ and $\vec{\Omega} = \Omega\vec{\beta}$. In the r.h.s. of the thermal equation, the contribution of the mean stratification has been neglected. N is the Brunt-Vaisala frequency. In Fourier space, the nonlinear (advection and pressure), gravity and Coriolis terms have now to be projected on the plane Π (since incompressibility $\vec{\nabla}.\vec{u} = 0$ still holds). One obtains:

$$(\frac{\partial}{\partial t} + \nu k^2)\, \hat{u}_i(\vec{k}, t) = -i k_m P_{ij}(\vec{k}) \int_{\vec{p}+\vec{q}=\vec{k}} \hat{u}_j(\vec{p}, t)\hat{u}_m(\vec{q}, t)d\vec{p}$$

$$- g\beta_j P_{ij}(\vec{k})\, \hat{\rho}'_*(\vec{k}, t)$$

$$- 2\Omega\, P_{ij}(\vec{k})\, \epsilon_{jab}\beta_a\, \hat{u}_b(\vec{k}, t) \qquad (IV-3-2)$$

$$(\frac{\partial}{\partial t} + \kappa k^2)\, \hat{\rho}'_*(\vec{k}, t) = -i k_j \int_{\vec{p}+\vec{q}=\vec{k}} \hat{u}_j(\vec{p}, t)\hat{\rho}'_*(\vec{q}, t)d\vec{p}$$

$$+ \frac{N^2}{g}\, \hat{u}_3(\vec{k}, t) \qquad (IV-3-3)$$

where ϵ_{ijl} is the antisymmetric tensor of order 3 , not equal to zero only if i, j and l are different, equal to 1 if the permutation i, j, l is even, and to -1 if the permutation is odd.

4 - Craya decomposition

The particular property that the velocity field is orthogonal to the wave vector $\vec{k}$ allows one to find other decompositions of the velocity field in Fourier space: the most common, often called the Craya decomposition (Craya, 1958), involves associating to $\vec{k}$ an orthonormal frame constructed in the following way: let $\vec{\alpha}$ be an arbitrary (but fixed) unit vector, and let us consider the reference frame

$$Z_{\vec{k}} \equiv \vec{i}(\vec{k}) = \frac{\vec{k} \times \vec{\alpha}}{\|\vec{k} \times \vec{\alpha}\|}, \quad \vec{j}(\vec{k}) = \frac{\vec{k} \times \vec{i}}{k}, \quad \frac{\vec{k}}{k} \quad .$$

The velocity field $\underline{\hat{u}}(\vec{k})$ is characterized by two components on[2] $\vec{i}(\vec{k})$ and $\vec{j}(\vec{k})$, for which it is possible to write evolution equations. This representation has been extensively used to study homogeneous strained or

---

[2] The complex number i such that $i^2 = -1$ has evidently no relation with the unit vector $\vec{i}$.

sheared turbulence (see Cambon et al., 1985). The same representation
is also known as the triad-interaction representation (Lee, J., 1979).

Let us now project $\hat{u}(\vec{k}, t)$ upon $\vec{i}$ and $\vec{j}$:

$$\hat{u}(\vec{k}, t) = u_V(\vec{k}, t)\, \vec{i}(\vec{k}) + u_W(\vec{k}, t)\, \vec{j}(\vec{k}) \quad . \qquad (IV - 4 - 1)$$

Back to physical space, we have

$$\vec{u}(\vec{x}, t) = \vec{u}_V(\vec{x}, t) + \vec{u}_W(\vec{x}, t) \quad , \qquad (IV - 4 - 2)$$

where $\vec{u}_V(\vec{x}, t)$ and $\vec{u}_W(\vec{x}, t)$ are respectively the inverse Fourier transforms of $u_V(\vec{k}, t)\, \vec{i}(\vec{k})$ and $u_W(\vec{k}, t)\, \vec{j}(\vec{k})$. Let us choose the axis of
coordinate $\vec{z}$ oriented in the direction of $\vec{\alpha}$. $\vec{u}_V(\vec{x}, t)$ is a horizontal[3]
non-divergent[4] velocity field, which can be put under the form:

$$\vec{u}_V(\vec{x}, t) = -\vec{\alpha} \times \vec{\nabla}_H\, \psi(\vec{x}, t) \quad , \qquad (IV - 4 - 3)$$

where ψ is an unknown function of $\vec{x}$ and t, $\vec{\nabla}_H$ being the gradient
taken in the direction perpendicular to $\vec{\alpha}$. Notice also that $(\vec{\nabla} \times \vec{u}_V).\vec{\alpha}$
is generally non zero, since its Fourier transform is $ik\, u_V(\vec{k}, t)\, [j(\vec{k}).\vec{\alpha}]$.
Therefore, $\vec{u}_V(\vec{x}, t)$ is vertically rotational, of stream function[5] ψ. On
the contrary, $\vec{u}_W(\vec{x}, t)$ is vertically irrotational, since the Fourier trans-
form of $\vec{\nabla} \times \vec{u}_W(\vec{x}, t)$ is $-ik u_W(\vec{k}, t)\, i(\vec{k})$. Since $\vec{k} \times \vec{i} = k\vec{j}$, $\vec{u}_W(\vec{x}, t)$
is the curl of a vector whose Fourier transform is parallel to $\vec{i}(\vec{k})$. Hence,
it may be put under the form:

$$\vec{u}_W(\vec{x}, t) = -\vec{\nabla} \times [\vec{\alpha} \times \vec{\nabla}_H \zeta(\vec{x}, t)] \quad , \qquad (IV - 4 - 4)$$

where ζ is another unknown function of $\vec{x}$ and t. this decomposition
was proposed, in a slightly different form than (IV-4-3) and (IV-4-4), by
Riley et al. (1981). $\vec{u}_W$ can also be written as

$$\vec{u}_W(\vec{x}, t) = \vec{\nabla}_H(\phi) - (\nabla_H^2 \zeta)\vec{\alpha} \quad , \qquad (IV - 4 - 5)$$

with $\phi = \partial\zeta/\partial z$.

The indices V and W stand respectively for *vortex* and *wave*
component. This terminology makes sense in the case of homogeneous
stably-stratified turbulence: indeed, taking $\vec{\alpha}$ equal to the vertical unit
vector $\vec{\beta}$, it was shown by Riley et al. (1981) that the field $\vec{u}_W$ satisfies,
for small amplitudes and when $\vec{\Omega} = \vec{0}$, the equation of propagation of

---

[3] since perpendicular to $\vec{\alpha}$

[4] since its Fourier transform is normal to $\vec{k}$

[5] for a given z

internal-gravity waves. The field $\vec{u}_V$ may in certain conditions, and for a given z, satisfy a two-dimensional Navier-Stokes equation in the limit of low Froude numbers: this would tend to justify the ideas of collapse to two-dimensional turbulence under stratification mentioned in Chapter III, although, as already stressed, the dissipative effects due to the vertical variability could be important. Notice finally that the "vortex-wave" decomposition still holds in the case of unstratified non-rotating fluid, but that there is in this case no physical significance attached to the words "wave" and "vortex".

Such a decomposition is a-priori different from Helmholtz's decomposition of a divergent field into a non-divergent part and an irrotational part. Both decompositions coincide when the horizontal velocity field is two-dimensional (i.e. independent of z). This is in particular the case for the Barré de Saint-Venant equations for a shallow fluid layer with a free surface introduced in Chapter II: in the case of a solid-body rotation $\vec{\Omega}$, the "vortex" field is then essentially a horizontal two dimensional turbulent field which interacts with a surface inertial-gravity waves field (Farge and Sadourny, 1989).

5 - Complex helical waves decomposition

Helicity in a flow is the scalar product of the velocity and the vorticity. It will be considered physically in detail in the following sections. Here we will introduce a particular decomposition of the velocity field that will be called the complex helical wave decomposition, introduced by Lesieur (1972).

Flows we will call helical waves have been introduced by Moffatt (1970): let $\vec{k}$ be a given vector, and $\vec{i}$ and $\vec{j}$ two orthogonal unit vectors perpendicular to $\vec{k}$ such that the frame formed by $\vec{i}, \vec{j}$ and $\vec{k}$ should be direct. One considers in the physical space the velocity field

$$\vec{V}_1^+(\vec{k}, \vec{x}) = (\cos \vec{k}.\vec{x})\, \vec{j} + (\sin \vec{k}.\vec{x})\, \vec{i} \qquad (IV-5-1)$$

which is such that

$$\vec{\nabla} \times \vec{V}_1^+ = k\, \vec{V}_1^+ \quad . \qquad (IV-5-2)$$

This flow is an eigenmode of the curl operator, and a particular case of what is called a Beltrami flow. Its vectorial product with its curl is zero, thus satisfying the vorticity equation associated with the Euler equation[6] with constant density and no rotation. Its helicity $\vec{V}_1^+.(\vec{\nabla} \times \vec{V}_1^+)$ is

---

[6] in an infinite domain

positive and equal to k. In the same way, one will consider the following flows

$$\vec{V}_2^+(\vec{k}, \vec{x}) = (\sin \vec{k}.\vec{x}) \, \vec{j} - (\cos \vec{k}.\vec{x}) \, \vec{i} \qquad (IV - 5 - 3)$$

$$\vec{V}_1^-(\vec{k}, \vec{x}) = (\cos \vec{k}.\vec{x}) \, \vec{j} - (\sin \vec{k}.\vec{x}) \, \vec{i} \qquad (IV - 5 - 4)$$

$$\vec{V}_2^-(\vec{k}, \vec{x}) = (\sin \vec{k}.\vec{x}) \, \vec{j} + (\cos \vec{k}.\vec{x}) \, \vec{i} \qquad (IV - 5 - 5)$$

with

$$\vec{\nabla} \times \vec{V}_2^+ = k \, \vec{V}_2^+, \vec{\nabla} \times \vec{V}_1^- = -k \, \vec{V}_1^-, \vec{\nabla} \times \vec{V}_2^- = -k \, \vec{V}_2^-,$$

and of respective helicity $k, -k, -k$. They too are Beltrami flows and solutions of the Euler equation. We will now introduce complex helical waves

$$\vec{V}^+(\vec{k}, \vec{x}) = \vec{V}_1^+(\vec{k}, \vec{x}) + i \, \vec{V}_2^+(\vec{k}, \vec{x}) = (\vec{j} - i \, \vec{i}) \exp i\vec{k}.\vec{x} \quad (IV - 5 - 6)$$

$$\vec{V}^-(\vec{k}, \vec{x}) = \vec{V}_1^-(\vec{k}, \vec{x}) + i \, \vec{V}_2^-(\vec{k}, \vec{x}) = (\vec{j} + i \, \vec{i}) \exp i\vec{k}.\vec{x} \quad (IV - 5 - 7)$$

which are still Beltrami flows and have a respective helicity k and $-k$.

Now, we assume that $\vec{k}$ is one of the wave vectors of the Fourier decomposition of $\vec{u}(\vec{x}, t)$, and that $\vec{i}(\vec{k})$ and $\vec{j}(\vec{k})$ are given as in section 4 , associated to $\vec{k}$ with the aid of the fixed arbitrary unit vector $\vec{\alpha}$. It is easy to show that

$$\frac{1}{2}[\hat{u}(\vec{k}, t) + i \, \frac{\vec{k} \times \hat{u}(k, t)}{k}] \exp i\vec{k}.\vec{x} = u^+(\vec{k}, t) \, \vec{V}^+(\vec{k}, \vec{x}) \quad (IV - 5 - 8)$$

$$\frac{1}{2}[\hat{u}(\vec{k}, t) - i \, \frac{\vec{k} \times \hat{u}(k, t)}{k}] \exp i\vec{k}.\vec{x} = u^-(\vec{k}, t) \, \vec{V}^-(\vec{k}, \vec{x}) \quad (IV - 5 - 9)$$

where $u^+(\vec{k}, t)$ and $u^-(\vec{k}, t)$ are two complex numbers. Eqs (IV-1-4), (IV-5-8) and (IV-5-9) thus permit one to write the following decomposition of any non-divergent field $\vec{u}(\vec{x}, t)$ into positive and negative complex helical waves:

$$\vec{u}(\vec{x}, t) = \int [u^+(\vec{k}, t) \, \vec{V}^+(\vec{k}, \vec{x}) + u^-(\vec{k}, t) \, \vec{V}^-(\vec{k}, \vec{x})]d\vec{k} \quad . \quad (IV - 5 - 10)$$

The orthogonality relations existing between the $\vec{V}^+$ and $\vec{V}^-$ allow one to invert (IV-5-10), showing that the decomposition of $\vec{u}(\vec{x}, t)$ is unique once the fixed vector $\vec{\alpha}$ determining the vectors $\vec{i}(\vec{k})$ and $\vec{j}(\vec{k})$ has been chosen. One obtains

$$2 \, u^+(\vec{k}, t) = (\frac{1}{2\pi})^3 \int \vec{u}(\vec{x}, t).\vec{V}^+(\vec{k}, \vec{x})^* d\vec{x} \qquad (IV - 5 - 11)$$

$$2\,u^-(\vec{k},t) = (\frac{1}{2\pi})^3 \int \vec{u}(\vec{x},t).\vec{V}^-(\vec{k},\vec{x})^* d\vec{x}\ , \qquad (IV-5-12)$$

where the symbol "*" refers to the complex conjugate. We have therefore obtained an orthogonal decomposition of the velocity field along complex helical waves, which could prove to be useful for the study of isotropic helical or non helical turbulence. Similar decompositions have been introduced by Moses (1971) and Cambon (1983). One can also note that the coordinates of $\vec{\nabla} \times \vec{u}(\vec{x},t)$ in this "helical space" are ku^+ and $-ku^-$, while those of $\nabla^2 \vec{u}(\vec{x},t)$ are $k^2 u^+$ and $-k^2 u^-$. Notice finally that the change of $\vec{k}$ into $-\vec{k}$ in $\vec{V}^+(\vec{k},\vec{x})$, $\vec{V}^-(\vec{k},\vec{x})$, $u^+(\vec{k},t)$, $u^-(\vec{k},t)$ is equivalent to taking their complex conjugate.

Equation (IV-2-7) can also be projected in this helical space: if $u_a(\vec{k},t)$ stands for u^+ or u^- according to the value +1 or -1 of the parameter a, one obtains (Lesieur, 1972):

$$(\frac{\partial}{\partial t} + \nu k^2)u_a(\vec{k},t) = -\frac{1}{2}\int_{\vec{p}+\vec{q}=\vec{k}} Q_{abc}(\vec{k},\vec{p},\vec{q})u_b(\vec{p},t)u_c(\vec{q},t)d\vec{p}$$

$$(IV-5-13)$$

with

$$Q_{abc}(\vec{k},\vec{p},\vec{q}) = -bp\,[\vec{j}(\vec{k}) + i\,a\,\vec{i}(\vec{k}), \vec{j}(\vec{p}) - i\,b\,\vec{i}(\vec{p}),$$
$$\vec{j}(\vec{q}) - i\,c\,\vec{i}(\vec{q})] \qquad (IV-5-14)$$

where [.,.,.] holds for a mixed product of three vectors, and the "indices" b and c take the values +1 or -1, with summation upon the repeated indices.

Finally, it is easy to relate the latter complex helical decomposition to the Craya decomposition: with the aid of (IV-4-1), (IV-5-11) and (IV-5-12), it is obtained (Lesieur, 1972, Cambon, 1983):

$$2u^+(\vec{k},t) = \hat{u}(\vec{k},t).(\vec{j} + i\,\vec{i}) = u_W(\vec{k},t) + i\,u_V(\vec{k},t)$$
$$2u^-(\vec{k},t) = \hat{u}(\vec{k},t).(\vec{j} - i\,\vec{i}) = u_W(\vec{k},t) - i\,u_V(\vec{k},t)$$

$$(IV-5-15)$$

or equivalently

$$u_V(\vec{k},t) = i\,[u^-(\vec{k},t) - u^+(\vec{k},t)]$$
$$u_W(\vec{k},t) = [u^-(\vec{k},t) + u^+(\vec{k},t)]$$

$$(IV-5-16)$$

Notice finally that $\vec{i}(-\vec{k}) = -\vec{i}(\vec{k})$ and $\vec{j}(-\vec{k}) = \vec{j}(\vec{k})$. Hence, using (IV-4-1) and the relation

$$\hat{u}(-\vec{k},t) = \hat{u}(\vec{k},t)^*\ , \qquad (IV-5-17)$$

due to the fact that $\vec{u}(\vec{x},t)$ is real, it is found:

$$u_V(-\vec{k},t) = -u_V(\vec{k},t)^*$$
$$u_W(-\vec{k},t) = u_W(\vec{k},t)^*\ .$$

$$(IV-5-18)$$

Chapter V

KINEMATICS OF HOMOGENEOUS TURBULENCE

1 - Utilization of random functions

From a mathematical standpoint, the velocity field $\vec{u}(\vec{x},t)$ will be assumed to be a random function defined on a sample space (see e.g. Papoulis, 1965). One can imagine for instance that we record the longitudinal air velocity at a given location in a wind tunnel: if the experiment is repeated N times in the same conditions, one obtains N realizations of the velocity evolution, each of them corresponding to a point in the sample space. For instance Figure V-1 represents four recordings of the u' velocity fluctuations obtained in such an experiment.

In a statistical description of the flow, we consider an "ensemble average", i.e. a statistical average performed on an infinite number of independent realizations. The ensemble average operator will be noted as $< \ . \ >$: for example in the above experiment, let $u^{(i)}(\vec{x},t)$ be any component of the velocity at location $\vec{x}$ and time t measured during the experiment "i" ("i"realization). The ensemble average of the product of n of these components at n locations $\vec{x}_1, \vec{x}_2, ...\vec{x}_n$ and n times $t_1, t_2, ..., t_n$ will be given by

$$< u(\vec{x}_1,t_1)u(\vec{x}_2,t_2)....u(\vec{x}_n,t_n) >= \lim_{N\to\infty} \frac{1}{N} \sum_{i=1}^{N}$$
$$u^{(i)}(\vec{x}_1,t_1)u^{(i)}(\vec{x}_2,t_2)...u^{(i)}(\vec{x}_n,t_n) \quad (V-1-1)$$

This ensemble average operator is analogous to the one used in statistical thermodynamics. It is not to be confused with temporal or spatial averages, except in certain conditions which will be specified below.

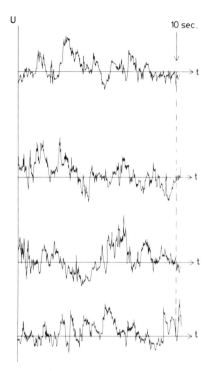

Figure V-1: four independent samples of the longitudinal velocity fluctuations recorded in the turbulent channel of the Institut de Mécanique de Grenoble (courtesy Y. Gagne)

In the following, the various functions characterizing the turbulent flow will be considered as random functions: this will in particular be the case for the velocity $\vec{u}(\vec{x}, t)$, the vorticity, the pressure, the temperature or the density fluctuations, etc. Notice also that these random functions are defined on the four-dimensional space $(\vec{x}, t)$.

2 - Moments of the velocity field, homogeneity and stationarity

Definition: The "n"th order moment of the velocity field is the ensemble average of any tensorial product of n components of the velocity field: $< \vec{u}(\vec{x}, t) >$ is the mean velocity at time t. $< u_i(\vec{x}_1, t_1) u_j(\vec{x}_2, t_2) >$ is the velocity correlation tensor at points $\vec{x}_1$ and $\vec{x}_2$ and at times t_1 and t_2. In the same way it is possible to define the moments of order $3, 4, ..., n$.

Homogeneity: the turbulence is said to be *homogeneous* if all the mean quantities built with a set of n points $\vec{x}_1, \vec{x}_2, ... \vec{x}_n$ (at times

$t_1, t_2, .., t_n)$ are invariant under any translation of the set $(\vec{x}_1, \vec{x}_2, ...\vec{x}_n$). One has in particular

$$< u_{\alpha_1}(\vec{x}_1, t_1)...u_{\alpha_n}(\vec{x}_n, t_n) >=$$
$$< u_{\alpha_1}(\vec{x}_1 + \vec{y}, t_1)...u_{\alpha_n}(\vec{x}_n + \vec{y}, t_n) > \quad (V - 2 - 1)$$

For instance the second order velocity correlation tensor is then given by

$$U_{ij}(\vec{r}, t_1, t_2) =< u_i(\vec{x}_1, t_1)u_j(\vec{x}_1 + \vec{r}, t_2) > \qquad (V - 2 - 2)$$

Notice also that

$$U_{ij}(-\vec{r}, t_1, t_2) = U_{ji}(\vec{r}, t_2, t_1) \qquad (V - 2 - 3)$$

For a homogeneous turbulence, the mean velocity field $< \vec{u}(\vec{x}, t) >$ is independent[1] of $\vec{x}$. Generally one studies such a turbulence in a frame moving with the mean flow, so that $< \vec{u} >$ is then taken equal to zero. When turbulence is homogeneous, an *ergodic hypothesis* allows one to calculate an ensemble average as a spatial average: for instance

$$U_{ij}(\vec{r}, t_1, t_2) = \lim_{V \to \infty} \frac{1}{V} \int_V u_i(\vec{x}_1, t_1)u_j(\vec{x}_1 + \vec{r}, t_2)d\vec{x}_1 \qquad (V - 2 - 4)$$

No proof of the ergodic theorem is known for the Navier-Stokes equations. There is however some numerical evidence that it is valid for the truncated Euler equations (that is where only a finite number of degrees of freedom are retained) both in two dimensions (Fox and Orszag, 1973; Basdevant and Sadourny, 1975) and in three dimensions (Lee, J., 1982).

   Stationarity: turbulence is *stationary* if all the mean quantities involving n times $t_1 t_2...t_n$ are invariant under any translation of $(t_1 t_2...t_n)$. In particular

$$< u_{\alpha_1}(\vec{x}_1, t_1)...u_{\alpha_n}(\vec{x}_n, t_n) >=$$
$$< u_{\alpha_1}(\vec{x}_1, t_1 + \tau)...u_{\alpha_n}(\vec{x}_n, t_n + \tau) > \quad (V - 2 - 5)$$

---

   [1] It must be stressed, however, that it is possible to consider a turbulence homogeneous with respect to the fluctuations of velocity with a non-constant mean velocity, provided the mean velocity gradients are constant (Craya, 1958; Townsend, 1967; Maréchal, 1972; Cambon et al., 1981; Rogallo, 1981; Rogers and Moin, 1987 a). This question will not be envisaged here, although, as stressed in the above reference, it presents analogies with wall flows: the direct-numerical simulations carried out by Rogers and Moin (1987 a) show for instance the existence of hairpin vortices, resulting from the straining due to the mean shear of perturbed spanwise vortex filaments.

If one assumes a zero mean velocity, $(1/2)U_{ii}(\vec{0},t,t)$ is the mean kinetic energy per unit mass. So in a stationary turbulence this quantity will be independent of time. This implies that a stationary turbulence needs to be sustained by external forces, otherwise the kinetic energy would decay with time, due to viscous dissipation. For a stationary turbulence, an ergodic hypothesis allows one to calculate an ensemble average as a time average. This monograph will mainly be devoted to homogeneous turbulence, but stationarity will not necessarily be assumed: indeed the latter assumption requires, as just seen above, the use of external forces which are generally only a mathematical trick to sustain the turbulence and can strongly modify its structure, specially in the scales where the energy is injected. On the contrary, the freely-evolving turbulence (also called decaying turbulence), as can be obtained for instance in a wind tunnel downstream of a grid, reorganizes according to its own dynamics, and might give more information about the nonlinear interactions between the various scales of motion.

3 - Isotropy

Definition: a homogeneous turbulence will be said to be "isotropic" if all the mean quantities concerning a set of n points $\vec{x}_1\vec{x}_2...\vec{x}_n$ (at times $t_1t_2...t_n$) are invariant under any simultaneous arbitrary rotation of the set of the n points and of the axis of coordinates. The first immediate consequences are

$$< \vec{u}(\vec{x},t) >= \vec{0} \qquad\qquad (V-3-1)$$

as can be seen in Figure V-2 : indeed, a rotation of angle π about an axis perpendicular to $x\vec{x}_i$ implies

$$< \vec{u}(\vec{x},t) >= - < \vec{u}(\vec{x},t) >= \vec{0} \quad ,$$

and for any scalar quantity $\vartheta(\vec{x},t)$:

$$< \vartheta(\vec{x},t)\vec{u}(\vec{x},t) >= 0 \quad . \qquad\qquad (V-3-2)$$

In fact, it will be shown later that the scalar-velocity correlation between two distinct points $\vec{x}$ and $\vec{y}$ is also zero. Notice also that an isotropic turbulence must be homogeneous, since a translation can be decomposed as the product of two rotations.

Longitudinal velocity correlation: let two points $\vec{x}$ and $\vec{y}$ be separated by $\vec{r}$, and $u(\vec{x},t)$ be the projection of the velocity $\vec{u}(\vec{x},t)$ on $\vec{r}$. The longitudinal correlation is defined by (Batchelor, 1953, Hinze, 1975)*

$$F(r,t,t') =< u(\vec{x},t)u(\vec{x}+\vec{r},t') > \quad . \qquad\qquad (V-3-3)$$

---

* See Figure V-3

It is independent of the direction of $\vec{r}$ because of the isotropy assumption.

Transverse velocity correlation: Let $v(\vec{x}, t)$ be the projection of the velocity on an axis $\vec{xv}$ normal to $\vec{r}$. The transverse velocity correlation is defined as

$$G(r, t, t') = < v(\vec{x}, t)v(\vec{x} + \vec{r}, t') > \quad . \qquad (V - 3 - 4)$$

Due to isotropy it is also independent of the direction of $\vec{r}$. Notice that a correlation like $< u(\vec{x}, t)v(\vec{x} + \vec{r}, t') >$ is always zero, as can be easily seen with a rotation of angle π around the axis $\vec{xu}$.

Cross velocity correlation: Let $w(\vec{x}, t)$ be the projection of the velocity on an axis $\vec{xw}$ normal to the $(\vec{r}, \vec{xv})$ plane. The cross velocity correlation is defined as:

$$H(r, t, t') = < v(\vec{x}, t)w(\vec{x} + \vec{r}, t') > \quad . \qquad (V - 3 - 5)$$

One can notice in particular that $H(0, t, t') = 0$.

Figure V-2: the mean velocity of isotropic turbulence is zero, since a rotation of angle π about an axis of origin $\vec{x}$ perpendicular to $\vec{xx_i}$ can change u_i into $-u_i$.

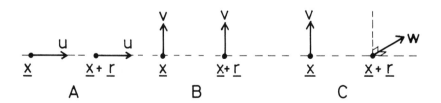

Figure V-3: schematic representation of longitudinal, transverse and cross velocity correlations.

In the present monograph, isotropic turbulence has been defined as statistically invariant under any rotation. Such a turbulence can possess or not the property of also being statistically invariant under any plane symmetry ("mirror symmetry"). In this case the cross velocity correlation $H(r, t, t')$ is zero, and there is no preferred helical tendency in the flow. In this case, one says that the turbulence possesses no helicity. The reader is warned that our definition of isotropy is different from Batchelor's (1953), which includes the mirror symmetry property in the isotropy definition.

Helicity: the quantity

$$H_e = \frac{1}{2} < \vec{u}(\vec{x}, t).[\vec{\nabla} \times \vec{u}(\vec{x}, t)] > \quad , \qquad (V-3-6)$$

is called the mean helicity of the flow. It is **evidently** zero if the turbulence possesses the mirror symmetry property. Like the mean kinetic energy and passive-scalar variance, it is conserved by the nonlinear terms of statistically homogeneous Navier-Stokes equations. The helicity seems to play an essential role in some magneto-hydrodynamic flows of electrically conducting fluids ($M.H.D.$ flows): indeed it can be shown that helicity enhances the development of magnetic fields, producing the so-called "dynamo effect". A review of this problem can be found in Moffatt (1978). It could be, for instance, that a dynamo effect in the turbulent outer earth core is responsible for the existence of the Earth's magnetic field.

At this point we recall how in the case of isotropic turbulence (with or without helicity), the second order velocity correlation tensor written in the physical space may be expressed only in terms of the quantities $F(r, t, t')$, $G(r, t, t')$, and $H(r, t, t')$. The derivation is not simple[2] and can be partially found in Batchelor (1953): let $\vec{a}$ and $\vec{b}$ be two arbitrary fixed vectors; the contracted tensorial product

$$a_i U_{ij}(\vec{r}, t, t') b_j$$

is a scalar, and consequently must be invariant under rotation of the three vectors $(\vec{r}, \vec{a}, \vec{b})$, without worrying about rotation of the axis of coordinates. It can thus be shown (Robertson, 1940) that this scalar is only a function of $\vec{r}.\vec{r}$ $\vec{r}.\vec{a}$ $\vec{r}.\vec{b}$ $\vec{a}.\vec{a}$ $\vec{b}.\vec{b}$ $\vec{a}.\vec{b}$ and $\vec{r}.(\vec{a} \times \vec{b})$, i.e. of the lengths, relative angles and orientation of this set of three vectors. Furthermore, it is linear with respect to the coordinates of $\vec{a}$ and $\vec{b}$. The only possible combination is then

$$a_i U_{ij}(\vec{r}, t, t') b_j = A(r, t, t') \vec{a}.\vec{b} + B(r, t, t')(\vec{r}.\vec{a})(\vec{r}.\vec{b})$$
$$+ C(r, t, t') \epsilon_{ijs} a_i b_j r_s \quad . \qquad (V-3-7)$$

---

[2] A simpler derivation in Fourier space will be given in Section 6.

If we choose the vectors $\vec{a}$ and $\vec{b}$ corresponding respectively to the unit vectors of axis x_i and x_j, we obtain:

$$U_{ij}(\vec{r},t,t') = A(r,t,t')\delta_{ij} + B(r,t,t')r_i r_j + C(r,t,t')\epsilon_{ijs}r_s .$$
$$(V-3-8)$$

In fact, such an expression can be obtained much more easily, using the notion of the spectral tensor in the local Craya space, which will be introduced in section 6.

One finally chooses to take a reference frame such that $\vec{r}$ has components $(r,0,0)$. This yields

$$U_{11}(r,t,t') = A(r,t,t') + r^2 B(r,t,t') = F(r,t,t')$$
$$U_{22}(r,t,t') = A(r,t,t') = G(r,t,t') \qquad\qquad (V-3-9)$$
$$U_{23}(r,t,t') = rC(r,t,t') = H(r,t,t')$$

and the final expression for $U_{ij}(\vec{r},t,t')$ is

$$U_{ij}(\vec{r},t,t') = G(r,t,t')\delta_{ij} + \frac{[F(r,t,t') - G(r,t,t')]}{r^2}r_i r_j$$
$$+ H(r,t,t')\epsilon_{ijs}\frac{r_s}{r} . \qquad\qquad (V-3-10)$$

Incompressibility condition: The incompressibility condition

$$\frac{\partial u_i}{\partial x_i} = 0 \qquad\qquad (V-3-11)$$

allows one to obtain a relationship between $F(r,t,t')$ and $G(r,t,t')$. Writing (V-3-11) under the form

$$\frac{\partial}{\partial r_i}U_{ij}(\vec{r},t,t') = 0 \qquad\qquad (V-3-12)$$

and replacing $U_{ij}(\vec{r},t,t')$ by its expression (V-3-10), it can be shown after some simple algebra (see Batchelor, 1953) that

$$G(r,t,t') = F(r,t,t') + \frac{r}{2}\frac{\partial F}{\partial r} . \qquad\qquad (V-3-13)$$

Let us now calculate the helicity of the flow, defined by (V-3-6): the following calculation is not a simple one, but enables one to become familiar with these tensorial calculations. From the definition of helicity, one has:

$$2H_e = \lim_{\vec{y}\to\vec{x}} <\vec{u}(\vec{x},t).[\vec{\nabla}\times\vec{u}(\vec{y},t)] >$$

$$= \lim_{\vec{y}\to\vec{x}} \epsilon_{ijl} < u_i(\vec{x},t)\frac{\partial u_l(\vec{y},t)}{\partial y_j} >$$

$$= \lim_{\vec{r}\to\vec{0}} \epsilon_{ijl}\frac{\partial U_{il}(\vec{r},t,t)}{\partial r_j}$$

$$= \lim_{\vec{r}\to\vec{0}} (\frac{\partial}{\partial r_j})\epsilon_{ijl}U_{il}(\vec{r},t,t)$$

and, using (V-3-10)

$$(\frac{\partial}{\partial r_j})\epsilon_{ijl}U_{il}(\vec{r},t,t) = -\epsilon_{ijl}\epsilon_{ial}\frac{\partial}{\partial r_j}[H(r,t,t)\frac{r_a}{r}]$$

$$= -\epsilon_{ijl}\epsilon_{ial}(\frac{r_a r_j}{r^2}\frac{\partial H}{\partial r} + \frac{H}{r}\delta_{ja} - \frac{H}{r^3}r_a r_j)$$

$$= -2\frac{\partial H}{\partial r} - 4\frac{H}{r}$$

When $\vec{r}$ goes to zero, $(\partial H/\partial r)$ is equivalent to (H/r), since $H(0) = 0$, and the mean helicity is equal to

$$H_e = -3\lim_{r \to 0}\frac{H(r,t,t)}{r} = -3\,C(0,t,t) \quad . \qquad (V-3-14)$$

Another interesting result is to show that the scalar-velocity correlation in two different points $\vec{x}$ and $\vec{x} + \vec{r}$ is always zero in isotropic three-dimensional turbulence (Hinze, 1975): let

$$S_i(\vec{r},t,t') = <\vartheta(\vec{x},t)u_i(\vec{x}+\vec{r},t')> \qquad (V-3-15)$$

be the correlation of any scalar ϑ (for instance the pressure or the temperature) with the velocity field. Let a_i be an arbitrary vector. The scalar a_iS_i must be invariant under any rotation. It must therefore depend upon r, a, and $\vec{a}.\vec{r}$; due to the linearity with respect to a_i, we must have

$$a_iS_i(\vec{r},t,t') = S(r,t,t')r_ia_i \qquad (V-3-16)$$

which yields

$$S_i(\vec{r},t,t') = S(r,t,t')r_i \quad . \qquad (V-3-17)$$

The incompressibility condition $\partial S_i/\partial r_i = 0$ implies

$$3S(r,t,t') + r\frac{\partial S}{\partial r} = 0 \quad . \qquad (V-3-18)$$

Thus, $S_i(\vec{r},t,t')$ is proportional to $r^{-3}r_i$. On the other hand it cannot diverge at $\vec{r} = 0$ since $S_i(\vec{0},t,t')$ is zero, as already seen above. Consequently, $S_i(\vec{r},t,t')$ must be equal to zero for arbitrary $\vec{r}$. One consequence is, for instance, that the pressure-velocity correlation is always zero in isotropic turbulence . This is also true for the temperature-velocity correlation, or the density-velocity correlation. Another consequence is that the scalar-vorticity correlation is also zero. As will be seen later, this is not automatically ensured in two-dimensional isotropic turbulence.

This zero scalar-velocity correlation property is valid only when the velocity and the scalar satisfy the isotropy conditions: indeed it might happen that an inhomogeneous scalar field is diffused by homogeneous isotropic turbulence, as for instance in a "thermal mixing layer" where two regions of a grid turbulence are initially at a different temperature (La Rue and Libby, 1981). The particular case of scalar-velocity correlation in two-dimensional isotropic turbulence will be considered later, as a special case of three-dimensional axisymmetric turbulence.

4 - The spectral tensor of an isotropic turbulence

definition: the spectral tensor of a homogeneous turbulence is the Fourier transform of the second order velocity correlation tensor

$$\hat{U}_{ij}(\vec{k},t,t') = (\frac{1}{2\pi})^3 \int [\exp -i\vec{k}.\vec{r}] \ U_{ij}(\vec{r},t,t') \ d\vec{r} \ . \qquad (V-4-1)$$

The spectral tensor can also be viewed as the velocity correlation tensor in Fourier space: indeed, let us calculate

$$< \hat{u}_i(\vec{k}',t)\hat{u}_j(\vec{k},t) >=$$

$$(\frac{1}{2\pi})^6 \int [\exp -i(\vec{k}'.\vec{x} + \vec{k}.\vec{x}')] \ \ < u_i(\vec{x},t)u_j(\vec{x}',t') > d\vec{x}d\vec{x}'$$

$$= (\frac{1}{2\pi})^6 \int [\exp -i\vec{k}.\vec{r}] \ [\exp -i(\vec{k} + \vec{k}').\vec{x}] \ U_{ij}(r,t,t')d\vec{r}$$

Then one obtains, using (IV-1-5):

$$< \hat{u}_i(\vec{k}',t)\hat{u}_j(\vec{k},t') >= (\frac{1}{2\pi})^3 \delta(\vec{k} + \vec{k}') \int [\exp -i\vec{k}.\vec{r}] \ U_{ij}(\vec{r},t,t')d\vec{r}$$

which yields

$$< \hat{u}_i(\vec{k}',t)\hat{u}_j(\vec{k},t') >= \hat{U}_{ij}(\vec{k},t,t') \ \delta(\vec{k} + \vec{k}') \ . \qquad (V-4-2)$$

We notice also from (V-4-1) and (V-2-3) that the tensor $\hat{U}_{ij}(\vec{k},t,t)$ is hermitian, that is

$$\hat{U}_{ji}(\vec{k},t,t) = \hat{U}_{ij}^*(\vec{k},t,t)$$

where the symbol * stands for the complex conjugate. Eq. (V-4-2) shows that for homogeneous turbulence there is no correlation in Fourier space between two wave vectors whose sum is different from zero. This result can be generalized to a set of N vectors, which must satisfy this zero-sum condition in order to allow a non-zero velocity correlation between them. Eq. (V-4-2) shows also that $\hat{U}_{ii}(\vec{k},t,t)$ is real and positive since it is "proportional" (modulo a Dirac distribution) to $< \hat{u}_i^*(\vec{k},t)\hat{u}_i(\vec{k},t) >$. In fact, it may be easier to consider these quantities from the point of view of the turbulence within a box introduced in (IV-1-1), since the Dirac distributions disappear, and $\hat{U}_{ii}(\vec{k},t,t)$ then becomes exactly equal to $< \hat{u}_i^*(\vec{k},t)\hat{u}_i(\vec{k},t) >$ (the symbol "B" has been omitted, and δk has been set equal to 1).

In the case of isotropic turbulence [3], the spectral tensor takes a simple form which can be obtained either by taking the Fourier transform of (V-3-8) or by working directly in Fourier space: indeed the isotropy hypothesis which leads to (V-3-8) is also valid in Fourier space, and the isotropic spectral tensor can then be written as

$$\hat{U}_{ij}(\vec{k},t,t') = \hat{A}(k,t,t')\delta_{ij} + \hat{B}(k,t,t')k_ik_j + \hat{C}(k,t,t')\epsilon_{ijl}k_l . \quad (V-4-3)$$

Therefore, the incompressibility condition in Fourier space implies

$$k_j\hat{U}_{ij}(\vec{k},t,t') = 0 \qquad\qquad (V-4-4)$$

and

$$\hat{B}(k,t,t') = -\frac{\hat{A}(k,t,t')}{k^2}$$

$$\hat{U}_{ij}(\vec{k},t,t') = \hat{A}(k,t,t')P_{ij}(\vec{k}) + \hat{C}(k,t,t')\epsilon_{ijs}\,k_s\ , \qquad (V-4-5)$$

where $P_{ij}(\vec{k})$ has been defined in (IV-2-4). A simpler derivation of this result will be given in section 6, when working in the local Craya frame.

5 - Energy, helicity, enstrophy and scalar spectra

Considering the case $t = t'$, one obtains

$$\hat{U}_{ii}(\vec{k},t,t) = \hat{U}(k,t) = 2\hat{A}(k,t) \qquad\qquad (V-5-1)$$

where $\hat{U}(k,t)$ is the trace (real and positive) of the tensor $\hat{U}_{ij}(\vec{k},t,t)$. Eq. (V-4-5) can then be written as

$$\hat{U}_{ij}(\vec{k},t,t) = \frac{1}{2}[\hat{U}(k,t)P_{ij}(\vec{k}) + i\tilde{U}(k,t)\epsilon_{ijs}k_s]\ . \qquad (V-5-2)$$

In two-dimensional isotropic turbulence, (V-5-2) has to be replaced by

$$\hat{U}_{ij}(\vec{k},t,t) = \hat{U}(k,t)P_{ij}(\vec{k})\ . \qquad\qquad (V-5-3)$$

---
[3] with or without helicity

It will be shown later that $\tilde{U}(k,t)$ is real. We first calculate the mean kinetic energy per unit mass

$$\frac{1}{2} < \vec{u}(\vec{x},t)^2 > = \frac{1}{2}U_{ii}(\vec{0},t) = \frac{1}{2}\int \hat{U}(k,t)d\vec{k}$$

$$= \int_0^{+\infty} 2\pi k^2 \hat{U}(k,t)dk \quad \text{(in three dimensions)}$$

$$= \int_0^{+\infty} \pi k \hat{U}(k,t)dk \quad \text{(in two dimensions)}$$

$$(V-5-4)$$

This determines the kinetic energy spectrum, density of kinetic energy at wave number k, and such that

$$E(k,t) = 2\pi k^2 \hat{U}(k,t) \quad \text{in three dimensions} , \qquad (V-5-5)$$

$$E(k,t) = \pi k \hat{U}(k,t) \quad \text{in two dimensions} . \qquad (V-5-6)$$

$E(k,t)$ corresponds to the kinetic energy in Fourier space integrated on a sphere (or a circle in two dimensions) of radius k. It is always real and positive.

As for the helicity, it can be shown to be equal to

$$H_e = -\frac{i}{2}\epsilon_{ijl}\int k_j \hat{U}_{il}(-\vec{k},t,t)d\vec{k}$$

which yields, using (V-5-2)

$$H_e = \int_0^{+\infty} 2\pi k^4 \tilde{U}(k,t)dk \quad . \qquad (V-5-7)$$

This determines the helicity spectrum,

$$H(k) = 2\pi k^4 \tilde{U}(k) \quad , \qquad (V-5-8)$$

density of helicity at wave number k. This quantity can only be defined in three dimensions, since it is zero in two dimensions. Finally, the spectral tensor (V-5-2) is written for $t' = t$

$$\hat{U}_{ij}(\vec{k},t,t) = \frac{1}{2}[\frac{E(k,t)}{2\pi k^2}P_{ij}(\vec{k}) + i\ \epsilon_{ijs}k_s \frac{H(k,t)}{2\pi k^4}] \qquad (V-5-9)$$

in three dimensions and

$$\hat{U}_{ij}(\vec{k},t,t) = \frac{E(k,t)}{\pi k}P_{ij}(\vec{k}) \qquad (V-5-10)$$

in two dimensions.

The enstrophy:

The enstrophy is the variance of the vorticity

$$D(t) = \frac{1}{2} <\vec{\omega}^2> \quad . \qquad\qquad (V-5-11)$$

One can show that for homogeneous turbulence the enstrophy is equal to

$$D(t) = -\frac{1}{2} < \vec{u}(\vec{x},t).\nabla^2\vec{u}(\vec{x},t) > \quad . \qquad\qquad (V-5-12)$$

Indeed, let $\vec{A}(\vec{x},t)$ and $\vec{B}(\vec{x},t)$ be two vector fields depending on the velocity field. Since, because of homogeneity,

$$< a_i\epsilon_{ijs}\frac{\partial b_s}{\partial x_j} >= -<\frac{\partial a_i}{\partial x_j}\epsilon_{ijs}b_s >$$

one has

$$< \vec{A}.\,(\vec{\nabla}\times\vec{B}) > \,=\, <\vec{B}.\,(\vec{\nabla}\times\vec{A}) > \quad . \qquad\qquad (V-5-13)$$

Now we use the property of a non divergent flow that

$$\vec{\nabla}\times(\vec{\nabla}\times\vec{u}) = -\nabla^2\vec{u} \quad , \qquad\qquad (V-5-14)$$

which shows (V-5-12). This allows calculation of the enstrophy: since the Fourier transform of $\nabla^2\vec{u}(\vec{x},t)$ is $-k^2\hat{u}(\vec{k},t)$, we have

$$D(t) = \frac{1}{2}\int d\vec{k}\,d\vec{k}'\,[\exp i(\vec{k}+\vec{k}').\vec{x}]k^2 < \hat{u}_i(\vec{k}',t)u_i(\vec{k},t) >$$

$$= \frac{1}{2}\int k^2\hat{U}_{ii}(\vec{k},t)d\vec{k} = \int_0^{+\infty} k^2 E(k,t)dk \quad .$$

$$(V-5-15)$$

It follows that the enstrophy is completely determined by the energy spectrum. This is not the case for the helicity spectrum.

The scalar spectrum

Let $\vartheta(\vec{x},t)$ be a scalar satisfying the isotropy conditions: the Fourier transform of the scalar $\hat{\vartheta}(\vec{k},t)$ can easily be shown in three dimensions to satisfy

$$< \hat{\vartheta}(\vec{k}',t)\hat{\vartheta}(\vec{k},t) >= \frac{E_\Theta(k,t)}{2\pi k^2}\,\delta(\vec{k}+\vec{k}') \qquad\qquad (V-5-16)$$

where $E_\Theta(k,t)/2\pi k^2$ is the Fourier transform of the spatial scalar correlation $< \vartheta(\vec{x},t)\vartheta(\vec{x}+\vec{r},t) > $. $E_\Theta(k,t)$ is the scalar spectrum, half the scalar variance density, and such that

$$\frac{1}{2} < \vartheta^2(\vec{x},t) >= \int_0^{+\infty} E_\Theta(k,t)dk \quad . \qquad\qquad (V-5-17)$$

Of particular interest will be the study of a passive scalar spectrum in isotropic turbulence, to which Chapter VIII will be devoted. Analogous expressions to (V-5-16) and (V-5-17) exist in two dimensions (see Chapter IX).

6 - Alternative expressions of the spectral tensor

Let us write eq. (V-4-2) as

$$< \underline{\hat{u}}(\vec{k}',t) \otimes \underline{\hat{u}}(\vec{k},t) >= \bar{\bar{U}}(\vec{k},t)\delta(\vec{k}+\vec{k}') \quad , \qquad (V-6-1)$$

where the tensor $\bar{\bar{U}}(\vec{k},t)$ is hermitian. When projected in the local Craya space associated to the wave vector $\vec{k}$ its coordinates become

$$\bar{\bar{U}}(\vec{k},t) = \begin{pmatrix} A(\vec{k},t) & C(\vec{k},t) \\ C^*(\vec{k},t) & B(\vec{k},t) \end{pmatrix} \qquad (V-6-2)$$

with

$$\begin{aligned}
< u_V(\vec{k}',t)u_V(\vec{k},t) > &= -A(\vec{k},t)\ \delta(\vec{k}+\vec{k}') \\
< u_V(\vec{k}',t)u_W(\vec{k},t) > &= -C(\vec{k},t)\ \delta(\vec{k}+\vec{k}') \\
< u_W(\vec{k}',t)u_V(\vec{k},t) > &= C^*(\vec{k},t)\ \delta(\vec{k}+\vec{k}') \\
< u_W(\vec{k}',t)u_W(\vec{k},t) > &= B(\vec{k},t)\ \delta(\vec{k}+\vec{k}')
\end{aligned} \qquad (V-6-3)$$

where u_V and u_W are defined by (IV-4-1) and where the functions $A(\vec{k},t)$ and $B(\vec{k},t)$ are real[4]. The minus signs in eq. (V-6-3) are due to the fact that the projection of $\underline{\hat{u}}(-\vec{k},t)$ on $\vec{i}(\vec{k})$ and $\vec{j}(\vec{k})$ is $-u_V(-\vec{k},t)$ and $u_W(-\vec{k},t)$. Such expressions are valid as soon as the turbulence is homogeneous, and isotropy is not required. It can then be easily shown that the mean kinetic energy and helicity are respectively equal to

$$\frac{1}{2}\int(A+B)d\vec{k} \quad \text{and} \quad \int k\ \Im(C)d\vec{k} \qquad (V-6-4)$$

where $\Im(C)$ stands for the imaginary part of $C(\vec{k},t)$: indeed, for a turbulence within a box, the coordinates of the spectral tensor in the local space are, with a proper choice of wave number units

$$\begin{pmatrix} < u_V^*(\vec{k},t)u_V(\vec{k},t) >= A(\vec{k},t) & < u_V^*(\vec{k},t)u_W(\vec{k},t) >= C(\vec{k},t) \\ < u_W^*(\vec{k},t)u_V(\vec{k},t) >= C^*(\vec{k},t) & < u_W^*(\vec{k},t)u_W(\vec{k},t) >= B(\vec{k},t) \end{pmatrix}$$

---

[4] These functions $A(\vec{k},t)$, $B(\vec{k},t)$ and $C(\vec{k},t)$ have no evident relation with other functions $\hat{A}$, $\hat{B}$, $\hat{C}$ introduced above in section 4.

and the mean kinetic energy is

$$\frac{1}{2}\sum_{\vec{k}} < \underline{\hat{u}}^*(\vec{k},t).\underline{\hat{u}}(\vec{k},t) > = \frac{1}{2}\sum_{\vec{k}} A(\vec{k},t) + B(\vec{k},t) \quad,$$

while the mean helicity is

$$\frac{1}{2}\sum_{\vec{k}} < \underline{\hat{u}}^*(\vec{k},t).[i\ \vec{k}\times\underline{\hat{u}}(\vec{k},t)] > = \frac{1}{2}\sum_{\vec{k}} -i\ \vec{k}. < \underline{\hat{u}}^*(\vec{k},t)\times\underline{\hat{u}}(\vec{k},t) >$$

$$= \sum_{\vec{k}} k\ \Im[C(\vec{k},t)] \quad.$$

The advantage of working in this local frame is that the isotropy consequences can be implemented extremely easily, since any rotation acting
simultaneously on $\vec{k}$ and $\vec{\alpha}$ will rotate also the local frame $(\vec{i},\vec{j},\vec{k}/k)$. Let
us for example envisage a rotation of angle $\pi/2$ about $\vec{k}$, which leaves
$\underline{\hat{u}}(\vec{k},t)$ unchanged, but transforms $\vec{i}$ into $\vec{j}$ and $\vec{j}$ into $-\vec{i}$. The coordinates
of the spectral tensor in the new local frame are

$$\begin{pmatrix} B(\vec{k},t) & -C^*(\vec{k},t) \\ -C(\vec{k},t) & A(\vec{k},t) \end{pmatrix}$$

and isotropy immediately implies that $A = B$ and C is imaginary.
Since any simultaneous rotation of $\vec{k}$ and $\vec{\alpha}$ leaves the coordinates of the
tensor unchanged, it follows that $A,B,$ and C depend on k and t only.
Finally, eq. (V-6-2) can be expressed under the form

$$\frac{1}{4\pi k^2}\begin{pmatrix} E(k,t) & i\ H(k,t)/k \\ -i\ H(k,t)/k & E(k,t) \end{pmatrix} \qquad (V-6-5)$$

where, because of (V-6-4), $E(k,t)$ and $H(k,t)$ are exactly the energy and
helicity spectra already defined. Such an expression has been proposed
by Cambon (1983). It allows recovery of $\hat{U}_{ij}(\vec{k},t)$ given by (V-5-9) with
the proper change of coordinates: indeed the spectral tensor can, from
(V-6-5), be written three-dimensionally in the local frame as

$$\frac{E(k,t)}{4\pi k^2}\begin{pmatrix} 1 & 0 & 0 \\ 0 & 1 & 0 \\ 0 & 0 & 0 \end{pmatrix} + i\frac{H(k,t)}{4\pi k^3}\begin{pmatrix} 0 & 1 & 0 \\ -1 & 0 & 0 \\ 0 & 0 & 0 \end{pmatrix}$$

where the first matrix, contracted with any three-dimensional vector,
projects it in the plane perpendicular to $\vec{k}$, while the second multiplies
it vectorially by $\vec{k}/k$. In the original reference frame, the projection in
the plane perpendicular to $\vec{k}$ is represented by the tensor $P_{ij}(\vec{k})$, while

the vectorial multiplication by $\vec{k}/k$ is $\epsilon_{ijs}k_s/k$. Such an expression, when transformed back into the physical space, yields a velocity correlation tensor of the form (V-3-8), which avoids the somewhat unsatisfying derivation given in section 3.

As for the spectral tensor in the complex helical wave decomposition, it is given by, using eqs. (IV-5-15), (V-6-3) and (V-6-5)

$$< u^+(\vec{k}',t)u^+(\vec{k},t) >= \frac{E^{++}(k,t)}{4\pi k^2}\delta(\vec{k}+\vec{k}') \qquad (V-6-6)$$

$$< u^+(\vec{k}',t)u^-(\vec{k},t) >=< u^-(\vec{k}',t)u^+(\vec{k},t) >= 0 \qquad (V-6-7)$$

$$< u^-(\vec{k}',t)u^-(\vec{k},t) >= \frac{E^{--}(k,t)}{4\pi k^2}\delta(\vec{k}+\vec{k}') \qquad (V-6-8)$$

with (Lesieur, 1972)

$$E^{++}(k,t) = \frac{1}{2}[E(k,t) + \frac{H(k,t)}{k}] \qquad (V-6-9)$$

$$E^{--}(k,t) = \frac{1}{2}[E(k,t) - \frac{H(k,t)}{k}] \ . \qquad (V-6-10)$$

Since E^{++} and E^{--} must be, by definition, positive, this demonstrates the inequality

$$\frac{|H(k,t)|}{k} < E(k,t) \ . \qquad (V-6-11)$$

It shows also that $H(k)$ (and hence $\tilde{U}(k)$ introduced in (V-5-2)) is real.

Finally the mean kinetic energy, helicity and enstrophy are given, using these complex helical coordinates, by

$$\frac{1}{2} < \vec{u}^2 >= \int_0^{+\infty} [E^{++}(k) + E^{--}(k)]dk \qquad (V-6-12)$$

$$\frac{1}{2} < \vec{u}.\,(\vec{\nabla} \times \vec{u}) >= \int_0^{+\infty} k[E^{++}(k) - E^{--}(k)]dk \qquad (V-6-13)$$

$$\frac{1}{2} < \vec{\omega}^2 >= \int_0^{+\infty} k^2[E^{++}(k) + E^{--}(k)]dk \ . \qquad (V-6-14)$$

The scalar-velocity correlation in three-dimensional isotropic turbulence can also be considered in spectral space: let $\hat{\vartheta}(\vec{k},t)$ be the Fourier transform of the scalar, any rotation of angle π about the vector $\vec{k}$ will change $\vec{i}$ into $-\vec{i}$ and $\vec{j}$ into $-\vec{j}$, with for consequence the nullifying of $< \hat{\vartheta}(\vec{k},t)\hat{u}_V(\vec{k},t) >$ and $< \hat{\vartheta}(\vec{k},t)\hat{u}_W(\vec{k},t) >$, and hence of $< \hat{\vartheta}(\vec{k},t)\hat{u}(\vec{k},t) >$. We recover the fact that the scalar-velocity correlation is zero, as already shown in the physical space.

7 - Axisymmetric turbulence

An important class of homogeneous non isotropic turbulent flows possesses the property of axisymmetry, that is statistical invariance under rotations about one particular axis $\vec{a}$. This type of turbulence may correspond for instance to rotating or stratified turbulence. The same kind of analysis as in the isotropic case can be performed, namely projecting the spectral tensor in the local Craya frame, in which it still has the form (V-6-2): A and B are real, but C is no more a pure imaginary. The three quantities depend now on k and $\cos \theta$, as can be easily checked by writing the invariance of the coordinates of the tensor under any rotation about $\vec{a}$ (θ is the angle between $\vec{a}$ and $\vec{k}$). The spectral tensor is thus of the form

$$\begin{pmatrix} e - Z_1 & Z_2 + i\,h/k \\ Z_2 - i\,h/k & e + Z_1 \end{pmatrix} \qquad (V-7-1)$$

already proposed in Cambon (1983), where the real functions e, Z_1, Z_2 and h depend only on k, $\cos \theta$ and t. The mean kinetic energy and helicity are, from (V-6-4), equal to $\int e\,d\vec{k}$ and $\int h\,d\vec{k}$. In the complex helical waves frame, the coordinates of the tensor are

$$\begin{pmatrix} e^{++} & Z^*/2 \\ Z/2 & e^{--} \end{pmatrix} \qquad (V-7-2)$$

where $Z = Z_1 + iZ_2$, e^{++} and e^{--} being related to e and h in the same way as E^{++} and E^{--} are related to E and H through (V-6-9) and (V-6-10).

As noticed by Cambon (1983), the statistical invariance with respect to a plane containing $\vec{a}$ implies that h is zero and Z is real ($Z_2 = 0$). This axisymmetric turbulence "without helicity" exists if initial and boundary conditions contain no helicity. Z_1 characterizes thus the degree of anisotropy of the flow. In the latter case, the spectral tensor is diagonal in the local Craya frame, and is characterized by the two scalar functions

$$\Phi_1(k, \cos\theta, t) = e - Z_1; \quad \Phi_2(k, \cos\theta, t) = e + Z_1$$

as was noticed by several authors (Chandrasekhar, 1950, Batchelor, 1953, Herring, 1974, and Capéran, 1982). We recall that Φ_1 characterizes the "vortex" spectrum, and Φ_2 the "wave" spectrum, in the terminology of Riley et al. (1981). In that case the spectral tensor can be written in this local frame as

$$\Phi_2 \begin{pmatrix} 1 & 0 & 0 \\ 0 & 1 & 0 \\ 0 & 0 & 0 \end{pmatrix} + (\Phi_1 - \Phi_2) \begin{pmatrix} 1 & 0 & 0 \\ 0 & 0 & 0 \\ 0 & 0 & 0 \end{pmatrix}$$

or equivalently

$$\hat{U}_{ij}(\vec{k},t) = \Phi_2 P_{ij}(\vec{k}) + (\Phi_1 - \Phi_2)Q_{ij}(\vec{k}) \qquad (V-7-3)$$

in the original frame. $Q_{ij}(\vec{k})$ is the tensor which projects on $\vec{i}$, the first unit vector of the local frame (the vortex mode): it is the equivalent of $P_{ij}(\vec{k})$ for isotropic two-dimensional turbulence. The relation (V-7-3) might be useful for the study of homogeneous stably-stratified turbulence.

Finally, let us write the scalar velocity correlation of axisymmetric turbulence, which we need for instance when considering the stratified turbulence problem on the basis of Boussinesq approximation. The same reasoning as above shows that, if $\hat{\vartheta}(\vec{k},t)$ is the Fourier transform of the scalar $\vartheta(\vec{x},t)$

$$< \hat{\vartheta}(\vec{k}',t)u_V(\vec{k},t) > = F_1(k,\cos\theta,t) \; \delta(\vec{k}+\vec{k}') \qquad (V-7-4)$$

$$< \hat{\vartheta}(\vec{k}',t)u_W(\vec{k},t) > = F_2(k,\cos\theta,t) \; \delta(\vec{k}+\vec{k}') \; , \qquad (V-7-5)$$

F_1 is zero if the turbulence has the mirror symmetry just mentioned above.

In the particular case of two-dimensional isotropic turbulence, where $\vec{k}$ lies in a plane perpendicular to $\vec{\alpha}$, $u_W(\vec{k},t)$ is evidently zero. A rotation of π about $\vec{\alpha}$ transforms $F_2(k,t)$ into $-F_2^*(\vec{k},t)$, and F_2 is then a pure imaginary. If this turbulence possesses the mirror symmetry property (with respect to planes containing $\vec{\alpha}$), the scalar-velocity correlation will thus become zero. This result shows why isotropic two-dimensional turbulence may differ from isotropic three-dimensional turbulence, and has not automatically zero scalar-velocity or zero scalar-vorticity correlations. As remarked in Lesieur and Herring (1985), the one-point scalar-vorticity correlation in two-dimensional turbulence, if not zero, is conserved by the non linear terms of the equations.

Chapter VI

PHENOMENOLOGICAL THEORIES

1 - Inhomogeneous turbulence

When one is interested in predicting the average quantities associated with a turbulent flow, the main difficulty comes from what is called the *closure problem*, which arises from the non-linearity of the Navier-Stokes equations: for instance, when averaging these equations, one obtains the well known Reynolds equations (previously derived by Barré de Saint-Venant and Boussinesq. These equations are analogous to the Navier-Stokes equations for the mean velocity and pressure, but with the supplementary turbulent Reynold stresses

$$\sigma_{ij}^t = -\rho <u_i'u_j'> \qquad (VI-1-1)$$

generally responsible for a loss of momentum in the mean motion. Physically, these Reynolds stresses can be understood as fictitious stresses which allow one to consider the mean motion as a real flow motion, submitted to the mean pressure gradients, the mean molecular stresses and the turbulent Reynolds stresses. But the Reynolds equation is not closed, in the sense that the Reynolds stresses are unknown.

Now, one may try to deduce the Reynolds stresses directly from the Navier-Stokes equations: one multiplies the latter by u_j, averages, and obtains an equation involving the third-order correlations. So, an evolution equation for these correlations will involve the fourth-order correlations, and so on: this is what is called the closure problem in turbulence, where a system describing the coupled evolution of the statistical moments will always contain one more unknown moment than the equations. The statistical theories of turbulence need therefore a closure hypothesis, which is a supplementary equation between the moments, generally quite arbitrary. For inhomogeneous turbulence, and when the average quantities involve only one point $\vec{x}$ of the physical space, one talks of *one-point closure modelling*.

1.1 The mixing-length theory

The simplest way of determining the Reynolds stresses for inhomogeneous turbulence is to make an eddy-viscosity assumption, that is, assume that the deviator of the corresponding tensor can be written as (suppose for simplicity in this section that ρ is uniform):

$$\sigma_{ij}^t + \frac{\rho}{3} <u_s'u_s'> \delta_{ij} = \mu_t(\frac{\partial <u_i>}{\partial x_j} + \frac{\partial <u_j>}{\partial x_i}) \,. \qquad (VI-1-2)$$

μ_t is an eddy-viscosity, given by a mixing length argument (Prandtl, 1925)

$$\frac{\mu_t}{\rho} = l_m v \quad , \qquad (VI-1-3)$$

where the mixing length l_m is a turbulent analogue of the mean free path of molecules in the kinetic theory of gases, and the velocity v characterizes the turbulent fluctuations. In turbulent free-shear flows for instance, it is assumed that l scales on the layer thickness δ, and v on the typical mean-velocity difference across the layer U.

1.2 Application of mixing-length to turbulent-shear flows

1.2.1 The plane jet

Let $<u>$, $<v>$ and $<w>$ be the components of the mean velocity $<\vec{u}(\vec{x},t)>$ respectively in the streamwise x, transverse y and spanwise z directions. The Reynolds equations, written for the first two components of the velocity, write:

$$<u> \frac{\partial}{\partial x} <u> + <v> \frac{\partial}{\partial y} <u> = -\frac{1}{\rho} \frac{\partial}{\partial x} <p>$$

$$-\frac{\partial}{\partial x} <u'^2> -\frac{\partial}{\partial y} <u'v'>$$

$$<u> \frac{\partial}{\partial x} <v> + <v> \frac{\partial}{\partial y} <v> = -\frac{1}{\rho} \frac{\partial}{\partial y} <p>$$

$$-\frac{\partial}{\partial x} <u'v'> -\frac{\partial}{\partial y} <v'^2>$$

$$(VI-1-4)$$

where the $\partial/\partial z$ terms have been discarded, for we assume spanwise homogeneity[1]. Let $u_0(x)$ be the velocity at the centre of the jet, and $\delta(x)$ a properly defined jet width: in the second equation of (VI-1-4), and due to continuity, the two terms of the l.h.s. are of same order $u_0\bar{v}/\delta$, $\bar{v}$ being a typical transverse mean velocity. Let $v_*(x)$ be a typical turbulent velocity such that $<u'^2> \sim \, | <u'v'> | \, \sim <v'^2> \sim v_*^2$. Since the average parameters of the jet vary much faster in the transverse than in the streamwise direction, the r.h.s. of the $<v>$ momentum equation reduces to $-(1/\rho)(\partial/\partial y) <p + \rho v'^2>$, where the pressure gradient dominates: indeed, the transverse pressure gradient term is of the order of u_0^2/δ, large in front of v_*^2/δ, since v_* will be assumed to be of the form $v_*(x) = a\, u_0(x)$, where $a \ll 1$ is a constant. Finally, and since $\bar{v} \ll u_0$, the $<v>$ equation reduces to

$$\frac{\partial}{\partial y} <p> = 0 \quad,$$

which shows that the mean pressure in the jet is uniform and equal to the pressure p_0 at infinity. Finally, the Reynolds equations for the plane jet yield:

$$<u> \frac{\partial}{\partial x} <u> + <v> \frac{\partial}{\partial y} <u> = -\frac{\partial}{\partial y} <u'v'> \quad, \qquad (VI-1-5)$$

or equivalently (because of the incompressibility of the mean flow)

$$\frac{\partial}{\partial x} <u>^2 + \frac{\partial}{\partial y} <u><v> = -\frac{\partial}{\partial y} <u'v'> \quad. \qquad (VI-1-6)$$

This equation, integrated on from $y = -\infty$ to $y = +\infty$ yields (assuming the Reynolds stresses are zero for y going to infinity)

$$\frac{d}{dx} \int_{-\infty}^{+\infty} <u>^2 (x)\, dy = 0 \quad, \qquad (VI-1-7)$$

which expresses the conservation downstream of the mean momentum flux across the jet[2]. Following Prandtl (1925) and Schlichting (1968), we

---

[1] We neglect also the molecular viscous terms, which are small in front of their eddy-viscous counterparts. However, this will not be valid for wall flows close to the boundary.

[2] It is well known that, due to the continuous turbulent entrainment of outer fluid by the jet, the mass is not conserved across the jet downstream. Such a conservation would correspond to the constancy with x of $\int_{-\infty}^{+\infty} <u> (x)\, dy$.

introduce an eddy-viscosity $\nu_t(x) \sim l v_* = b\, \delta(x) u_0(x)$, where $b << 1$ is a constant and l the integral scale of turbulence. Since

$$| \frac{\partial}{\partial y} <u> | >> | \frac{\partial}{\partial y} <v> | \quad ,$$

the momentum equation is written:

$$<u> \frac{\partial}{\partial x} <u> + <v> \frac{\partial}{\partial y} <u> = \nu_t(x) \frac{\partial^2}{\partial y^2} <u> \quad . \quad (VI-1-8)$$

We look for self-similar solutions of the form

$$<u> (x,y) = u_0(x)\, f(\eta)$$
$$\eta = \frac{y}{\delta(x)} \qquad \qquad (VI-1-9)$$

Notice first that this implies, from eq. (VI-1-7),

$$\frac{d}{dx} u_0^2\, \delta = 0 \quad . \qquad (VI-1-10)$$

Since the mean flow is non divergent, one may introduce a stream function which is necessarily of the form

$$<\psi> (x,y) = u_0 \delta\, F(\eta) \quad , \qquad (VI-1-11)$$

with $f(\eta) = dF/d\eta = F'$. Substituting into eq. (VI-1-8) yields finally:

$$F'^2 - \frac{(u_0\delta)'}{u_0'\delta} FF'' = b \frac{u_0}{u_0'\delta} F''' \quad , \qquad (VI-1-12)$$

(where the " ' " symbol refers to a derivative with respect to x or η) with the boundary conditions

$$<u> (x,0) = u_0(x) \, , \quad <u> (x,\infty) = 0 \, ,$$
$$\frac{\partial}{\partial y} <u> (x,0) = 0 \, , \quad \frac{\partial}{\partial y} <u> (x,\infty) = 0 \, ,$$

which write:

$$F'(0) = 1 \, , \quad F'(\infty) = 0 \, , \quad F''(0) = 0 \, , \quad F''(\infty) = 0 \quad . \quad (VI-1-13)$$

For $\eta = 0$, it is found that $u_0/u_0'\delta$ is a constant, which is negative since $u_0' < 0$. Using (VI-1-10) in order to eliminate δ, we finally obtain

$$\frac{d}{dx}(\frac{1}{u_0^2}) = \text{constant} \quad , \qquad (VI-1-14)$$

which proves that

$$u_0(x) \propto x^{-1/2}$$
$$\delta(x) \sim x \quad .$$

$$(VI-1-15)$$

This allows to write eq. (VI-1-12) as

$$F'^2 + FF'' + \nu_t^+ F''' = 0 \quad , \tag{VI-1-16}$$

where ν_t^+ is a positive number, which will be chosen equal to 1/2. This fixes the choice of δ. Integrating twice leads finally to

$$F(\frac{y}{\delta}) = \tanh \frac{y}{\delta} \quad , \tag{VI-1-17}$$

and

$$<u> (x,y) \propto x^{-1/2} \frac{1}{\cosh^2(y/\delta)} \quad , \tag{VI-1-18}$$

which is the velocity profile of the turbulent Bickley jet. Notice also that the $<v>$ component is not zero for $y \to \infty$.

1.2.2 The round jet

With the same assumptions as above, the mean pressure is uniform, and the $<u>$ equation writes:

$$<u> \frac{\partial}{\partial x} <u> + <v> \frac{\partial}{\partial y} <u> + <w> \frac{\partial}{\partial z} <u> = - \frac{\partial}{\partial y} <u'v'>$$
$$- \frac{\partial}{\partial z} <u'w'> \quad .$$

$$(VI-1-19)$$

Again, the momentum flux is an invariant:

$$\frac{d}{dx} \iint <u>^2 \, dydz = 0 \quad .$$

One assumes that the longitudinal velocity is of the form

$$<u> (x,y) = u_0(x) \, f(\eta)$$
$$\eta = \frac{r}{\delta(x)} \tag{VI-1-20}$$
$$r = \sqrt{y^2 + z^2} \quad .$$

Hence, eq. (VI-1-19) yields

$$\frac{d}{dx}(u_0 \, \delta) = 0 \quad , \tag{VI-1-21}$$

which shows that the eddy-viscosity is now a constant ν_t. Finally, the momentum equation reduces to:

$$<u> \frac{\partial}{\partial x} <u> + <v> \frac{\partial}{\partial y} <u> + <w> \frac{\partial}{\partial z} <u> = \nu_t(\frac{\partial^2}{\partial y^2} <u>$$

$$+ \frac{\partial^2}{\partial z^2} <u>) \quad .$$

$$(VI-1-22)$$

We now assume that the jet is axisymmetric, and work using cylindrical coordinates x and r. Let $u(x,r) =<u>$ and $v(x,r) =<v>$ be respectively the longitudinal and radial mean velocity. The momentum and continuity equations write:

$$u\frac{\partial u}{\partial x} + v\frac{\partial u}{\partial r} = \nu_t \frac{1}{r}\frac{\partial}{\partial r}(r\frac{\partial u}{\partial r})$$

$$\frac{\partial ru}{\partial x} + \frac{\partial rv}{\partial r} = 0 \quad .$$

$$(VI-1-23)$$

The complete calculation may be done by introducing a generalized stream function $\psi(x,r)$ such that

$$u = \frac{1}{r}\frac{\partial \psi}{\partial r} \quad , \quad v = -\frac{1}{r}\frac{\partial \psi}{\partial x} \quad .$$

Taking into account the boundary conditions, it is finally found that $u_0(x) \propto x^{-1}$, $\delta(x) \sim x$, and:

$$<u> (x,r) = u_0(x) \frac{1}{[1+(\eta^2/8)]^2}$$

$$<v> (x,r) = \frac{1}{2}u_0(x)\delta' \frac{\eta[1-(\eta^2/8)]}{[1+(\eta^2/8)]^2} \quad .$$

$$(VI-1-24)$$

1.2.3 The plane wake

In the far region of the wake, one assumes that

$$<u(x,y)> = U - \bar{u}_d(x,y) \quad ,$$

where $\bar{u}_d(x,y)$ is the deficit-velocity profile, assumed to be much smaller than the velocity at infinity U. Estimates of the orders of magnitude in the Reynolds equations allow one to show that $<p>$ is uniform. The momentum equation reduces to

$$U\frac{\partial \bar{u}_d}{\partial x} = \frac{\partial}{\partial y} <u'v'> \quad ,$$

$$(VI-1-25)$$

whose integration across the wake yields

$$\frac{d}{dx} \int_{-\infty}^{+\infty} \bar{u}_d(x,y)\, dy = 0 \quad . \qquad (VI-1-26)$$

Assuming that the deficit velocity is of the form

$$u_d(x,y) = u_0(x)\, f\left(\frac{y}{\delta}\right) \quad , \qquad (VI-1-27)$$

eq. (VI-1-26) yields

$$\frac{d}{dx}\, u_0\, \delta = 0 \quad , \qquad (VI-1-28)$$

which expresses the conservation of the mass in the wake. Again, we suppose that $\nu_t = b\, u_0 \delta$, where b is a constant number. It is obtained

$$U\frac{\partial u_d}{\partial x} = \nu_t \frac{\partial^2 u_d}{\partial y^2} \quad , \qquad (VI-1-29)$$

or equivalently

$$f - \frac{u_0\, \delta'}{u_0'\, \delta}\, \eta f' = b\, \frac{u_0^2}{U u_0' \delta}\, f'' \quad , \qquad (VI-1-30)$$

with the boundary conditions

$$f(0) = 1 \ , \ f'(0) = 0 \ , \ f(\infty) = 0 \ , \ f'(\infty) = 0 \quad . \qquad (VI-1-31)$$

For $\eta = 0$, the coefficient of f'' in the r.h.s. of (VI-1-30) must be x independent: this, together with eq. (VI-1-28), yields

$$\begin{array}{c} u_0(x) \propto x^{-1/2} \\ \delta(x) \propto x^{1/2} \end{array} \quad , \qquad (VI-1-32)$$

and eq. (VI-1-30) leads, with a proper definition of δ, to

$$f + \eta f' + f'' = 0 \quad , \qquad (VI-1-33)$$

which is also valid in the laminar-viscous case. It may easily be solved to give $f = \exp -(\eta^2/2)$, and the deficit velocity is:

$$u_d(x,y) \propto x^{-1/2}\, \exp -\frac{y^2}{2\delta^2} \quad . \qquad (VI-1-33)$$

1.2.4 The round wake

We will just summarize the results. In cylindrical coordinates, we take

$$<u> = U - u_0(x)\, f(\frac{r}{\delta}) \quad . \qquad (VI-1-34)$$

the mass conservation now implies that

$$\frac{d}{dx}\, u_0\, \delta^2 = 0 \quad . \qquad (VI-1-35)$$

The Reynolds equation is written

$$U\frac{\partial u_d}{\partial x} = \nu_t(\frac{\partial^2 u_d}{\partial r^2} + \frac{1}{r}\frac{\partial u_d}{\partial r}) \quad , \qquad (VI-1-36)$$

whose solution is:

$$u_0(x) \propto x^{-2/3}$$
$$\delta(x) \propto x^{1/3} \qquad (VI-1-37)$$
$$f = \exp -\frac{\eta^2}{2} \quad .$$

1.2.5 The plane mixing layer

The analysis is more difficult here in the spatially-growing case, since no analytic solution is found: if U_1 and U_2 are the respective velocities of the two streams ($U_1 > U_2$), with

$$\bar{U} = \frac{U_1 + U_2}{2} \quad , \quad U = \frac{U_1 - U_2}{2} \quad ,$$

one looks for self-similar solutions of the form

$$<u> (x,y) = \bar{U} + U\, F'(\eta)$$
$$\eta = \frac{y}{\delta} \quad , \qquad (VI-1-38)$$

$\delta(x)$ being a properly-defined thickness of the layer. The stream function of the mean flow is

$$<\psi> (x,y) = U\delta\, F(\eta) + \bar{U}\, y \quad ,$$

and the transverse mean velocity

$$<v> (x,y) = U\delta' (\eta F' - F) \quad .$$

The boundary conditions for F' are $F'(+\infty) = 1, F'(-\infty) = -1$. As for the above flows, the mean pressure is found to be uniform. The Reynolds equation for $<u>$ is the same as (VI-1-8), but the eddy-viscosity is now $\nu_t(x) = b\,U\delta(x)$. One obtains:

$$bF''' + \delta'F''(F + \eta\frac{\bar{U}}{U}) = 0 \quad . \tag{VI-1-39}$$

This shows that δ' is a constant, and hence $\delta(x) \sim x$. A numerical solution of this differential equation is shown in Schlichting (1968). It differs from the hyperbolic-tangent velocity profile which, however, is not far from the experimental results. In the case $U/\bar{U} << 1$ (or in the temporally-growing case) the equation reduces to

$$bF''' + \delta'\frac{\bar{U}}{U}\eta F'' = 0 \quad ,$$

and admits for solution an error function, with

$$F''(0) = \frac{2\sqrt{C}}{\sqrt{2\pi}} \quad , \quad C = \frac{\delta'\bar{U}}{bU} \quad .$$

Notice that, if the layer thickness is defined as the mean vorticity thickness:

$$\delta(x) = 2U/\frac{\partial}{\partial y} <u>(x,0) = \frac{2\delta}{F''(0)} \quad ,$$

this yields $F''(0) = 2$, and implies

$$\frac{\bar{U}}{U}\delta' = 2\pi\,b \quad .$$

Experiments where the Reynolds stresses can be measured and the eddy-viscosity determined (see e.g. Wygnanski and Fiedler, 1970), show that $b \approx 0.02$. This leads to a non-dimensional vorticity spreading rate of 0.126, to be compared with the more commonly accepted[3] experimental value of 0.17 for a *natural mixing layer* without external forcing.

It has to be stressed finally that these statistical predictions based on the mixing-length theory do in no way contradict the existence of coherent structures in the turbulent flows considered above: the coherent structures have a self-similar evolution which is part of the global self-similarity of turbulence.

---

[3] Note, however, that some controversies exist concerning the universality of this parameter.

1.2.6 The boundary layer

In the turbulent boundary layer, we assume for simplicity a statistical homogeneity in the x and z directions, and a parallel mean flow. The Reynolds equation writes now:

$$0 = \frac{1}{\rho} \frac{\partial}{\partial y} \left(\mu \frac{\partial <u>}{\partial y} - \rho <u'v'> \right) \quad , \qquad (VI-1-40)$$

and expresses the independence of the total[4] stress σ with respect to y (the distance to the boundary). The friction velocity v_*, such that $\sigma = -\rho v_*^2$, characterizes the velocity fluctuations far from the boundary. The viscous sublayer δ_v is defined as a critical layer where the viscous and turbulent stresses are of the same order, which yields

$$\mu \frac{v_*}{\delta_v} = \rho v_*^2$$

$$\delta_v = \frac{\nu}{v_*} \quad .$$

Above this layer, one assumes that

$$\nu_t(y) = A \, y \, v_* \quad , \qquad (VI-1-41)$$

A being called the Karman constant. Hence, the Reynolds equation reduces in this range to

$$\frac{d <u>}{dy} = A^{-1} \, v_* \, y^{-1} \quad ,$$

which leads to the famous logarithmic mean velocity profile:

$$\frac{<u>}{v_*} = A^{-1} \ln\left(\frac{y v_*}{\nu}\right) + B \quad . \qquad (VI-1-42)$$

This has been very well verified experimentally, since the observations of Nikuradse (1932). The values found from the experiments for the constants are $A = 0.42$ and $B = 5.84$.

2 - Triad interactions and detailed conservation

We return now to homogeneous turbulence, where the mean velocity is zero[5], and our interest lies in correlations between *different* points of

---

[4] turbulent and viscous
[5] provided one works in a frame moving with the mean velocity

the space: indeed these quantities give access to energy transfers between different scales of motion. Even in this case the closure problem arises, and the closures we will try to develop are sometimes referred to as *two-point closures*. They involve in particular the velocity correlations between two wave vectors in Fourier space, namely the spectral tensor.

The simplest of these spectral theories are obtained via phenomenological or dimensional arguments which will be presented in this chapter. Some of them are based on conservation properties of the Navier-Stokes equations, and on particular behaviour due to the triadic character of the energy exchanges. Let us first consider the exact evolution equation of the energy spectrum in terms of the triple-velocity moments:

For homogeneous isotropic turbulence, it is possible to obtain from the Navier-Stokes equations an exact equation relating the longitudinal second order velocity correlation to the third order velocity correlations which result from the "advection" term $u_j \partial u_i / \partial x_j$ (we recall that the pressure-velocity correlation, as already mentioned, is then zero). The resulting equation is known as the Karman-Howarth equation, and its derivation can be found (for instance) in Hinze (1975). An equivalent equation for the trace of the spectral tensor of homogeneous turbulence (with constant uniform density) can be found by working in Fourier-space (see Rose and Sulem, 1978): one writes the time evolution equations of $\hat{u}_i(\vec{k}',t)$ and $\hat{u}_j(\vec{k},t)$, multiplies respectively by $\hat{u}_j(\vec{k},t)$ and $\hat{u}_i(\vec{k}',t)$, adds the two equations, and obtains the time derivative of $< \hat{u}_i(\vec{k}',t)\hat{u}_i(\vec{k},t) >$. When integrating on $\vec{k}'$ and taking the trace of the tensor, one finally obtains after some algebra

$$(\frac{\partial}{\partial t} + 2\nu k^2)\hat{U}_{ii}(\vec{k},t) = -P_{ijm}(\vec{k})$$

$$\int \Im[< \hat{u}_i(\vec{k})\hat{u}_j(\vec{p})\hat{u}_m(\vec{q}) >]d\vec{p}d\vec{q} \quad (VI-2-1)$$

with

$$P_{ijm}(\vec{k}) = k_m P_{ij}(\vec{k}) + k_j P_{im}(\vec{k}) \quad (VI-2-2)$$

and where $\Im$ stands for the imaginary part. In the integrand of (VI-2-1), the $< uuu >$ term is proportional to a $\delta(\vec{k}+\vec{p}+\vec{q})$ Dirac function, so that only the triads $\vec{k}+\vec{p}+\vec{q} = \vec{0}$ are involved in the integration. It allows one to show a theorem of "detailed conservation" of kinetic energy: indeed, let eq. (VI-2-1) be written under a symmetrized form

$$(\frac{\partial}{\partial t} + 2\nu k^2)\hat{U}_{ii}(\vec{k},t) = \int s(\vec{k},\vec{p},\vec{q})\delta(\vec{k}+\vec{p}+\vec{q})d\vec{p}\,d\vec{q} \quad (VI-2-3)$$

with

$$s(\vec{k},\vec{p},\vec{q}) = s(\vec{k},\vec{q},\vec{p}) \quad . \quad (VI-2-4)$$

From (VI-2-1) one obtains (Rose and Sulem, 1978)

$$s(\vec{k},\vec{p},\vec{q})\delta(\vec{k}+\vec{p}+\vec{q}) =$$
$$-\Im[<(\vec{k}.\vec{\hat{u}}(\vec{q}))(\vec{\hat{u}}(\vec{k}).\vec{\hat{u}}(\vec{p}))> + <(\vec{k}.\vec{\hat{u}}(\vec{p}))(\vec{\hat{u}}(\vec{k}).\vec{\hat{u}}(\vec{q}))>]. \qquad (VI-2-5)$$

Using the incompressibility condition $\vec{k}.\hat{\underline{u}}(\hat{k}) = 0$, one finally obtains

$$s(\vec{k},\vec{p},\vec{q}) + s(\vec{p},\vec{q},\vec{k}) + s(\vec{q},\vec{k},\vec{p}) = 0 \qquad (VI-2-6)$$

for triads such that $\vec{k}+\vec{p}+\vec{q}=\vec{0}$. The interpretation of this result is that, if only three modes (such that $\vec{k}+\vec{p}+\vec{q}=0$) were interacting, non-linear exchanges of kinetic energy between these modes would conserve the energy. Also, one can derive a mean kinetic energy conservation result: by integrating (VI-2-3) on wave vector $\vec{k}$, one obtains

$$\frac{1}{2}\int(\frac{\partial}{\partial t} + 2\nu k^2)\hat{U}_{ii}(\vec{k},t)d\vec{k} = \frac{1}{2}\int s(\vec{k},\vec{p},\vec{q})\delta(\vec{k}+\vec{p}+\vec{q})d\vec{k}d\vec{p}d\vec{q}$$
$$= \frac{1}{6}\int[s(\vec{k},\vec{p},\vec{q}) + s(\vec{p},\vec{q},\vec{k}) + s(\vec{q},\vec{k},\vec{p})]\delta(\vec{k}+\vec{p}+\vec{q})d\vec{k}d\vec{p}d\vec{q}$$
$$= 0 \qquad (VI-2-7)$$

If the turbulence is isotropic, (VI-2-7) is equivalent to

$$\frac{d}{dt}\int_0^{+\infty} E(k,t)dk + 2\nu\int_0^{+\infty} k^2 E(k,t)dk = 0 \quad . \qquad (VI-2-8)$$

This result shows that the mean kinetic energy is conserved by the non-linear terms of the Navier-Stokes equations, and dissipated by molecular viscosity at a rate $2\nu\int_0^{+\infty} k^2 E(k,t)dk$. Eq (VI 2-8) may of course be shown much more easily in physical space, as will be seen below. But the detailed conservation property gives significant information about the way non-linear interactions redistribute the energy between the modes.

This theorem of detailed conservation can actually be generalized to any quadratic quantity conserved by the non-linear terms of the Navier-Stokes equations, i.e. the mean kinetic energy, the helicity (in three dimensions), the enstrophy (in two dimensions) and the passive scalar variance: Let $\hat{U}_c(\vec{k},t)$ be the density of this conserved quantity at the wave vector $\vec{k}$, satisfying the evolution equation

$$(\frac{\partial}{\partial t} + 2\kappa_c k^2)\hat{U}_c(\vec{k},t) = \int s_c(\vec{k},\vec{p},\vec{q})\delta(\vec{k}+\vec{p}+\vec{q})d\vec{p}d\vec{q} \qquad (VI-2-9)$$

with

$$s_c(\vec{k},\vec{p},\vec{q}) = s_c(\vec{k},\vec{q},\vec{p}) \qquad (VI-2-10)$$

where κ_c arising in the l.h.s. of (VI-2-9) is the molecular diffusivity of the quantity, which will be either the molecular viscosity or the conductivity. A result from Kraichnan (1975) states that $s_c(\vec{k}, \vec{p}, \vec{q})$ satisfies a detailed conservation relation analogous to (VI-2-6). This will have profound consequences for the dynamics of two-dimensional isotropic turbulence, since the enstrophy transfer density at wave vector $\vec{k}$ is $k^2 s(\vec{k}, \vec{p}, \vec{q})$, thus implying both the detailed energy and enstrophy conservation:

$$k^2 s(\vec{k}, \vec{p}, \vec{q}) + p^2 s(\vec{p}, \vec{q}, \vec{k}) + q^2 s(\vec{q}, \vec{k}, \vec{p}) = 0 \quad . \qquad (VI - 2 - 11)$$

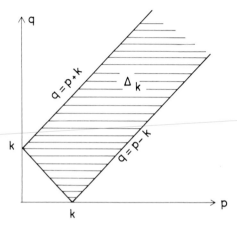

Figure VI-1: domain in the (p, q) plane such that (k, p, q) should be the sides of a triangle, and allowing triad interactions with the wave number k.

2.1 Quadratic invariants in physical space

Let us write directly in physical space the evolution equations for homogeneous turbulence of the following quadratic quantities: kinetic energy, helicity and passive scalar variance.

2.1.1 Kinetic energy

We start with an incompressible Navier-Stokes equation[6] under the form

$$\frac{\partial u_i}{\partial t} + u_j \frac{\partial u_i}{\partial x_j} = -\frac{1}{\rho}\frac{\partial p}{\partial x_i} + \nu \nabla^2 u_i \quad , \qquad (VI - 2 - 12)$$

that we multiply by u_i, for fixed i. It yields

$$\frac{1}{2}\frac{\partial}{\partial t}u_i^2 + \frac{1}{2}u_j\frac{\partial}{\partial x_j}u_i^2 = -\frac{1}{\rho}u_i\frac{\partial p}{\partial x_i} + \nu\, u_i\nabla^2 u_i \quad , \qquad (VI - 2 - 13)$$

---

[6] with constant density

or equivalently, because of incompressibility:

$$\frac{1}{2}\frac{\partial}{\partial t}u_i^2 + \frac{1}{2}\frac{\partial}{\partial x_j}u_j u_i^2 = -\frac{1}{\rho}u_i\frac{\partial p}{\partial x_i} + \nu\, u_i\nabla^2 u_i \quad .$$

After summation upon the i, one obtains

$$\frac{1}{2}\frac{\partial}{\partial t}\vec{u}^2 + \frac{1}{2}\frac{\partial}{\partial x_j}u_j\vec{u}^2 = -\frac{1}{\rho}\frac{\partial p u_i}{\partial x_i} + \nu\vec{u}\,.\,\nabla^2\vec{u} \quad .$$

Making use of the homogeneity property and averaging, one has:

$$\frac{d}{dt}\frac{1}{2}<\vec{u}^2> = \nu\,<\vec{u}\,.\,\nabla^2\vec{u}> \quad .$$

Using eq.(V-5-12), the mean kinetic energy evolution equation is:

$$\frac{d}{dt}\frac{1}{2}<\vec{u}^2> = -2\nu\,D(t) \quad , \qquad\qquad (VI-2-14)$$

which allows to recover eq. (VI-2-8) when turbulence is isotropic.

2.1.2 Helicity

We consider both eq. (VI-2-12) as well as the corresponding vorticity equation:

$$\frac{\partial \omega_i}{\partial t} + u_j\frac{\partial \omega_i}{\partial x_j} = \omega_j\frac{\partial u_i}{\partial x_j} + \nu\nabla^2\omega_i \quad , \qquad\qquad (VI-2-15)$$

multiply them respectively by ω_i and u_i, add and average. Making use of the incompressibility and homogeneity conditions, non-linear terms vanish, and the mean helicity evolution equation is (with the aid of (V-5-13) and (V-5-14):

$$\frac{dH_e}{dt} = -\nu <\vec{\omega}.(\vec{\nabla}\times\vec{\omega})>$$
$$H_e = \frac{1}{2}<\vec{u}\,.\,\vec{\omega}> \qquad\qquad (VI-2-16)$$

It is an easy exercise to calculate the helicity dissipation spectrum, using the Craya representation, and the final mean helicity equation is for isotropic turbulence:

$$\frac{d}{dt}\int_0^{+\infty} H(k,t)dk + 2\nu\int_0^{+\infty} k^2 H(k,t)dk = 0 \quad . \qquad (VI-2-17)$$

2.1.3 Passive scalar

We consider a passive scalar $\vartheta(\vec{x}, t)$ satisfying the equation

$$\frac{\partial \vartheta}{\partial t} + u_j \frac{\partial \vartheta}{\partial x_j} = \kappa \nabla^2 \vartheta \quad . \qquad (VI-2-18)$$

Multiplying by ϑ and averaging yields, for homogeneous turbulence

$$\frac{1}{2}\frac{d}{dt} <\vartheta^2> = \kappa <\vartheta \; \nabla^2 \vartheta> = -\kappa <(\vec{\nabla}\vartheta)^2> \quad , \qquad (VI-2-19)$$

which may be written for isotropic turbulence as

$$\frac{d}{dt} \int_0^{+\infty} E_\Theta(k, t)dk + 2\kappa \int_0^{+\infty} k^2 E_\Theta(k, t)dk = 0 \quad . \qquad (VI-2-20)$$

Therefore, the above three quantities are conserved by the non-linear terms of the equations. We will call them *quadratic invariants of turbulence*, although they are not invariant, since the viscous dissipation will be seen to be of prior importance in three-dimensional turbulence.

3 - Transfer and Flux

For isotropic turbulence, and after multiplication of eq. (VI-2-3) by $2\pi k^2$ (or πk in two-dimensional turbulence), one obtains

$$\left(\frac{\partial}{\partial t} + 2\nu k^2\right) E(k, t) = T(k, t) \qquad (VI-3-1)$$

where $T(k, t)$ corresponds to the triple-velocity correlations coming from non-linear interactions of the Navier-Stokes equations:

$$T(k, t) = \int_{\vec{k}+\vec{p}+\vec{q}=\vec{0}} 2\pi k^2 s(k, p, q)d\vec{p} = \iint_{\Delta_k} dp \; dq \; S(k, p, q)$$
$$(VI-3-2)$$

with

$$S(k, p, q) = 4\pi^2 kpq \; s(k, p, q) \qquad (VI-3-3)$$

$s(k, p, q)$ being defined in (VI-2-3). In (VI-3-2), the double integral is performed on a domain Δ_k of the plane (p, q) such that the positive numbers $p = |\vec{p}|$ and $q = |\vec{q}|$ should be the sides of a triangle of a third side $k = |\vec{k}|$. This domain is shown in Figure VI-1. $S(k, p, q)$ is symmetric in p and q, and possesses the detailed conservation property (VI-2-6) . When considering forced turbulence, one needs to add a forcing term $F_o(k)$ to the r.h.s. of (VI-3-1). This forcing term is a mathematical expedient which allows the supplying of the viscous loss of kinetic energy and convergence towards a stationary energy spectrum $E(k)$. It has generally no physical reality, since the forces acting on real flows induce anisotropy and inhomogeneity in the large scales. $T(k,t)$ will be called the kinetic energy transfer. The kinetic energy flux through wave number k is defined as

$$\Pi(k,t) = \int_k^{+\infty} T(k',t)\,dk' \qquad\qquad (VI-3-4)$$

which is equivalent to

$$T(k,t) = -\frac{\partial \Pi(k,t)}{\partial k} \quad . \qquad\qquad (VI-3-5)$$

The kinetic energy dissipation result (VI-2-8) implies that

$$\int_0^{+\infty} T(k,t)\,dk = 0 \quad . \qquad\qquad (VI-3-6)$$

This result can also be deduced from (VI-3-2), given that $S(k,p,q)$ satisfies the detailed conservation property and that the domain of the $[k,p,q]$ space, such that (k,p,q) are the sides of a triangle, is invariant by circular permutation of (k,p,q). Thus, (VI-3-6) is also valid for forced turbulence. One also has

$$\Pi(k,t) = -\int_0^k T(k',t)\,dk' \quad . \qquad\qquad (VI-3-7)$$

A transfer function $T(k,t)$ calculated in a direct-numerical simulation of isotropic turbulence is shown on Figure VI-2: it is negative in the large energy-containing eddies, and positive at high wave numbers, indicating a tendency for the energy to cascade from large to small scales. At higher Reynolds numbers, we will see later on that the transfer function presents a plateau at zero value, indicating a constant flux of kinetic energy.

Kraichnan (1971 a) has shown, using the symmetry and detailed conservation properties of $S(k,p,q)$, that the energy flux through a wave number k could be written as

$$\Pi(k,t) = \Pi^+(k,t) - \Pi^-(k,t) \qquad\qquad (VI-3-8)$$

$$\Pi^+(k,t) = \int_k^{\infty} dk' \int_0^k\int_0^k S(k',p,q)dpdq \qquad\qquad (VI-3-9)$$

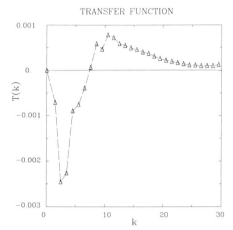

TRANSFER FUNCTION

Figure VI-2: transfer function of three-dimensional isotropic turbulence, computed in a direct numerical simulation (courtesy O. Métais, Institut de Mécanique de Grenoble).

$$\Pi^-(k,t) = \int_0^k dk' \int_k^\infty \int_k^\infty S(k',p,q)dpdq \quad . \qquad (VI-3-10)$$

This formalism will be useful when considering the notion of enstrophy cascade in two-dimensional turbulence (see Chapter IX), and the eddy-viscosity concept in spectral space for three-dimensional isotropic turbulence with a separation of scales (turbulence with a "spectral gap", Pouquet et. al., 1983): in this last case $-\Pi^-$ is the flux of kinetic energy from the "large" scales (low wave numbers) to the "small" scales (large wave numbers), and the corresponding energy transfer $\partial \Pi^-/\partial k$ can be approximated (via the stochastic models discussed in the next chapter) as $-2\nu_t k^2 E(k)$, ν_t depending on the kinetic energy in the small scales. Since the molecular viscous energy transfer is $-2\nu k^2 E(k)$, ν_t can be interpreted as an eddy-viscosity.

Then an exact result can be obtained directly on the Navier-Stokes equations without any approximation: let us assume that there exists a forcing $F_o(k)$ concentrated on a narrow spectral band in the vicinity of a wave number k_i, which will thus be characteristic of the "large energy-containing eddies". We assume also that the turbulence is stationary, so that the energy spectrum and the transfer are independent of t (see Figure VI-3). Eq. (VI-3-1) with the forcing term writes

$$2\nu k^2\ E(k) = T(k) + F_o(k) \quad . \qquad (VI-3-11)$$

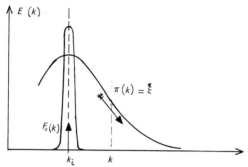

Figure VI-3: schematic stationary kinetic energy spectrum forced, within the Navier-Stokes equation in the limit of zero viscosity, by a narrow forcing spectrum $F_o(k)$ concentrated at k_i. The kinetic energy flux is, for $k > k_i$, equal to the injection rate ϵ.

For $k \neq k_i$, it yields

$$T(k) = 2\nu k^2 E(k) \qquad\qquad (VI-3-12)$$

and for k fixed

$$\lim_{\nu \to 0} T(k) = 0 \quad . \qquad\qquad (VI-3-13)$$

Now let

$$\epsilon = \int_0^{+\infty} F_o(k)dk \qquad\qquad (VI-3-14)$$

be the rate of injection of kinetic energy. Integrating eq. (VI-3-11) from 0 to ∞, and using eq. (VI-3-6), leads to

$$\epsilon = 2\nu \int_0^{+\infty} k^2 E(k)dk \quad , \qquad\qquad (VI-3-15)$$

and the injection of kinetic energy is balanced by molecular viscous dissipation, which is quite obvious intuitively. If in eq. (VI-3-15) one lets ν go to zero (ϵ being imposed by the stationary injection), the enstrophy will go to infinity. Finally integration of eq. (VI-3-11) from 0 to $k \neq k_i$ yields

$$\lim_{\nu \to 0} \Pi(k) = 0, k < k_i \qquad\qquad (VI-3-16)$$

$$\lim_{\nu \to 0} \Pi(k) = \epsilon, k > k_i \qquad\qquad (VI-3-17)$$

(for a fixed k). The conclusion of this very simple discussion is that there exists a spectral range extending beyond the injection wave number k_i where, at vanishing viscosity, the kinetic energy transfer is identically zero, the energy flux is constant and equal to the injection rate; furthermore the enstrophy of such a turbulence blows up. We anticipate that these results are the major ingredients of the Kolmogorov theory, which will be looked at in the next section.

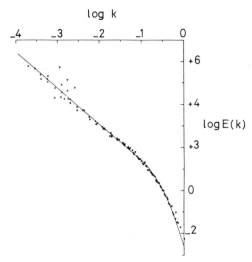

log k

log E(k)

Figure VI-4: longitudinal kinetic energy spectrum measured in the three-dimensional turbulence generated in a tidal channel in the ocean (Knight inlet, British Columbia): turbulence is decaying with time, and the spectra at various stages of the evolution are normalized by the Kolmogorov dissipative scale. They display the self-similar behaviour proposed in (VI-6-10). The spectra exhibit a three-decade Kolmogorov $k^{-5/3}$ inertial range (from Gargett et al., 1984, courtesy J. Fluid Mech.).

4 - The Kolmogorov theory

The Kolmogorov theory (Kolmogorov, 1941), certainly the most famous theory of isotropic turbulence, can be explained in several ways. We have just seen that for stationary isotropic turbulence forced at a rate ϵ in a narrow spectral range around k_i, and in the limit of an infinite Reynolds number (or equivalently zero viscosity), the energy flux $\Pi(k)$ is independent of k and equal to ϵ for $k > k_i$, ϵ being also the viscous dissipation rate. This shows that ϵ is an extremely important parameter which controls the energy flux from the large scales where it is injected to the small scales where it will be dissipated by viscosity: this scheme of progressive energy cascade from large to smaller-size eddies has been immortalized by Richardson[7] (1922) with his parody of Jonathan Swift's *fleas sonnet* [8], and this infinite hierarchy of eddies/fleas sucking the en-

---

[7] *Big whirls have little whirls, which feed on their velocity, and little whirls have lesser whirls, and so on to viscosity.*

[8] *A flea hath smaller fleas that on him prey; and these have smaller yet*

ergy/blood of the bigger ones on which they ride, while they are being sucked by smaller eddies/fleas riding on them. So Kolmogorov's theory assumes that the energy spectrum at wave numbers greater than k_i depends only on ϵ and k. A dimensional analysis, based on the Vaschy-Buckingham π-theorem, enables one to easily show that

$$E(k) = C_K \epsilon^{2/3} k^{-5/3} \qquad (VI-4-1)$$

where C_K is a universal constant called the Kolmogorov constant. When the turbulence is freely decaying, eq. (VI-4-1) can be generalized with

$$\epsilon(t) = -\frac{d}{dt} \int_0^{+\infty} E(k,t) dk \quad ,$$

the dissipation rate of kinetic energy, given by (VI-2-8) and formally identical to (VI-3-15). k_i is now a function of t, and will later on be seen to decrease with time. The Richardson-Kolmogorov cascade scheme is certainly questionable since it does not correspond physically to well identified instabilities arising in the fluid. But, and for whatever may be the reason, the law (VI-4-1) is remarkably well verified experimentally in the small scales of a flow when the Reynolds number is sufficiently high: this is, for instance the case in the ocean (Grant et al., 1962; Gargett et al., 1984), the atmosphere (Champagne et al., 1977), or for laboratory experiments such as grid turbulence (Dumas, 1962; Gagne, 1987)[9] mixing layers (Browand and Ho, 1983), or jets (Gibson, 1963; Giger et al., 1985). As an example, the spectrum measured in the ocean in a tidal channel in British Columbia (the "Knight Inlet", see Gargett et al., 1984) displays a Kolmogorov law extending to nearly three decades. This spectrum is presented in Figure VI-4. The value of the universal Kolmogorov constant C_K, found experimentally, is of the order of 1.5.

Kolmogorov's law is not, of course, valid for any scale of motion: under k_i, the spectrum will be influenced by forcing (if any) and by long-range (*non-local*) interactions which will be studied in the following chapters. Above a sufficiently high wave number k_d, called the Kolmogorov wave number, the viscous dissipation will damp the velocity perturbations. The order of magnitude of this wave number can for instance be obtained by taking a schematic energy spectrum equal to zero for $k < k_i$ and $k > k_d$ and given by (VI-4-1) in between, and by calculating ϵ with the aid of eq. (VI-3-15). One finds

$$k_d = \left(\frac{\epsilon}{\nu^3}\right)^{1/4} \qquad (VI-4-2)$$

---

to bite'em, and so proceed ad infinitum, quoted from Frisch and Orszag (1990).

[9] These two experiments were carried out in the Modane ONERA wind tunnel, where the Reynolds numbers are huge.

For $k > k_d$, the energy spectrum will rapidly (possibly exponentially) drop to negligible values. This range is called the dissipation range. The inverse of k_d is the Kolmogorov dissipative scale l_d, while the inverse of k_i is of the order of the integral scale l obtained after integration from $r = 0$ to infinity and normalization by $< u^2 >$ of the simultaneous longitudinal correlation coefficient obtained from (V-3-3). Notice finally that when the molecular viscosity goes to zero, k_d goes to infinity, the enstrophy diverges as $k_d^{4/3}$, and the rate of dissipation ϵ is finite and independent of viscosity. This is, as stressed by Orszag (1977), one of the main properties characterizing three-dimensional isotropic turbulence, namely a *finite* viscous dissipation of energy when the viscosity vanishes. As already seen above, this is physically due to the stretching of the vortex filaments by the turbulence, which dramatically increases the enstrophy, thus compensating the low molecular viscosity in the energy-dissipation rate.

4.1 Oboukhov's theory

Kolmogorov's law can also be derived, using the so-called Oboukhov theory, by introducing a constant flux of kinetic energy ϵ proportional to the "available cascading kinetic energy" in the vicinity of k, divided by a characteristic local time of the cascade $\tau(k)$, assumed to depend only on k and $E(k)$, and thus equal to

$$\tau(k) = [k^3 E(k)]^{-1/2} \quad . \tag{VI-4-3}$$

It will be seen in Chapter VII that $\tau(k)$ is a non-linear time characteristic of the relaxation of triple-velocity correlations towards a quasi-equilibrium state. The available kinetic energy in the vicinity of k can be obtained by integration of $E(k)$ on a logarithmic spectral vicinity of k, and is of the order of $kE(k)$. This again yields Kolmogorov's law (VI-4-1).

We recall that this law is valid in the so-called "inertial-range", when molecular-viscous and external-forcing effects can be neglected. An equivalent analysis can be performed in the physical space (see Rose and Sulem, 1978), by considering an eddy (the word "eddy" is not associated here with a particular structure of the flow) with typical rotational velocity v_r and radius r. The inertial time (or "turnover time") of this eddy is r/v_r . If one assumes that this eddy loses an appreciable part of its energy during a turnover time, the energy dissipation rate ϵ is proportional to $v_r^2/(r/v_r)$. We obtain

$$v_r \approx (\epsilon r)^{1/3} \quad . \tag{VI-4-4}$$

Let us associate to r a wave number $k = r^{-1}$. The kinetic energy of eddies in a spectral vicinity of k is proportional e.g. to $\int_{k/10}^{k} E(p)dp$, and therefore (as already stressed) to $kE(k)$, if $E(k)$ decreases following a power law of k. Thus v_r^2 has to be associated to $kE(k)$, and eq (VI-4-4) is equivalent to the Kolmogorov law. v_r is a typical velocity difference between two points whose distance r corresponds to inertial-range eddies, filtered of the velocity fluctuations due to eddies of size smaller than r. Now, let us consider the second-order velocity structure function

$$F_2(r,t) = < [\vec{u}(\vec{x},t) - \vec{u}(\vec{x}+\vec{r},t)]^2 > \quad . \qquad (VI-4-5)$$

Within the same approximation, it should be of the order of $\int_k^\infty E(p)dp$, and still proportional to v_r^2 in the inertial range. It is therefore found

$$F_2(r,t) \sim (\epsilon r)^{2/3} \quad , \qquad (VI-4-6)$$

for $l_d < r < l$. This can be checked directly in the following manner, as shown by Orszag (1977): an exact expression of the structure function in terms of $E(k)$ is (Batchelor, 1953):

$$F_2(r,t) = 2 \int \hat{U}_{ii}(\vec{k},t) (1 - \exp i\vec{k}.\vec{r}) \, d\vec{k}$$
$$= 4 \int_0^{+\infty} E(k,t) (1 - \frac{\sin kr}{kr}) \, dk \quad , \qquad (VI-4-7)$$

and the replacement of $E(k)$ in (VI-4-7) by an infinite inertial range extending from 0 to ∞ yields

$$F_2(r,t) = 4.82 \, C_K \, (\epsilon r)^{2/3} \quad . \qquad (VI-4-8)$$

However, such an inertial range will exist in physical space only if the extension of the Kolmogorov range in spectral space is wide enough. Otherwise, side effects coming from both the dissipative range and the energy-containing range will contaminate the law (VI-4-8).

Notice finally that, within this presentation of the Kolmogorov law in physical space, one may build with v_r and r a local Reynolds number

$$R(r) = \frac{r \, v_r}{\nu} \quad , \qquad (VI-4-9)$$

which is proportional to $\epsilon^{1/3} r^{4/3}/\nu$ when using eq. (VI-4-4). Hence, for

$$r < l_d = (\frac{\nu^3}{\epsilon})^{1/4} \quad ,$$

this local Reynolds number falls under 1, and viscous effects become preponderant. This allows us to understand why motions are damped by viscosity in the dissipation range, under the Kolmogorov dissipative scale.

5 - The Richardson law

Let us now consider a pair of Lagrangian tracers with a r.m.s. separation of $R(t)$, $R(t)$ still lying in the inertial range (l_d, l). They would disperse in the mean with a separation velocity given by eq. (VI-4-4) with $r = R(t)$. The dispersion coefficient σ will be given by

$$\sigma = \frac{1}{2}\frac{d}{dt}R^2 = R\frac{dR}{dt} \sim Rv_R \sim \epsilon^{1/3}R^{4/3} \quad . \qquad (VI-5-1)$$

Thus a $k^{-5/3}$ isotropic energy spectrum is equivalent to a $R^{4/3}$ turbulent dispersion coefficient. The latter law is known as the Richardson law (Richardson, 1926): Indeed a $R^{4/3}$ dispersion law was proposed by Richardson fifteen years before Kolmogorov's prediction, on the basis of atmospheric experimental diffusion data. Although Richardson's measurements ignored the dependence of σ upon ϵ, he certainly can be considered as the precursor of the Kolmogorov law, as stressed by Leith (private communication). Let us remark finally that, within the classical phenomenology of three-dimensional isotropic turbulence, the Richardson law (VI-5-1) will be shown to govern the evolution of the velocity and temperature integral scales in various situations (stationary or freely-decaying turbulence, different velocity and scalar integral scales).

6 - Characteristic scales of turbulence

We have already introduced the integral scale and the dissipative scale, between which the inertial-range eddies see their kinetic energy organize along the Kolmogorov energy cascade. As shown above, the dissipative scale can be understood with the aid of eq. (VI-4-4). This is also the case for the integral scale: indeed, let $v^2 = <\vec{u}^2>$ be the velocity variance, which is characteristic of the kinetic energy of the energy-containing eddies of length scale l. Therefore the integral scale l is such that $v \approx (\epsilon\, l)^{1/3}$, which yields

$$l \sim \frac{v^3}{\epsilon} \quad . \qquad (VI-6-1)$$

The value of the dissipative scale l_d measured in the atmosphere is about 1 mm, while it is 0.1 mm in a laboratory grid turbulence in the air. The reason being that, in the latter case, the Reynolds numbers are moderate, and the viscous dissipative rates of energy are stronger than their asymptotic value at high Reynolds numbers.

6.1 The degrees of freedom of turbulence

Preceeding relations allow one to write

$$\frac{l}{l_d} = \frac{k_d}{k_i} \sim \left(\frac{vl}{\nu}\right)^{3/4} \quad , \qquad (VI - 6 - 2)$$

which shows that the extension of the inertial range in Fourier space goes to infinity with the three-fourth power of the large-scale turbulent Reynolds number R_l. This result is extremely important, for it gives an upper bound for the number of degrees of freedom which are needed to describe the motion (from dissipative scales under which the motion is quickly damped by viscosity, to large-scale energy-containing eddies) in each direction of the space: indeed, let us work for instance within the discreet Fourier representation of a periodic flow in a box of size l. The velocity field may be expanded as an infinite series

$$\vec{u}(\vec{x}, t) = \sum_{n_1, n_2, n_3} \hat{\underline{u}}(\vec{k}, t) \exp i\vec{k}.\vec{x} \quad , \qquad (VI - 6 - 3)$$

where the components k_i of $\vec{k}$ are

$$k_1 = \frac{2\pi}{l} n_1 , \ k_2 = \frac{2\pi}{l} n_2 , \ k_3 = \frac{2\pi}{l} n_3 \quad , \qquad (VI - 6 - 4)$$

n_1, n_2 and n_3 being relative integers. It is clear that this expansion may be truncated for $|k_i| > A k_d$, where A is a constant of the order of unity. This gives an upper bound for n_i of the order of $k_d l = k_d/k_i$. Hence, one expects that the total number of degrees of freedom of the flow is *a-priori* of the order of k_d/k_i in each direction of space. This yields a maximum total number for the degrees of freedom equal to $(k_d/k_i)^3$ in three-dimensional turbulence, and $(k_d/k_i)^2$ in two-dimensional turbulence[10]. This number has been interpreted by Constantin et al. (1985) as an upper bound on the dimension of the Navier-Stokes attractor. It follows from eq. (VI-6-2) that the total number of degrees of freedom is of the order of $R_l^{9/4}$ in three dimensions. Notice that in two-dimensional turbulence, the phenomenology of the enstrophy cascade (see Chapter IX) leads to a total number of degrees of freedom equal to R_l.

---

[10] In two-dimensional turbulence, it will be shown in Chapter IX that the equivalent Kolmogorov wave number k_d characterizes the enstrophy dissipation.

6.1.1 The dimension of the attractor

Several experimental or numerical attempts, still based on the concept of dimension of the attractor, have recently been carried out in order to determine whether the *actual* number of degrees of freedom of the flow could not be smaller than the upper bounds presented above. These attempts have not, up to now, improved significantly these bounds (see Sreenevasan and Strykowski, 1984; Atten et al., 1984; Lafon, 1985). The question is still open to know whether, in flows where spatially organized large structures exist (such as atmospheric planetary scale motions, mixing layers, wakes, jets, boundary layers, thermal convective turbulence, rotating flows, etc), the dynamics of these large scales may be modelled, with a proper parameterization of the exchanges with smaller scales, by a dynamical system involving a relatively low number of degrees of freedom, and possibly displaying a chaotic behaviour and strange-attractor solutions. Such an approach has been proposed by Lumley (1967) and applied to the boundary layer by Aubry et al. (1988). This could be a way of bridging the *chaos* and statistical fully-developed points of view of turbulence, the latter being used to understand and model the energy exchanges between the small-scale turbulence and the large organized scales analysed by the former. But a serious obstacle to such an analysis lies in the fact that, in developed turbulence, there is no spectral separation of scales between the coherent structures and the rest of the turbulence. It is doubtful in particular that the Kolmogorov cascade could be described in terms of dynamical systems and strange attractors.

To end this discussion, the Kolmogorov viewpoint of energy cascade from large to small scales has often been opposed to the experimental evidence that large scales could pair and amalgamate, leading to the formation of larger structures. In fact, as already mentioned in Chapter III, there does not seem to be any contradiction at all between both mechanisms, which certainly occur simultaneously: the large scales of the flows may be quasi-two-dimensional (in the mixing layer for instance) and obey the two-dimensional vorticity conservation constraint which implies strong inverse transfers of energy. On the other hand, they will simultaneously degenerate, through successive instabilities due to the three-dimensional perturbations they are submitted to, towards small-scale Kolmogorov fully developed three-dimensional turbulence which will dissipate the kinetic energy of the large scales or of the mean flow.

6.2 The Taylor microscale

A third characteristic length scale is often used in turbulence, mainly by experimentalists. It is the Taylor microscale, characteristic of the mean spatial extension of the velocity gradients, and defined by

$$\lambda^2 = \frac{<\vec{u}^2>}{<(\vec{\nabla} \times \vec{u})^2>} \qquad \text{(VI} - 6 - 5)$$

Since $\epsilon = \nu <(\vec{\nabla} \times \vec{u})^2>$, eq. (VI-6-5) is equivalent to

$$\lambda^2 = \frac{<\vec{u}^2> \nu}{\epsilon} \quad . \qquad \text{(VI} - 6 - 6)$$

In a typical grid turbulence laboratory experiment, the integral scale l is of the order of 4 cm (the grid mesh), the **Kolmogorov** scale is 0.1 mm, and the Taylor microscale is 2 mm .

The Reynolds number $R_\lambda = v\lambda/\nu$ satisfies

$$R_\lambda \sim R_l^{\frac{1}{2}} \sim (\frac{l}{\lambda}) \quad . \qquad \text{(VI} - 6 - 7)$$

The law (VI-6-7) is very well verified experimentally, even at moderate Reynolds numbers. The proportionality constant in (VI-6-7) is such that $R_\lambda = 4R_l^{1/2}$, and an experimental value of R_l equal to 300 corresponds to $R_\lambda \approx 70$ (see Hinze, 1975).

6.3 Self-similar decay

The integral and dissipative scales allow one to propose self-similar expressions for the energy spectrum equivalent to Karman-Howarth (1938) solutions for the spatial velocity correlations in the physical space, i.e.

$$E(k,t) = EL \ F(k \ L) \qquad \text{(VI} - 6 - 8)$$

where E and L are respectively a typical kinetic energy and a typical scale, and $F(x)$ is a dimensionless function of the dimensionless argument x. If E and L are chosen as v^2 and l, one obtains a self similar solution

$$E(k,t) = \ v^2(t) \ l(t) \ F[k \ l(t)] \qquad \text{(VI} - 6 - 9)$$

which can be shown to be valid in describing the energy-containing and inertial ranges. In particular, the assumption that F is a power law of kl and $E(k,t)$ is independent of v leads to the Kolmogorov law. If E and L are chosen to characterize the dissipation range, the self similar solution is

$$E(k,t) = \frac{\nu^2}{l_d} \ G(k \ l_d) \qquad \text{(VI} - 6 - 10)$$

where G is another dimensionless function. Eq. (VI-6-10) effectively predicts the behaviour of the energy spectrum in the inertial and dissipative ranges. The assumption that G is a power law of kl_d and $E(k,t)$ does not depend on ν yields again the Kolmogorov law.

7 - Skewness factor and enstrophy divergence

7.1 The skewness factor

Let $X(\omega)$ be a random variable of zero mean. Its skewness factor is defined as $<X^3>/<X^2>^{3/2}$. The skewness factor of isotropic turbulence will be defined in the present monograph as the skewness factor of $-\partial u_1/\partial x_1$, u_1 being any component of the velocity:

$$s = -<(\frac{\partial u_1}{\partial x_1})^3>/<(\frac{\partial u_1}{\partial x_1})^2>^{3/2} \quad . \qquad (VI-7-1)$$

This factor would be equal to zero if the random function $\vec{u}$ were gaussian. Experimental values of s found in grid turbulence (Batchelor and Townsend, 1949) are of the order of 0.4, and direct numerical simulations at moderate Reynolds numbers done by Orszag and Patterson (1972) give a value of 0.5. This shows that turbulence cannot certainly be considered as gaussian. Actually the skewness characterizes the rate at which the enstrophy increases by vortex stretching: indeed, one obtains from eq. (VI-2-15), after multiplication by ω_i, averaging, and making use of the incompressibility condition:

$$\frac{d}{dt}D(t) = <\omega_i\omega_j\frac{\partial u_i}{\partial x_j}> -2\nu\, P(t) \qquad (VI-7-2)$$

where the enstrophy $D(t)$ is given by (V-5-11) and (V-5-15), $P(t)$ being the "palinstrophy"[11] defined by

$$P(t) = \frac{1}{2}<(\vec{\nabla}\times\vec{\omega})^2> = \int_0^{+\infty} k^4 E(k,t)\, dk \quad . \qquad (VI-7-3)$$

It has been shown by Batchelor and Townsend (1947) that, for isotropic three-dimensional turbulence, the skewness factor s is equal to

$$s = (\frac{135}{98})^{1/2} <\omega_i\omega_j\frac{\partial u_i}{\partial x_j}> /D(t)^{3/2} \quad . \qquad (VI-7-4)$$

---

[11] see footnote at the beginning of Chapter IX

Therefore eq. (VI 7-2) is written

$$\frac{d}{dt} D(t) = (\frac{98}{135})^{1/2} \, s \, D^{3/2} - 2\nu \, P(t) \quad . \qquad (VI - 7 - 5)$$

As noticed by Orszag (1970 a), who wrote this equation, a positive skewness factor is needed in order to increase the enstrophy by the stretching of vortex filaments. We notice also that the first term of the r.h.s. of (VI-7-5) is, from (VI-3-1), equal to $\int_0^{+\infty} k^2 T(k,t) dk$: therefore, the skewness factor is also equal to (Orszag, 1977)

$$s(t) = (\frac{135}{98})^{1/2} \int_0^{+\infty} k^2 T(k,t) \, dk / D(t)^{3/2} \quad . \qquad (VI - 7 - 4')$$

Hence it will certainly be positive, since the k^2 factor will enhance the high wave number positive contribution of $T(k,t)$ (we recall that $T(k,t)$ satisfies eq. (VI-3-6) in the Navier-Stokes case). At high times, and if turbulence is unforced and decays self-similarly, it has been noticed by Orszag (1977) that the term dD/dt in eq. (VI-7-5) is now much smaller than the two terms of the r.h.s., so that we have approximately:

$$s = 2.35 \frac{\nu \, P(t)}{D(t)^{3/2}} \quad . \qquad (VI - 7 - 6)$$

Then the use of (VI-6-10) to calculate $P(t)$ and $D(t)$, which are dominated by viscous-range contributions, yields the constancy of $s(t)$ with time. This property seems to be characteristic of decaying three-dimensional isotropic turbulence, as well as the finiteness of the kinetic energy dissipation.

Now we are going to consider the question of the enstrophy and skewness factor time evolution, for an unforced initial-value problem, where we assume an initial energy spectrum sharply peaked at $k_i(0)$, with a finite enstrophy $D(0)$ (hereafter called *type S* initial conditions. The problem will be looked at both in the case of Euler and Navier-Stokes equations.

7.2 Does enstrophy blow up at a finite time in a perfect fluid?

We will show in Chapter VII stochastic models of the Euler equations predicting the blow up of the enstrophy at a finite time. As will be seen below, this is an extremely controversial question which we are going to examine here on the basis of the phenomenology. If one accepts the idea that a Kolmogorov spectrum extending to infinity[12] will eventually form,

---

[12] since $k_D \to \infty$ when $\nu \to 0$ if ϵ is finite

yielding an infinite enstrophy, the questions which arise are: "will the enstrophy-divergence time be finite or not"?, or "will it take a finite time to build up a Kolmogorov spectrum extending to infinity"? To answer these questions, several phenomenological models will be considered. We will use the Euler counterpart of (VI-7-5)

$$\frac{d}{dt}D(t) = (\frac{98}{135})^{1/2} \, s \, D^{3/2} \quad . \qquad (VI-7-7)$$

7.2.1 The constant skewness model

Let us suppose a skewness factor s constant with time. The enstrophy will thus be equal to

$$D(t) = \frac{D(0)}{[1-(t/t_c)]^2} \qquad (VI-7-8)$$

with

$$t_c = \frac{1}{0.425 \, s} \, D(0)^{-1/2} \quad . \qquad (VI-7-9)$$

Therefore the enstrophy blows up at the critical time t_c (equal to $5.9 \, D(0)^{-1/2}$ for $s = 0.4$).

But the skewness factor certainly does not fulfill this assumption: direct-numerical simulations of Euler equations carried out with gaussian[13] *type S* initial conditions[14] show that the skewness factor grows initially from zero to finite values[15] (Wray, 1988). In fact, one can show the following results, which are valid for Euler equations as long as the enstrophy remains finite:

7.2.2 Positiveness of the skewness

Theorem: for solutions[16] of Euler equations which are limits of Navier-Stokes solutions when $\nu \to 0$, the skewness factor is positive.

---

[13] see details on gaussian functions in Chapter VII

[14] These simulations are valid as soon as the characteristic maximum wave number $k_E(t)$ of the kinetic-energy spectrum does not exceed the computation cutoff wave number k_c .

[15] These calculations, using pseudo-spectral methods, were carried out with the same initial velocity field, and a maximum cutoff wave number k_c respectively equal to 32, 64 and 128. The skewness factor evolution consisted in two stages: a growth up to a maximum corresponding to the time when the ultra-violet kinetic-energy cascade starts being affected by the cutoff k_c, then a decay afterwards, due to the build up of gaussian absolute-equilibrium ensemble solutions whith a kinetic-energy spectrum proportional to k^2 (see Chapter X). The calculation seems to indicate that the maximum of the skewness grows with k_c .

[16] we work in Fourier space, and these solutions will be characterized by their kinetic energy spectrum

The derivation proceeds as follows: due to (VI-2-14), the kinetic energy of the Euler solutions we consider will be conserved with time as long as the enstrophy remains finite. Therefore, the transfer $T(k,t)$ will satisfy eq. (VI-3-6). The same argument as used above in the case of a viscous fluid shows then that $\int_0^{+\infty} k^2 T(k,t)dk$ will be positive if the transfer has the form displayed in Figure VI-2, which all the numerical simulations show: the transfer is negative in the vivinity of k_i and positive in the vicinity of k_E. Hence the skewness factor will be, from eq. (VI-7-4'), positive and the enstrophy will grow, due to eq. (VI-7-7).

For instance, if the initial velocity is gaussian, the skewness factor will grow from zero, and either keep on growing, or reach a maximum and decay. Then, another theorem can be shown:

7.2.3 Enstrophy blow up theorem

Theorem: if the skewness $s(t)$ has a strictly positive lower bound s_0, the enstrophy will blow up at a finite time

$$t_* < \frac{1}{0.425\ s_0}\ D(0)^{-1/2}\ . \tag{VI-7-10}$$

The derivation of this result is staighforward: one considers eq. (VI-7-7), which writes

$$\frac{d}{dt}D(t)^{-1/2} = -0.425\ s\ . \tag{VI-7-11}$$

In the same way, the function $D_{s_0}(t)$, solution of

$$\frac{d}{dt}D_{s_0}(t) = (\frac{98}{135})^{1/2}s_0\ D_{s_0}^{3/2}\ ,$$

with $D_{s_0}(0) = D(0)$ satisfies

$$\frac{d}{dt}D_{s_0}(t)^{-1/2} = -0.425\ s_0\ . \tag{VI-7-12}$$

Since $s \geq s_0$, eqs. (VI-7-11) and (VI-7-12) yield

$$\frac{d}{dt}D_{s_0}(t)^{-1/2} \geq \frac{d}{dt}D(t)^{-1/2}\ ,$$

and the function $D_{s_0}(t)^{-1/2} - D(t)^{-1/2}$, initially zero, is increasing with time. Hence $D(t) \geq D_{s_0}(t)$. Since $D_{s_0}(t)$ blows up at t_c given by (VI-7-7) with $s = s_0$, this demonstrates the theorem.

In the above example where the velocity is initially gaussian, for instance, a continuous growth of the skewness (which might possibly be extrapolated from Wray's (1988) calculations) implies the enstrophy blow up. Even a subsequent decay of the skewness from a maximum to a strictly positive value leads also to the enstrophy divergence at a finite time[17]. It may be easily shown as well by solving eq. (VI-7-11) under the form

$$D(t) = \frac{D(0)}{[1 - 0.425 D(0)^{1/2} \int_0^t s(\tau) d\tau]^2} \quad,$$

that a blow up at finite time will occur if the skewness does not decrease to 0 at large times faster than t^{-1} .

7.2.4 A self-similar model

Here, we give arguments based on self-similarity, which predict also an inviscid blow up of the enstrophy at a finite time. We still start at $t = 0$ with a kinetic-energy spectrum peaking at $k_i(0)$, and assume that the spectrum starts developing a $k^{-5/3}$ Kolmogorov range between $k_i(t)$ and $k_E(t)$. The latter will tend to spread out towards higher wave numbers, under the action of the global deformation due to the larger scales, which create smaller scales than k_E^{-1} at the rate:

$$\frac{1}{k_E} \frac{d}{dt} k_E = \sqrt{D}(t) \quad . \qquad (VI-7-13)$$

This equation is a statistical equivalent of the passive-vector equation (II-6-7). One assumes also that, in the neighbourhood of $k_E(t)$, the kinetic-energy spectrum is wholly determined by $E(k_E, t)$ and k/k_E , and may be written as:

$$E(k, t) = E(k_E, t) \, F(\frac{k}{k_E})$$

$$E(k_E, t) = (\frac{k_E}{k_i})^{-5/3} \, E(k_i, t) \qquad (VI-7-14)$$

$$E(k_i, t) \sim v_0^2 \, k_i(t)^{-1} \quad .$$

In this analysis, k_E plays the same role as the Kolmogorov wave number in eq. (VI-6-10). $F(x)$ is a non-dimensional function. v_0 is the r.m.s. velocity, independent of time since we still assume that the enstrophy is finite and has not yet diverged. Now, one may use eq. (VI-7-14) to calculate the enstrophy

$$D(t) \sim v_0^2 k_i^{-1} \, (\frac{k_E}{k_i})^{-5/3} \, k_E^3 = v_0^2 k_i^{2/3} k_E^{4/3} \quad, \qquad (VI-7-15)$$

---

[17] indeed, just take as initial time the time where the skewness is maximum

where it has been assumed that the integral $\int_0^{+\infty} x^2\, F(x)\, dx$ converges[18]. Substituting into (VI-7-13) leads to

$$\frac{d}{dt}k_E^{-2/3} \sim -v_0 k_i^{1/3} \quad ,$$

which yields finally:

$$k_E(t)^{-2/3} = k_E(0)^{-2/3} - Av_0 \int_0^t k_i(\tau)^{1/3}\, d\tau \quad , \qquad (VI-7-16)$$

where A is a non-dimensional constant. The exact dependance of $k_i(t)$
upon t is not known, but direct-numerical simulations show that k_i does
not vary very much during the build up of the ultra-violet cascade. Hence
$k_E(t)$ will blow up at a finite time t_c proportional to $1/v_0 k_i$. Again,
the enstrophy within this model is proportional to $1/(t_c - t)^2$, and the
skewness $\sim dD^{-1/2}/dt$ is constant. This model shows that it takes a
finite time to build up a Kolmogorov spectrum extending to infinity.

One may of course in this model question the validity of the ultra-
violet cascade law (VI-7-13) for $k_E(t)$. But one can get rid of this assumption in the following way: one still assumes that a Kolmogorov
spectrum has developed from k_i to k_E, with an energy spectrum given
by (VII-7-14). The transfer $T(k, t)$ is supposed to behave in the following
manner:

$$
\begin{aligned}
T(k,t) &= -T(k_i)\, f(k/k_i) \quad , \quad k \sim k_i \\
T(k,t) &= 0 \quad \text{in the Kolmogorov range} \qquad (VI-7-17) \\
T(k,t) &= T(k_E)\, g(k/k_E) \quad ,
\end{aligned}
$$

where $f(x)$ and $g(x)$ are non-dimensional positive functions of the non-
dimensional variable x, $-T(k_i)$ and $T(k_E)$ characteristic values of $T(k,t)$
in the neighbourhood of k_i and k_E respectively. $T(k_i)$ and $T(k_E)$ are
supposed to be positive. Due to the energy conservation (VI-3-6), it
is easy to show that $T(k_E, t) = k_i k_E^{-1} T(k_i)$. Hence, with the aid of
(VI-7-4') and (VI-7-15), the skewness factor is equal to

$$s = \frac{k_i T(k_i)(k_E^2 - k_i^2)}{v_0^3 k_i k_E^2} = v_0^{-3} T(k_i)[1 - (\frac{k_i}{k_E})^2] \,. \qquad (VI-7-18)$$

As already stressed, v_0 is fixed and k_i and $T(k_i)$ will not vary very much,
so that one again recovers a constant skewness model if $k_E >> k_i$.

From all this information, one may build up two scenarii where,
starting from an initial gaussian velocity field of kinetic-energy spectrum
peaking at $k_i(0)$, the skewness factor would:

---

[18] otherwise, the scaling is meaningless

a) either grow continuously with time, until the enstrophy blows up at a finite time t_c with formation of a Kolmogorov spectrum extending from $k_i(t)$ to infinity.

b) or grow up and saturate at a plateau, with the formation of a Kolmogorov range of limited extension from k_i to k_E. In this case also, the enstrophy will afterwards blow up at a finite time t_c, with simultaneously (from eq. VI-7-15) the rejection of k_E to infinity.

7.2.5 Oboukhov's enstrophy blow up model

To finish with these inviscid enstrophy blow up models, we recall a calculation done by Brissaud et al. (1973) showing an exact result of enstrophy blow up in the case of a modified time-dependent Oboukhov theory proposed by Panchev (1971): this model is written in the inviscid case

$$\frac{\partial}{\partial t} \int_0^k E(p,t)dp = -\frac{1}{\tau(k)} \int_k^\infty E(p,t)dp \qquad (VI-7-19)$$

with $\tau(k) = [\int_0^k p^2 E(p,t)dp]^{-1/2}$. The time $\tau(k)$, which will also be used in the next chapter, characterizes the shearing action of the "large scales" $< k$ upon the "small scales" $> k$. The l.h.s. of (VI-7-19) is the loss of kinetic energy of the large scales, equal to the "available" cascading kinetic energy $\int_k^\infty E(p)dp$ divided by the time $\tau(k)$. Starting with an initially rapidly decreasing spectrum with a finite initial enstrophy $D(0)$, and by differentiation of (VI-7-19) with respect to k, multiplication by k^2 and integration over k, one obtains

$$\frac{dD(t)}{dt} = \int_0^{+\infty} k^2 E(k)\tau(k)dk[\int_0^k p^2 E(p)dp - \frac{k^2}{2} \int_k^\infty E(p)dp] \quad .$$

By successive majorations and minorations of the integrals, described in Brissaud et al. (1973) , one obtains the following inequality

$$\frac{dD(t)}{dt} > \frac{1}{4}D(t)^{3/2} \qquad (VI-7-20)$$

showing that the enstrophy blows up before the time $8\,D(0)^{-1/2}$.

7.2.6 Discussion

Up to now, and since the pioneering work of Leray (1934) on Navier-Stokes equations, only regularity results for finite times of the order of $D(0)^{-1/2}$ have been rigorously demonstrated for Navier-Stokes or Euler equations (see for a review Temam, 1977 and Lions, 1989). Direct numerical simulations of Euler equations, starting with the Taylor-Green (1937) vortex

$$
\begin{aligned}
u(\vec{x}) &= \cos x_1\ \sin x_2\ \sin x_3 \\
v(\vec{x}) &= -\sin x_1\ \cos x_2\ \sin x_3 \\
w(\vec{x}) &= 0
\end{aligned}
\qquad (VI-7-21)
$$

have been performed, using both the original perturbation expansion of Taylor and Green (1937) in powers of time (Orszag, 1977, Morf et al., 1980), and a truncated spectral method (Brachet et al., 1983). But the results concerning the appearance of a possible singularity at a finite time are difficult to interpret: indeed conclusions of Morf et al. (1980) are at variance with those of Brachet et al. (1983), depending upon the numerical method chosen. The more recent opinions expressed by the majority of the authors of these references are that there is no singularity. This statement concerns nevertheless only the inviscid Taylor-Green vortex. Returning to isotropic Euler turbulence starting with *type S* initial conditions, an important result shown above is that:

• any Kolmogorov $k^{-5/3}$ range of finite extension forming at a finite time will spread out to infinity ($k \to \infty$) in a finite time.

Therefore, conjectures ruling out the possibility of an enstrophy blow up at a finite time, have to rule out the possibility of any such Kolmogorov range forming at a finite time within the Euler equations.

Let us stress finally that the existence or not of a singularity at a finite time within unforced Euler equations may depend also upon the existence and the nature of a perturbation of small amplitude superposed on the initial velocity field. Let us consider for instance at some initial time t_0 a two-dimensional temporal mixing layer[19] consisting in a row of rolling up Kelvin-Helmholtz vortices. If no perturbation is brought, the further evolution of the flow will certainly not lead to any enstrophy blow up, since the vorticity cannot in two dimensions exceed its initial value. On the contrary, a three-dimensional perturbation of the type considered in Chapter III (section 3-1-1)[20], might lead in the Euler case to a blow up of vorticity in certain regions of the flow. A last remark concerns the possibility that Euler equations could lead to local blow-ups of vorticity, without divergence of the enstrophy.

7.3 The viscous case

Eq. (VI-7-7) is valid for a viscous flow[21] during the early period of the evolution when the kinetic-energy spectrum $E(k,t)$ decreases faster than k^{-5} for $k \to \infty$[22]. Meanwhile the kinetic energy, which is dissipated at the rate $2\nu D(t)$, is conserved with time in the limit of zero viscosity as long as the inviscid enstrophy remains finite, that is, for $t < t_c$ if one

---

[19] in an infinite domain

[20] that is, a hairpin vortex filament stretched by Kelvin-Helmholtz billows

[21] when the viscosity is small enough

[22] This assumption allows the palinstrophy $P(t)$ to remain finite, so that the viscous term $\nu P(t)$ in (VI-7-5) goes to zero with ν.

accepts the above ideas that the inviscid enstrophy blows up at a finite time.

For $t > t_c$, (VI-7-6) is no more valid, since the dissipative term in (VI-7-5) cannot be neglected anymore. One now uses the fact that the kinetic energy is, due to viscous dissipation, going to decay as

$$\frac{1}{2} <\vec{u}^2> \propto t^{-\alpha_E} \qquad\qquad (VI-7-22)$$

α_E being an exponent of the order of $1 \sim 1.5$, as will be seen in Chapter VII . Therefore, the enstrophy $\epsilon/2\nu$ will decay as

$$D(t) \propto \frac{t^{-\alpha_E-1}}{\nu} \qquad\qquad (VI-7-23)$$

as indicated in Figure VI-5. Due to the finiteness of ϵ in the limit of zero viscosity, the enstrophy will thus be infinite for $t > t_c$.

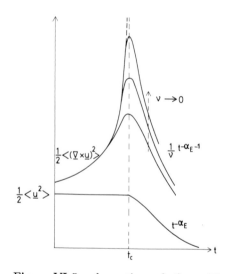

Figure VI-5: schematic evolution with time of the kinetic energy and the enstrophy in a freely-decaying three-dimensional isotropic turbulence when the viscosity goes to zero.

Statistical models of the Navier-Stokes equations displaying the same catastrophic behaviour will be introduced in Chapter VII.

One may finally also wonder about the behaviour of the skewness factor for $t > t_c$, both within Euler equations and within Navier-Stokes equations in the limit of zero viscosity: in this last case, we have seen above, using eqs (VI-6-10) and (VI-7-6) that the skewness saturates at a constant value. Things are different in the Euler case: indeed, if one assumes an infinite Kolmogorov range extending for $k > k_i$ forming at $t = t_c$ with a self-similar decaying energy spectrum given by (VI-4-1) ($\epsilon(t)$ decays with time now), the transfer will not be energy conservative anymore. $T(k,t)$ will be equal to $T(k_i)f(k/k_i)$ in the vicinity of k_i, and to $\partial E(k,t)/\partial t \sim -E(k,t)/t$ in the inertial range[23]. Hence

$$s \sim \frac{-k_i^3 T(k_i) - t^{-1}D}{D^{3/2}} \quad , \tag{VI-7-24}$$

and the skewness will fall to zero by negative values[24]. This is therefore an example of quantity related to turbulence which is different in the Euler case from the Navier-Stokes ($\nu \to 0$) case.

8 - The internal intermittency

We have already emphasized the intermittent character of the small scales of turbulence, as a result of the process of stretching of vortex filaments. Thus in a homogeneous three-dimensional isotropic turbulent flow, the intensity of the velocity fluctuations is not distributed in a uniform manner in space, and presents what one will call "internal intermittency". This intermittency is of a different nature from the "external intermittency" which characterizes the large coherent structures of a turbulent flow at the frontier with the outer irrotational flow, in turbulent boundary layers or jets for instance. The existence of internal intermittency is not in contradiction with the assumption of homogeneity, which is an average property of the flow. So the "local" kinetic energy dissipation rate $\epsilon = \nu(\vec{\nabla} \times \vec{u})^2$ displays important fluctuations about its mean value[25] $< \epsilon >$. A consequence is that the Kolmogorov $k^{-5/3}$ theory, which does not involve these fluctuations of ϵ, must certainly be corrected in order to take into account this intermittent character. This has been noticed by Kolmogorov himself, who proposed a theory in 1962, based on a lognormality assumption, which corrected his original theory (Kolmogorov, 1962). The same ideas were simultaneously expressed by Oboukhov (1962), and developed by Yaglom (1966). A different class of models was proposed by Novikov and Stewart (1964), and worked out by Frisch et al. (1978). Mandelbrot (1976) proposed a very interesting

---

[23] We assume that v_0^2 decay as a power of time. Remark also that the transfer is not exactly zero in the inertial range for decaying turbulence.

[24] here, k_i is a decreasing function of time, see Chapter VII

[25] In the rest of the book (except in VII-11), the notation ϵ is used for the quantity called $< \epsilon >$ in this section only.

unifying synthesis of these various theories, based on a weighted or absolute "curdling" principle (see also Mandelbrot, 1975). In this section, we will summarize the main results of both the Kolmogorov-Oboukhov-Yaglom theory and the Novikov-Stewart theory. A useful reference for this purpose is Gagne (1987).

8.1 The Kolmogorov-Oboukhov-Yaglom theory

This theory is developed in detail in Monin and Yaglom's (1975) textbook of turbulence. It assumes that the local kinetic-energy dissipation ϵ possesses a log-normal distribution (i.e. that the random function $\ln \epsilon$ is gaussian): as shown by Yaglom (1966), such a result comes from a model called "self-similar breakdown of turbulent eddies". More specifically, this theory introduces another local kinetic energy dissipation rate, $\epsilon_r(\vec{x}, t)$, which is the average of ϵ on a sphere of center $\vec{x}$ and radius $r/2$. Within the model, the variance σ_r^2 of $\ln \epsilon_r$ is of the form

$$\sigma_r^2 = Q(\vec{x}, t) + \chi \ln \frac{l}{r} \qquad (VI-8-1)$$

where l is the integral scale of turbulence, $Q(\vec{x}, t)$ a function depending on the large scales, and χ a universal constant. Eq. (VI-8-1) is only valid for $r \ll l$. From these assumptions it can be shown that the order q moments of ϵ are given by

$$< \epsilon_r^q > = < \epsilon >^q \exp \frac{1}{2} q(q-1)\sigma_r^2 \qquad (VI-8-2)$$

which yields, using (VI-8-1)

$$< \epsilon_r^q > = D_q < \epsilon >^q \left(\frac{l}{r}\right)^{\chi q(q-1)/2} \qquad (VI-8-3)$$

where $D_q(\vec{x}, t)$ is a coefficient which depends on the large scales of the flow. In the particular case $q = 2$, and using the statistical homogeneity condition, one has (see Monin and Yaglom, 1975)

$$< \epsilon(\vec{x}, t)\epsilon(\vec{x} + \vec{r}, t) > = \frac{1}{2}\left(\frac{\partial^2}{\partial r^2}\right)[r^2 < \epsilon_r(\vec{x}, t)^2 >]$$

$$= D_2 < \epsilon >^2 \left(\frac{l}{r}\right)^{\chi} \qquad (VI-8-4)$$

which shows that the parameter χ characterizes the spatial correlation of ϵ. At this point the classical Kolmogorov 1941 theory can be shown again locally, assuming that the crucial parameter is ϵ_r. This yields for the local

structure function given by (VI-4-6) the proportionality to $\epsilon_r^{2/3} r^{2/3}$.
After an ensemble averaging and use of (VI-8-3) for $q = 2/3$, we obtain

$$F_2(\vec{r}, t) \sim < \epsilon >^{2/3} r^{2/3} (\frac{r}{l})^{\chi/9} \qquad (VI - 8 - 5)$$

which, translated in the Fourier space, yields

$$E(k) \sim < \epsilon >^{2/3} k^{-5/3} (kl)^{-\chi/9} \quad . \qquad (VI - 8 - 6)$$

The same analysis for a structure function of order n would give with
the aid of (VI-8-3) with $q = n/3$

$$< v_r^n > \sim < \epsilon_r^{n/3} > r^{n/3} \sim < \epsilon >^{n/3} r^{n/3} (\frac{l}{r})^{\chi n(n-3)/18} \qquad (VI - 8 - 7)$$

as recalled in Anselmet et al. (1984). Eqs. (VI-8-5), (VI-8-6) and
(VI-8-7) show the departure from the Kolmogorov 1941 theory due to
the intermittency in the frame of the lognormal theory. In particular
(VI-8-7) shows that the structure function of order 6 is proportional to
$r^2(l/r)^\chi$.

8.2 The Novikov-Stewart (1964) model

This model, or a dynamical equivalent one known as the β-model (Frisch
et al., 1975), assumes that during the cascade process only a fraction β
of the volume occupied by the cascading eddies will be filled by turbulence: more specifically, let us consider an "eddy" of size l_p and volume
l_p^3 which gives rise to N eddies of size $l_p/2$: the fraction β of the volume
occupied by turbulence is

$$\beta = \frac{N}{2^3} \quad . \qquad (VI - 8 - 8)$$

N can be defined with the aid of the concept of fractal dimension of
Hausdorff (see Mandelbrot, 1975, 1976): the turbulent structures are
assumed to be self similar, and to lie on a fractal set of dimension D
such as $N = 2^D$, and

$$\beta = 2^{D-3} \quad . \qquad (VI - 8 - 9)$$

Thus the essence of the theory is to assume that the standard Kolmogorov phenomenology is valid only in the active regions: at step p of
the cascade corresponding to eddies of size l_p of typical velocity v_{l_p}, the
kinetic energy dissipation rate will be equal to $v_{l_p}^3/l_p$ in the active regions and to zero in the "non turbulent" regions: thus the mean kinetic
energy dissipation rate $< \epsilon >$ on the whole volume will be

$$< \epsilon > \sim \beta^p v_{l_p}^3/l_p \qquad (VI - 8 - 10)$$

since the eddies l_p occupy a fraction β^p of the initial volume. Hence the assumption of a constant mean kinetic energy flux along the cascade yields

$$v_{l_p} \sim (<\epsilon> l_p)^{1/3} \beta^{-p/3} \quad . \qquad (VI-8-11)$$

The kinetic energy spectrum associated to the wave number $k = l_p^{-1}$ is $v_{l_p}^2 k^{-1}$ in the active regions and zero elsewhere. Therefore the energy spectrum of the mean kinetic energy is

$$E(k) \sim \beta^p v_{l_p}^2 k^{-1} \sim \beta^{p/3} <\epsilon>^{2/3} k^{-5/3} \qquad (VI-8-12)$$

which, with the aid of (VI-8-9), can be written as

$$E(k) \sim (kl)^{-(3-D)/3} <\epsilon>^{2/3} k^{-5/3} \qquad (VI-8-13)$$

since $kl = l/l_p = 2^p$. Eq. (VI-8-13) shows the departure from the 1941 Kolmogorov law within the Novikov-Stewart theory.

The structure functions of order n of the velocity can then be evaluated in the same way, assuming the velocity differences to be zero in the non active regions:

$$<v_r^n> \sim \beta^s (<\epsilon> r)^{n/3} \beta^{-ns/3} \qquad (VI-8-14)$$

with $(l/r) = 2^s$. This yields

$$<v_r^n> \sim <\epsilon>^{n/3} r^{n/3} (\frac{l}{r})^{\chi(n-3)/3} \qquad (VI-8-15)$$

as has been recalled by Anselmet et al. (1984). In (VI-8-15) the coefficient χ is now defined as

$$\chi = 3 - D \quad . \qquad (VI-8-16)$$

Finally the kinetic energy dissipation product between two points separated by r is equal to v_r^6/r^2 if $r < l_s$ and zero if $r > l_s$. The spatial ϵ-correlation is thus

$$<\epsilon(\vec{x},t)\epsilon(\vec{x}+\vec{r},t)> \sim \frac{\beta^s v_r^6}{r^2} \sim <\epsilon>^2 (\frac{l}{r})^{\chi} \quad . \qquad (VI-8-17)$$

8.3 Experimental and numerical results

Eqs. (VI-8-4), (VI-8-7), (VI-8-15) and (VI-8-17) show that the coefficients χ introduced in both lognormal and β-model theories characterize the behaviour of the ϵ-correlation or of $<v_r^6>/r^2$, which decay as $(l/r)^{\chi}$ when r goes to zero (and in the limit of zero viscosity). This allows experimental determinations of χ, by measuring either the ϵ-correlation (or equivalently their Fourier spectrum) or the sixth-order velocity structure function. Values of the order of 0.5 , using the ϵ-correlation method in a wind tunnel (Gagne, 1978), the ϵ-spectrum method in a jet (Friehe et al. (1971), or the structure function method in the atmosphere (Van Atta and Chen, 1970), were thus determined. A numerical simulation done by Brachet (1982) tends to confirm this value. It seems however, as shown by Anselmet et al. (1984) in a turbulent jet, that a more reliable value of χ (determined with the aid of the sixth-order structure function) is of 0.20. The same work shows that, with this value of χ, the velocity structure functions up to the order 12 follow the quadratic dependence in n of the lognormal theory displayed in eq. (VI-8-7), and not the linear dependence of eq. (VI-8-15). So the lognormal theory seems to be valid for $n \leq 12$. For $12 < n \leq 18$ however, departures from the lognormal distribution have been observed by Anselmet et al. (1984). These departures could possibly be explained by what is called a multi-fractal approach (see Schertzer and Lovejoy, 1984, Argoul et al.[26], 1989).

Another point to mention is that, at the level of the kinetic-energy spectrum, the intermittency corrections which steepen the $k^{-5/3}$ Kolmogorov law are (with this value of χ) respectively in both theories of $\chi/9 \simeq 0.02$ (from eq. (8-6)) and $\chi/3 \simeq 0.06$ (from eqs. (VI-8-13) and (VI-8-16)). This is too small to allow an experimental verification at that level. This could also suggest that the closure theories envisaged in the next chapter, and which deal with energy spectra, could give satisfactory results, even if they cannot deal with the spatial intermittency envisaged here.

8.4 Temperature and velocity intermittency

The same experimental study as that carried out for the velocity by Anselmet et al. (1984), was done by Antonia et al. (1984) for a passive temperature. In these two studies (see also Gagne, 1987), the probability density functions[27] of respectively the velocity difference Δu and the temperature difference[28] $\Delta\Theta$, were found to have an *exponential tail*, that is, proportional to $\exp -|X|$ instead of the gaussian distribution $\exp -X^2$. This indicates small-scale intermittency. But the departure from gaussianity is much more marked for the temperature than for the velocity. On the other hand, Métais and Lesieur (1989, 1990) have determined various probability distribution functions in direct and large-eddy

---

[26] This work analyzes the turbulent signal coming from Gagne's (1987) data with the aid of the *wavelet analysis* technique.

[27] referred to as p.d.f.

[28] between two points a given distance apart

simulations of decaying isotropic turbulence: the velocity and temperature derivatives exhibit the same behaviour as for the above experimental
velocity and temperature differences, but the temperature itself has an
exponential distribution, contrary to the velocity which is very close to
gaussian. This shows that the passive temperature is also intermittent in
the large scales. On the other hand, the same calculations show that the
temperature becomes gaussian again when coupled with the velocity[29] in
stably-stratified turbulence[30]. This behaviour could be associated with
the regimes of respectively *hard* and *soft* turbulence found for the temperature by Castaing et al. (1989) in an experiment involving a heated
boundary layer: in this work, hard turbulence corresponds to an exponential temperature p.d.f., while soft turbulence refers to a gaussian
temperature.

Finally, it has to be stressed that the evolution towards an intermittent state is a natural tendency for a turbulent flow. When applied
to the universe, assumed to be fluid, this concept allows one to understand how the initially homogeneous universe of the "big bang" has lost
its homogeneity, and has now developed such an intermittent distribution of galaxies. More generally, intermittency seems to characterize any
dissipative non-linear system.

---

[29] through Boussinesq approximation

[30] More about these calculations (in the isotropic or stratified cases)
will be given throughout the following chapters.

Chapter VII

ANALYTICAL THEORIES AND STOCHASTIC MODELS

1 - Introduction

Our objective in this chapter is to provide the reader with a good understanding of the analytical theories and stochastic models of turbulence sometimes referred to as *two-point closures* since, as will be seen, they deal with correlations in two different points of the space (or two different wave numbers $\vec{k}$ and $\vec{k}'$ such that $\vec{k} + \vec{k}' = \vec{0}$ in the Fourier space). A whole book would not be sufficient to contain all the details of the algebra which is involved, and the reader will be referred to the quoted references for further details: of particular interest for that purpose are Orszag (1970 a, 1977), Leslie[1] (1973), and Rose and Sulem (1978). Here, we will mainly focus on the so-called *E.D.Q.N.M.* approximation (Eddy-Damped Quasi-Normal Markovian approximation), and will situate it among other theories of the same type. These theories can generally be presented from two different points of view, the stochastic model point of view, and the closure point of view. Some of these theories, as will be seen, do not exactly correspond to these points of view, but they lead to spectral equations of the same family, which can be solved with the same methods. We will not use too much energy deriving the "best" analytical theory, for it seems that they all have qualitatively the same defects and qualities, and differ essentially in the values of the inertial-range exponents. We will concentrate principally on the *E.D.Q.N.M.*, because in the case of isotropy it can be solved numerically at a much cheaper cost than the direct simulations, and al-

---

[1] Leslie's work reviews in detail Kraichnan's *Direct-Interaction Approximation*, see below

lows one to reach extremely high Reynolds numbers[2]. We will discuss to what extent the results can be relied upon for "real" turbulence (that is turbulence governed by the Navier-Stokes equation). The confidence we can have in these theories is based on some of the results of the numerical Large-Eddy Simulations of turbulence which will be discussed in Chapter XII. The results concerning a passive scalar diffusion will be given in Chapter VIII, those on two-dimensional turbulence in Chapter IX, and those on the predictability problem in Chapter XI. These closures will also, through the concept of non-local interactions (the equivalent of long-range interactions in physics), allow us to propose statistical parameterizations of the "subgrid-scales" useful for the large eddy simulations, as will be seen in Chapter XII. Our feeling about these closures is that, notwithstanding their inability to deal with spatial intermittency or situations with strong departures from gaussianity, they are a unique tool in studying the strong nonlinearities of developed turbulence, and allow handling of situations inaccessible to the non-linear stability analysis or to the so-called Renormalization Group techniques (see Forster, Nelson and Stephen, 1977, Fournier, 1977, Yakhot and Orszag, 1986, and Dannevik et al., 1987). Some details on these techniques will be given at the end of this chapter). Coupled with the large-eddy-simulations, they could contribute to a decisive advance of our understanding of "real world" anisotropic and inhomogeneous turbulence. Finally, they give a good description of the inverse cascading tendency of two-dimensional turbulence, and provide valuable information on the diffusion and dispersion problems in isotropic turbulence.

The closure problem inherent to a statistical description of turbulence has already been discussed in Chapter VI. Here we reformulate it for homogeneous turbulence in Fourier space: the following formal analysis can be found in a lot of works, for instance in Sulem et al., (1975): let $\hat{u}(\vec{k})$ represent the velocity field, and let the Navier-Stokes equation be written formally as

$$\frac{\partial \hat{u}(\vec{k})}{\partial t} = \hat{u}\hat{u} - \nu k^2 \hat{u}(\vec{k}) \qquad (VII-1-1)$$

which states for eq (IV-2-7), $\hat{u}\hat{u}$ representing the non linear convolution term. Since $<\vec{\hat{u}}>= \vec{0}$, the averaging of eq (VII-1-1) yields the trivial identity $\vec{0} = \vec{0}$. To obtain an evolution equation for the spectral tensor $\hat{U}_{ij}(\vec{k},t)$, one has to write the evolution equation for $\hat{u}(\vec{k'})$,

$$\frac{\partial \hat{u}(\vec{k'})}{\partial t} = \hat{u}\hat{u} - \nu k'^2 \hat{u}(\vec{k'}) \qquad (VII-1-2)$$

---

[2] with several-decade Kolmogorov inertial-range spectra

multiply it tensorially by $\hat{u}(\vec{k})$, multiply (VII-1-1) by $\hat{u}(\vec{k}')$, add the resulting equations, and average. We obtain

$$[\frac{\partial}{\partial t} + \nu(k'^2 + k^2)] <\hat{u}(\vec{k}')\hat{u}(\vec{k})> = <\hat{u}\hat{u}\hat{u}> \qquad (VII-1-3)$$

which, after a further integration on $\vec{k}'$ and because of (V-4-2), gives the desired equation for $\hat{U}_{ij}(\vec{k})$. The formal term $<\hat{u}\hat{u}\hat{u}>$, which is a linear combination of third-order moments of the velocity components, involves integrations on triads of wave numbers, and corresponds to the non-linear transfer between various scales of motion. The problem is evidently not closed, since the triple-velocity correlations are unknown. Then the use of a third evolution equation for $\hat{u}(\vec{k}'')$ analogous to (VII-1-1) and (VII-1-2) allows to write

$$[\frac{\partial}{\partial t} + \nu(k^2 + p^2 + q^2)] < \hat{u}(\vec{k})\hat{u}(\vec{p})\hat{u}(\vec{q}) > = < \hat{u}\hat{u}\hat{u}\hat{u} > \quad . \quad (VII-1-4)$$

One could go to higher orders, and still have one more unknown moment than equations. The hierarchy of the moments is for instance written in Tatsumi (1980) and André (1974), but this hierarchy is not closed, and the only way of solving the problem using this method is to introduce an arbitrary further relation between the velocity moments, called a closure hypothesis. We notice however that here there is no closure needed at the level of the pressure (which has been eliminated) or of the viscous dissipation, contrary to what happens in the "one-point closure modelling". In this monograph, we will focus on closures which assume that turbulence is close to gaussianity, which is not totally unphysical if we consider turbulence as the result of a superposition of independent Brownian-like chaotic motions to which the central limit theorem could apply. But we have already seen examples of strong departures from Gaussianity in turbulence, for instance when considering the p.d.f. in Chapter VI. However, kinetic-energy spectra and energy transfers do not seem to be much affected by these departures.

This chapter will be divided into two parts: the method and the results.

CHAPTER VII- PART A: THE METHOD

2 - The Quasi-Normal approximation

The most known of the closure hypotheses is the famous Quasi-Normal approximation, proposed by Millionshtchikov (1941) and independently by Chou (1940), as emphasized in Tatsumi (1980). The resulting spectral equations for isotropic turbulence were obtained independently by Proudman and Reid (1954) and Tatsumi (1957). Before developing this analysis, it is useful to recall the main results concerning the Gaussian random functions (see e.g. Blanc-Lapierre and Picinbono (1981):

Let X be a four-dimensional variable (three dimensions of space and one of time), and let $g(X)$ be a random function of X (g might also be a vector) of zero mean. We recall that $g(X)$ is a *gaussian random function* if, given N arbitrary numbers α_i and N values X_i of X, the linear combination $\sum \alpha_i g(X_i)$ is a *gaussian random variable*. The consequences of this definition are in particular that

-for any $X, g(X)$ is a random gaussian variable

-the odd moments of g are zero

-the even moments can be expressed in terms of the second order moments. In particular, fourth-order moments in four points $X_1...X_4$ satisfy

$$\begin{aligned}
<g(X_1)g(X_2)g(X_3)g(X_4)> = &<g(X_1)g(X_2)><g(X_3)g(X_4)> \\
&+ <g(X_1)g(X_3)><g(X_2)g(X_4)> \\
&+ <g(X_1)g(X_4)><g(X_2)g(X_3)> \quad . \quad \text{(VII} - 2 - 1)
\end{aligned}$$

Given any (non gaussian) random function[3] whose second-order moments are known, it is then possible to calculate the fictitious moments of order n that this function would have if it were a gaussian function: the difference between the actual n-th order moment of the function and the corresponding gaussian value is called the n-th order *cumulant*. In particular the odd cumulants are equal to the moments. For a gaussian function, all the cumulants are zero by definition.

Although the velocity probability densities measured experimentally in turbulence are not far from a normal distribution, they display an assymmetry about the mean value which is more and more marked on the successive derivatives of the turbulent signal (Gagne, 1978). The

---

[3] of zero mean

non-zero value of the skewness factor introduced in Chapter VI is a manifestation of this fact. Also, as already seen in Chapter VI, the p.d.f. of the velocity derivatives are closer to an exponential than to a gaussian. It is therefore irrealistic to approximate the velocity field of turbulence by a gaussian random function, since such a turbulence would have no energy transfer between wave numbers (we recall from (VII-1-3) that the transfer is proportional to third-order moments in Fourier space). The idea of the Quasi-Normal approximation (Q.N.) is to simply assume that the fourth-order cumulants are zero, without any assumption on the third-order moments. This allows one to close the problem at the level of eq. (VII-1-4), by replacing the fourth-order moment by the gaussian value obtained from (VII-2-1). It leads to

$$[\frac{\partial}{\partial t}+\nu(k^2+p^2+q^2)] <\hat{u}(\vec{k})\hat{u}(\vec{p})\hat{u}(\vec{q})>=\sum <\hat{u}\hat{u}><\hat{u}\hat{u}> \quad (\text{VII}-2-2)$$

where the sum $\sum$ corresponds to three terms coming from (VII-2-1), and to the various terms included in the $<\hat{u}\hat{u}\hat{u}>$ transfer of the r.h.s. of (VII-1-3). Again the terms $<\hat{u}\hat{u}><\hat{u}\hat{u}>$ are a complicated integral involving various wave numbers and components of the velocity. Then eq. (VII-1-3), where (because of homogeneity) $\vec{k}'$ has been taken equal to $-\vec{k}$, together with eq. (VII-2-2), yield

$$(\frac{\partial}{\partial t} + 2\nu k^2)\hat{U}_{ij}(\vec{k},t) = \int_0^t d\tau \int_{\vec{p}+\vec{q}=\vec{k}}[\exp -\nu(k^2 + p^2 + q^2)(t - \tau)]$$

$$\sum <\hat{u}\hat{u}><\hat{u}\hat{u}> \ d\vec{p} \quad (\text{VII}-2-3)$$

with τ as time argument in the products $<\hat{u}\hat{u}><\hat{u}\hat{u}>$. The calculation of these products involves a lengthy algebra which will not be given in this book, but is at hand for any reader who wants to use these theories. The exact set of equations (VII-1-3) and (VII-2-2) can be found for instance in Tatsumi (1980).

The calculation is simpler for isotropic turbulence without helicity: taking the trace of $\hat{U}_{ij}(\vec{k},t)$ in (VII-2-3) and replacing the spectral tensors by their isotropic values (V-5-9) leads to

$$(\frac{\partial}{\partial t} + 2\nu k^2)E(k,t) = \int_0^t d\tau \iint_{\Delta_k} dp \ dq$$

$$[\exp -\nu(k^2 + p^2 + q^2)(t - \tau)]S(k,p,q,\tau) \quad (\text{VII}-2-4)$$

with

$$S(k,p,q,\tau) = \frac{k^3}{pq} a(k,p,q) \ E(p,\tau)E(q,\tau)$$

$$(\text{VII}-2-5)$$

$$-\frac{1}{2}\frac{k}{pq} E(k,\tau)[p^2 b(k,p,q)E(q,\tau) + q^2 b(k,q,p)E(p,\tau)].$$

The integration in (VII-2-4) is done on the domain Δ_k in the (p,q) plane such that (k,p,q) should be the sides of a triangle, already defined in Chapter VI. The geometrical coefficients $a(k,p,q)$ and $b(k,p,q)$ are evaluated in Leslie (1973). They are equal to, using Leslie's (1973) and Kraichnan's (1959) notations

$$a(k,p,q) = \frac{1}{4k^2} P_{ijm}(\vec{k})\ P_{ibc}(\vec{k})P_{jb}(\vec{p})\ P_{mc}(\vec{q})$$

$$= \frac{1}{2}(1 - xyz - 2y^2z^2) \qquad (VII-2-6)$$

$$b(k,p,q) = \frac{1}{2}k^2 P_{cjm}(\vec{k})\ P_{jbc}(\vec{p})\ P_{mb}(\vec{q})$$

$$= \frac{p}{k}\ (xy + z^3) \qquad (VII-2-7)$$

with summation on the repeated indices ($P_{ijm}(\vec{k})$ has been defined previously in (VI-2-2)). x, y, and z are the cosines of the interior angles of the triangle (k,p,q) facing respectively the sides k, p, q. Actually $a(k,p,q)$ can be obtained by symmetrization of $b(k,p,q)$

$$a(k,p,q) = \frac{b(k,p,q) + b(k,q,p)}{2} \qquad (VII-2-8)$$

and eqs (VII-2-4) (VII-2-5) can be simplified using the above symmetry properties to

$$(\frac{\partial}{\partial t}+2\nu k^2)E(k,t) = \int_0^t d\tau \iint_{\Delta_k} dp\ dq\ [\exp-\nu(k^2 + p^2 + q^2)(t - \tau)]$$

$$\frac{k}{pq}\ b(k,p,q)\ [k^2 E(p,\tau) - p^2 E(k,\tau)]\ E(q,\tau)\quad . \qquad (VII-2-9)$$

It was only in the early sixties that the development of computers allowed a numerical resolution of the isotropic $Q.N.$ equation (VII-2-4). This was done by Ogura (1963) who showed that this approximation eventually led to the appearance of negative energy spectra in the energy-containing eddies range, and checked that this phenomenon was not a numerical artefact. Such a behaviour is of course unacceptable, since the energy spectrum, proportional to $< |\hat{u}(\vec{k})|^2 >$, is a positive quantity by nature. It shows that the $Q.N.$ approximation is incompatible with the dynamics of the Navier-Stokes equation. The same result was independently obtained by O'Brien and Francis (1963) who used the $Q.N.$ approximation to study the spectral evolution of a passive scalar in isotropic turbulence: it was thus the passive scalar spectrum which developed negative values. The reason for this anomalous behaviour of the $Q.N.$ theory was identified by Orszag (1970 b, 1977) who showed that the r.h.s. of (VII-2-2) was responsible for the build-up of too high third order moments. These moments saturate in reality, as is shown for instance experimentally by the not excessive values of the skewness factor. Then the role of the fourth order cumulants (discarded in the *Q.N.* theory) is to provide a damping action leading to a saturation of the third order moments. This is the motivation of the *E.D.Q.N.* and *E.D.Q.N.M.* theories:

3 - The Eddy-Damped Quasi-Normal type theories

Orszag (1970 b, 1977) proposed then to approximate the fourth-order cumulants neglected in (2-2) by a linear damping term, and to replace (VII-2-2) by

$$[\frac{\partial}{\partial t} + \nu(k^2 + p^2 + q^2) + \mu_{kpq}] < \hat{u}(\vec{k})\hat{u}(\vec{p})\hat{u}(\vec{q}) >$$
$$= \sum < \hat{u}\hat{u} >< \hat{u}\hat{u} > \qquad (VII - 3 - 1)$$

where μ_{kpq}, which has the dimension of the inverse of a time, is a characteristic "eddy-damping rate" of the third order moments associated to the triad (k, p, q). This parameter is arbitrary in the theory, and its choice is essential, at least if one wants to use the theory for quantitative predictions. For isotropic turbulence, the following expression

$$\mu_{kpq} = \mu_k + \mu_p + \mu_q \qquad (VII - 3 - 2)$$

where

$$\mu_k \sim [k^3 E(k)]^{1/2} \qquad (VII - 3 - 3)$$

is the inverse of the local non-linear time defined in (VI-4-3), has been proposed by Orszag (1970). The value of the numerical constant in front of the r.h.s. of (VII-3-3) can be shown to be related to the value of the Kolmogorov constant (we anticipate here that these Eddy-Damped theories will lead to a Kolmogorov $k^{-5/3}$ inertial-range spectrum for isotropic three-dimensional turbulence). Actually the choice (VII-3-3) is less satisfactory for a rapidly decreasing spectrum at high k (in the initial stage of evolution of decaying turbulence for instance), where μ_k, given by (VII-3-3), becomes a decreasing function of k: this is fairly irrealistic, since μ_k is a kind of characteristic frequency of turbulence, and one might expect an increase of the frequencies with k. It was thus proposed by Frisch (1974) and Pouquet et al. (1975) to modify μ_k as:

$$\mu_k = a_1 [\int_0^k p^2 E(p, t) dp]^{1/2} \quad , \qquad (VII - 3 - 4)$$

which is growing with k in any situation, and represents the average deformation rate of eddies $\sim k^{-1}$ by larger eddies. a_1 is a numerical constant whose choice will be specified later on. As a matter of fact, both expressions collapse in the inertial-range with a proper choice of the numerical constant in the r.h.s. of (VII-3-3) and (VII-3-4).

The choice of μ_k is more difficult in non-isotropic situations, for instance for problems where waves (Rossby waves, inertial or gravity waves) interact with turbulence (see Legras, 1978, Cambon et al., 1985), and this is still an open question.

This approximation is known as the Eddy-Damped Quasi-Normal approximation ($E.D.Q.N.$). It has to be stressed that the eddy-damping procedure concerns the third-order moments, and that there is no eddy-damping of the energy in the theory, where the kinetic energy is still conserved by non-linear interactions. The resulting evolution equation for the spectral tensor is formally identical to (VII-2-3), provided the viscous damping term arising in the exponential term of the r.h.s. be modified to take into account the eddy-damping. One obtains

$$(\frac{\partial}{\partial t} + 2\nu k^2)\hat{U}_{ij}(\vec{k},t) = \int_0^t d\tau \int_{\vec{p}+\vec{q}=\vec{k}} \exp -[\mu_{kpq} + \nu(k^2 + p^2 + q^2)](t - \tau)$$

$$\sum <\hat{u}\hat{u}><\hat{u}\hat{u}>(\tau)\, d\vec{p} \quad . \qquad (VII-3-5)$$

This $E.D.Q.N.$ approximation, though physically more acceptable, does not nevertheless guarantee the *realizability* (positiveness of the energy spectrum) in all the situations. As shown by Orszag (1977), this can be ensured with a minor modification, called the *markovianization*: this consists in assuming that the exponential term in the integrand of (VII-3-5) varies with a characteristic time $[\mu_{kpq} + \nu(k^2 + p^2 + q^2)]^{-1}$ much smaller than the characteristic evolution time of $\sum <\hat{u}\hat{u}><\hat{u}\hat{u}>$; the latter is of the order of the large-eddy turnover time of the turbulence. This assumption is valid in the inertial and dissipative range, but questionable in the energy-containing range where both times are of the same order. Nevertheless the markovianization allows a considerable simplification of the resulting spectral equations, while ensuring the realizability. Eq. (VII-3-5) is therefore changed into

$$(\frac{\partial}{\partial t} + 2\nu k^2)\hat{U}_{ij}(\vec{k},t) = \int_{\vec{p}+\vec{q}=\vec{k}} \theta_{kpq} \sum <\hat{u}\hat{u}><\hat{u}\hat{u}>(t)\, d\vec{p}$$

$$(VII-3-6)$$

with

$$\theta_{kpq} = \int_0^t \exp -[\mu_{kpq} + \nu(k^2 + p^2 + q^2)](t - \tau)\, d\tau \quad .$$

A last assumption consists in neglecting the time variation of μ_{kpq} in the above determination of θ. This yields for θ_{kpq} an expression proposed by Leith (1971)

$$\theta_{kpq} = \frac{1 - \exp-[\mu_{kpq} + \nu(k^2 + p^2 + q^2)]t}{\mu_{kpq} + \nu(k^2 + p^2 + q^2)} \quad . \qquad (VII - 3 - 7)$$

One can also notice that when t goes to infinity, θ_{kpq} is equivalent to $1/[\mu_{kpq} + \nu(k^2 + p^2 + q^2)]$, and allows one to recover eq. (VII-3-6) by simply neglecting the $\partial/\partial t$ term in (VII-3-1), which allows determination of the $<\hat{u}\hat{u}\hat{u}>$ term arising in eq. (VII-1-3).

This approximation is the Eddy-Damped Quasi-Normal Markovian approximation ($E.D.Q.N.M$). The time θ_{kpq} is characteristic of the relaxation (towards a quasi-equilibrium) by non-linear transfers and molecular viscosity of $<\hat{u}(\vec{k})\hat{u}(\vec{p})\hat{u}(\vec{q})>$. The final $E.D.Q.N.M.$ equation for the spectral tensor is for homogeneous turbulence (Lesieur, 1973)

$$(\frac{\partial}{\partial t} + 2\nu k^2)\hat{U}_{in}(\vec{k}, t) = \int_{\vec{p}+\vec{q}=\vec{k}} d\vec{p}\, \theta_{kpq}(t)P_{ijm}(\vec{k})$$

$$[P_{nab}(\vec{k})\hat{U}_{jb}(\vec{p}, t)\hat{U}_{ma}(\vec{q}, t) - 2P_{mab}(\vec{p})\hat{U}_{ja}(\vec{q}, t)\hat{U}_{bn}(\vec{k}, t)] \quad .$$
$$(VII - 3 - 8)$$

Let us write for instance the spectral $E.D.Q.N.M.$ equation in the case of three-dimensional isotropic turbulence without helicity: taking the trace of (VII-3-8), using (VI-3-3), (VII-2-6) and (VII-2-7), and noticing that

$$P_{ijm}(\vec{k})\, P_{bi}(\vec{k}) = P_{bjm}(\vec{k}),$$

one obtains

$$(\frac{\partial}{\partial t} + 2\nu k^2)E(k, t) =$$

$$\iint_{\Delta_k} dp\, dq\theta_{kpq}(t)\frac{k}{pq}b(k, p, q)E(q, t)[k^2 E(p, t) - p^2 E(k, t)] \quad .$$
$$(VII - 3 - 9)$$

This is, as expected, the equation obtained from (VII-2-9) after the eddy-damping and markovianization procedures. The realizability of eq. (VII-3-9) can be easily obtained by noticing that the coefficient $a(k, p, q)$ is positive (see Orszag, 1977).

4 - The stochastic models

This is another point of view, due to Kraichnan (1961) (see also Herring and Kraichnan, 1972), which consists in replacing the Navier-Stokes equation (for which we cannot solve exactly the closure problem),

by a set of modified equations having the same basic structural properties as Navier-Stokes (quadratic non linearity, non-linear quadratic invariants, existence of truncated inviscid equipartition solutions[4]), and
for which the closure problem can be solved. Thus, instead of considering eq. (IV-2-7), one replaces it by a set of N equations coupling N
fictitious random velocity fields $\hat{\underline{u}}^\alpha(\vec{k},t)$, in the following way[5]

$$(\frac{\partial}{\partial t} + \nu k^2)\hat{u}_i^\alpha(\vec{k},t) = -\frac{i}{N} k_m P_{ij}(\vec{k}) \int_{\vec{p}+\vec{q}=\vec{k}} \Phi_{\alpha\beta\sigma} \hat{u}_j^\beta(\vec{p},t)\hat{u}_m^\sigma(\vec{q},t)d\vec{p}$$

$$(VII-4-1)$$

where the $\Phi_{\alpha\beta\sigma}$ are N^3 gaussian random functions of time, symmetric
with respect to the permutations of α,β,σ, of zero mean, having the
same statistical properties, and which may also depend on the wave
numbers triad $\vec{k},\vec{p},\vec{q}$. It may be shown (Frisch et al., 1974) that in the
limit $N \to \infty$, the various $\hat{\underline{u}}^\alpha$ become independent and gaussian, and
that their energy spectrum satisfies a closed integro-differential equation
which depends on the choice of $\Phi_{\alpha\beta\sigma}$: if $\Phi_{\alpha\beta\sigma}(t)$ is a white-noise gaussian
function of temporal correlation given by

$$< \Phi_{\alpha\beta\sigma}(t)\Phi_{\alpha\beta\sigma}(t') >= 2\,\theta_{kpq}(t)\,\delta(t-t')   , \qquad (VII-4-2)$$

one recovers the same spectral equation as (VII-3-8). Thus, though the
"closure" and the "stochastic model" philosophies are different, they
finally correspond to the same spectral evolution equations: the first
method insists more on the physics of the departures from gaussianity,
while the second permits to keep the model along the same structural
parapets as the Navier-Stokes equation. Both philosophies are com-
plementary, and allow an understanding of which kind of performances
can be expected (and which cannot) from the theory. It is evident for
instance that small-scale intermittency will certainly be badly treated
by the spatial stirring due to the random phases $\Phi_{\alpha\beta\sigma}$, and that the
small-scale departures from the Kolmogorov law which can be predicted
with the aid of intermittency theories (see Chapter VI) escape to the
E.D.Q.N.M. Such a disability seems very light in comparison with the
power of this theory to make non-trivial predictions when applied to
the important mathematical questions (on the existence of singularities
for instance) or physical problems (turbulent diffusion, predictability,
two-dimensional turbulence for instance) envisaged below.

---

[4] see Chapter X

[5] An alternative stochastic-model philosophy, based on a stochastic
Langevin equation for the velocity, has also been introduced (see Leith,
1971, and Herring and Kraichnan, 1972). It yields the same final spectral
equations than the theories presented here.

When θ_{kpq} is taken equal to a constant θ_0, the particular *E.D.-Q.N.M.* equation obtained is called the Markovian Random Coupling Model (*M.R.C.M.*, proposed by Frisch et al., 1974) for it corresponds to a *markovian* version of an earlier model due to Kraichnan (1961), the Random Coupling Model (*R.C.M.*), obtained by taking $\Phi_{\alpha\beta\sigma}$ constant (that is independent of time and wave number, but random) in the stochastic model (VII-4-1). The *R.C.M.* gives the same resulting equations as the Direct-Interaction Approximation (*D.I.A.*, see Kraichnan, 1959), introduced earlier by Kraichnan through formal diagrammatic expansions of the velocity field[6]. It is not a markovian theory, and its spectral equations have a "time-memory", contrary to the *E.D.Q.N.M.* which has forgotten the past and is unable to give information on two-times velocity covariances. The *R.C.M.* (resp *D.I.A.*) introduces a Green's function formalism, in the following way: assume that Navier-Stokes equation possesses a forcing term $\hat{f}(\vec{k})$, and writes:

$$(\frac{\partial}{\partial t} + \nu k^2)\hat{u}(\vec{k},t) = l(\hat{u},\hat{u}) + \hat{f}(\vec{k},t) \quad , \qquad (VII-4-3)$$

where l is a bilinear operator of $\hat{u}$. The Green's function operator g represents the linearized response $\delta\hat{u}$ of $\hat{u}$ to a perturbation $\delta\hat{f}$ of the force $\hat{f}$, and is solution of the linearized equation:

$$(\frac{\partial}{\partial t} + \nu k^2)g = l(g,\hat{u}) + l(\hat{u},g) + I \quad , \qquad (VII-4-4)$$

where I is the identity operator, with

$$\delta\hat{u}_i(\vec{k},t) = \int_{-\infty}^{t} g_{ij}(\vec{k},t,t')\,\delta\hat{f}_j(\vec{k},t')\,dt' \quad . \qquad (VII-4-5)$$

We will assume furthermore that the stirring force is a white noise in time. Hence, for homogeneous turbulence, the correlation of the forcing writes:

$$<\hat{f}_i(\vec{k'},t)\hat{f}_j(\vec{k},t')> = F_{ij}(\vec{k},t)\,\delta(\vec{k}+\vec{k'})\,\delta(t-t') \quad . \qquad (VII-4-6)$$

Therefore the *D.I.A.* spectral equations involve two tensors, the two-time spectral tensor and the average Green's function. They write (see

---

[6] Leslie (1973, p 51) summarizes the various steps of the *D.I.A.*

Lesieur et al., 1971, Leslie 1973)[7]:

$$(\frac{\partial}{\partial t} + \nu k^2)\hat{U}_{in}(\vec{k}, t, t') =$$

$$\frac{1}{2}\int_{\vec{p}+\vec{q}=\vec{k}} d\vec{p} \int d\tau P_{ijm}(\vec{k})P_{lab}(\vec{k}) <g_{nl}> (\vec{k}, t', \tau)\hat{U}_{jb}(\vec{p}, t, \tau)\hat{U}_{ma}(\vec{q}, t, \tau)$$

$$-\int_{\vec{p}+\vec{q}=\vec{k}} d\vec{p} \int d\tau P_{ijm}(\vec{k})P_{lab}(\vec{p}) <g_{lm}> (\vec{p}, t, \tau)\hat{U}_{ja}(\vec{q}, t, \tau)\hat{U}_{bn}(\vec{k}, t, \tau)$$

$$+ <g_{nl}> (\vec{k}, t', t)\ F_{il}(\vec{k}, t)\quad .$$

$$(VII - 4 - 7)$$

$$(\frac{\partial}{\partial t} + \nu k^2) <g_{in}> (\vec{k}, t, t') =$$

$$-\int_{\vec{p}+\vec{q}=\vec{k}} d\vec{p} \int d\tau P_{ijm}(\vec{k})P_{lab}(\vec{p}) <g_{ml}> (\vec{p}, t, \tau) <g_{bn}> (\vec{k}, \tau), t')$$

$$\hat{U}_{aj}(\vec{q}, t, \tau) + P_{in}(\vec{k})\ \delta(t - t')\quad .$$

$$(VII - 4 - 8)$$

Now, the following operations are performed: a) "markovianize" the above *D.I.A.* equations in the sense that the average Green's function $<g_{in}(\vec{k}, t, \tau>$ is replaced by $\theta_0 P_{in}(\vec{k})\delta(t - \tau)$; b) remark that

$$\frac{\partial}{\partial t}U_{in}(\vec{k}, t, t) = \frac{\partial}{\partial t}[U_{in}(\vec{k}, t, t') + U_{ni}(-\vec{k}, t, t')]_{t'=t}\quad .$$

One recovers the *M.R.C.M.* spectral evolution equation (VII-3-8) (with $\theta_{kpq} = \theta_0$), with a forcing term $P_{nl}(\vec{k})F_{il}(\vec{k}, t)$.

As we will see below, the *D.I.A.* does not yield a Kolmogorov $k^{-5/3}$ inertial range, but a $k^{-3/2}$ range instead. One of the reasons for that is the non invariance of the theory under random Galilean transformations. Kraichnan (1965) proposed heuristic Lagrangian modifications of the *D.I.A.*, which restore the Galilean invariance, and are known as Lagrangian History Direct Interaction Approximations (*L.H.D.I.A.*). These theories are not widely utilized presently, and it is not certain that they improve other theories like the *E.D.Q.N.M.* for instance. More details on that can be found in Leslie (1973).

A last model we will briefly present is Kraichnan's Test-Field Model (*T.F.M.*, see Kraichnan, 1971 b). It is a markovian model of the *E.D.Q.-N.M.* type, that is satisfying eq. (VII-3-9), but with a relaxation time θ_{kpq} determined in a more sophisticated way than the simple relation

---

[7] In these two references, the two-time velocity spectral tensor is defined in such a way that it corresponds to the complex conjugate $\hat{U}_{ij}^*(\vec{k}, t, t') = \hat{U}_{ij}(-\vec{k}, t, t')$ of the spectral tensor $\hat{U}_{ij}(\vec{k}, t, t')$ used here and defined with the aid of (V-4-2)

(VII-3-7): the *T.F.M.* evaluates θ_{kpq} by studying the triple correlations of an auxiliary velocity field, the "test-field", transported by the turbulence itself. Such an analysis poses difficulties due to the loss of incompressibility of the test-field, which has to be separated into a solenoidal and compressible component. After various "ad hoc" approximations, one finally obtains coupled equations for the evolution of θ_{kpq}. As shown by Herring et al. (1982), this time, as given by the *T.F.M.*, reduces to the simple *E.D.Q.N.M.* time (VII-3-7) in the inertial range. Moderate Reynolds number direct-numerical simulations of turbulence seem to indicate a very good agreement with the *T.F.M.* calculations (Herring et al., 1973). It is nevertheless not evident that the further "price" paid for the use of the *T.F.M.* instead of the *E.D.Q.N.M.* is really worthwhile.

Figure VII-1 is an attempt to draw a map of this complicated closure world and of the routes which link the various theories: the double arrows indicate an exact result, while the simple lines correspond to an approximation or a modification. We will propose three possible trips starting from the Navier-Stokes equation, trips employing circuits which will be shown to communicate:

● **trip A**
The hiker takes the "moments equations" way, and from there goes either to the *Q.N.* approximation (by discarding the fourth order cumulants $< uuuu >_c$), or to the *E.D.Q.N.* approximation (by setting these cumulants equal to $-\mu < uuu >$). In any case he can commute easily from the *Q.N.* to the *E.D.Q.N.*, by changing νk^2 into $\mu_k + \nu k^2$. With the aid of a "markovianization", he will go from the *E.D.Q.N.* to the *E.D.Q.N.M.* The same operation would have led him from the *Q.N.* to the *Q.N.M.* (Quasi-Normal Markovian approximation), that he could also have reached from the *E.D.Q.N.M.*, by simply setting $\mu_k = 0$. Notice that the *Q.N.M.* was extensively studied by Tatsumi and colleagues (1978, 1980). We will discuss it by comparison with the *E.D.Q.N.M.* later on. Notice also that attached to the *E.D.Q.N.M.* is the *T.F.M.*, which has a special status for the time θ_{kpq}.

● **trip B**
The hiker takes the difficult path of the Feynman formal diagrammatic expansions: then he has the choice between the already mentioned Renormalization-Group Techniques (*R.N.G.*) (see below), or the *D.I.A.*. From there he can go to the *L.H.D.I.A.*, by restoring the random Galilean invariance.

● **trip C**
The hiker modifies the non-linear terms of the Navier-Stokes equation

with the aid of gaussian random phases $\Phi_{\alpha\beta\sigma}$, and goes to the Stochastic Models. Then he has several choices: taking the coupling phases satisfying (VII-4-2) enables him to go to the $E.D.Q.N.M.$ (and then rejoin the trip A). Taking the phases as random *variables* (and not *functions*) will permit him to go to the $R.C.M.$ which, as already stressed, is identical to the $D.I.A.$ This indicates a communication with the circuit B. A "markovianization" of the $R.C.M.$ at the level of the Green's function leads to the $M.R.C.M.$, which could also have been reached from the $E.D.Q.N.M.$ by letting $\theta_{kpq} \equiv \theta_0$. This provides a bridge to trip A. Finally, as already mentioned, the $M.R.C.M.$ can also be obtained from the stochastic models by letting the random phases be a white noise with respect to time, but independent of the wave numbers.

It is of course difficult to give a definitive answer to the question of what is the "best" analytical theory to use. From a personal experience of these tools, which will be developed in the following sections and chapters, I would suggest that if the turbulence is isotropic (in three- or two-dimensions), the $E.D.Q.N.M.$ is certainly an excellent tool to predict the energy transfers between various modes and the high Reynolds number behaviours, at low numerical cost (compared with the numerical direct or large-eddy-simulations). The $M.R.C.M.$ will be its faithful ally, allowing qualitative analytical predictions about the existence of inertial ranges, and the possible loss of regularity for vanishing viscosity or diffusivity. Some problems where we need information on two-point correlations will require the use of the $D.I.A.$ In non-isotropic situations such as the collapse of stably-stratified turbulence, the anisotropic $E.D.Q.N.M.$ with the isotropic form of the time θ_{kpq} could also allow a first step towards the understanding of the problem. This is nevertheless not true for rotating turbulence, where, as shown in Cambon et al. (1985), the only effect of the rotation within the $E.D.Q.N.M.$ is to modify θ in a non-trivial way. Such modifications require looking to higher order moments in the hierarchy, following methods developed e.g. by André (1974) and Legras (1978). However, it seems now that the increased computing possibilities offered by the present super-computers will render possible a direct numerical calculation of the large scales of the flows and of the inhomogeneities and anisotropies they contain, the small isotropic scales being modelled with the aid of the closures, for instance the $E.D.Q.N.M.$ For these reasons there does not seem to be an urgency to find a new closure of the type presented above: such a new closure would behave essentially in the same way as its predecessors. What seems more important is to learn how to handle the closures in the simplest way, analytically for instance when possible: indeed, one cannot imagine using complicated closure calculations in situations where direct-numerical simulations are at hand.

The rest of this Chapter will be devoted to analytical and numerical results concerning the closure spectral equations in the case of isotropic three-dimensional turbulence. Other situations will be envisaged in the following Chapters.

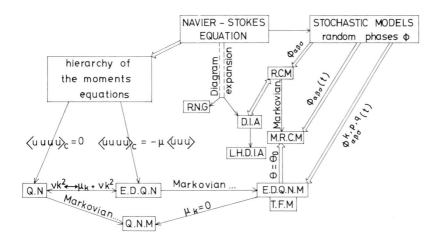

Figure VII-1: Map of the analytical statistical theories and stochastic models land (see text for comments)

CHAPTER VII-PART B: THE RESULTS

5 - Phenomenology of the closures

It is possible to derive a simple phenomenological analysis allowing one to predict the inertial-range exponents obtained in the frame of the closures. These predictions will of course be checked numerically later on, but they constitute a first step towards an analytical understanding of these theories, which in turn will enable us to understand more deeply the phenomenological analysis of turbulence presented in Chapter VI. We will concentrate on isotropic turbulence without helicity. In this case, eq. (VII-3-9) permits to calculate the $E.D.Q.N.M.$ energy flux

$$\Pi(K) = \int_K^\infty dk \iint_{\Delta_k} dp\ dq\ \theta_{kpq}(t)$$

$$\frac{k}{pq}\ b(k,p,q)E(q,t)[k^2 E(p,t) - p^2 E(k,t)] \ . (VII-5-1)$$

We are looking for an inertial-range, where viscous effects can be neglected, and where $\Pi(K)$ will be independent of K and equal to ϵ. It has been shown by Kraichnan (1971 a, see also André and Lesieur, 1977) that, within the $E.D.Q.N.M.$ or the $T.F.M.$ theories, and in the Kolmogorov inertial-range, wave numbers in a spectral vicinity of one decade about K participate in more than 80% of the energy flux across K. This allows us to assume that, to a first approximation, the integral (VII-5-1) is dominated by wave numbers k, p, q, of order K (i.e. comprised for instance between $K/10$ and $10K$). Assuming also that the quantities under the integral (VII-5-1) vary as powers of k, p, q, and remembering a remark already made in Chapter VI that $\int_{K/10}^K E(k)dk$ is of the order of $KE(K)$, we finally obtain

$$\Pi(k) \sim \theta(k)\ k^4\ E(k)^2 \qquad\qquad (VII-5-2)$$

where $\theta(k)$ is the value taken by θ_{kpq} for $k = p = q$. We notice also that if k is smaller than the Kolmogorov dissipative wave number, and for large times, $\theta(k)$ is of the order of $1/[k^3 E(k)]^{1/2}$. Inserting this value in (VII-5-2) and looking for solutions such that $\Pi(k) \equiv \epsilon$ yields

$$E(k) \sim \epsilon^{2/3} k^{-5/3} \qquad . \qquad\qquad (VII-5-3)$$

This shows that the $E.D.Q.N.M.$ leads to a Kolmogorov $k^{-5/3}$ energy cascade, which was to be expected as soon as $\theta(k)$ was a function of k and $E(k)$ only. The Kolmogorov constant in the r.h.s. of (VII-5-3) can easily be shown to be proportional to the two-third power of the adjustable constant of the theory, arising in (VII-3-3) or (VII-3-4)[8].

In the case of the $M.R.C.M.$, the time $\theta(k)$ is equal to θ_0, and the above analysis gives

$$E(k) \sim (\frac{\epsilon}{\theta_0})^{1/2} k^{-2} \qquad . \qquad\qquad (VII-5-4)$$

It is not the Kolmogorov law, which is not surprising since θ_0 enters now as an independent parameter. The -2 power is here equivalent to the

---

[8] see e.g. André and Lesieur (1977)

actual $-5/3$ power of Kolmogorov's theory. It will be seen later on that the *M.R.C.M.* is nevertheless a very good tool which permits analytical calculations and predicts the existence and direction of the cascades.

For the *R.C.M.* (resp. *D.I.A.*), and though eq. (VII-5-1) is not exact, eq. (VII-5-2) can still be shown to be valid, $\theta(k)$ being now the relaxation time of the Green's function. In the inertial range, this time is of order $1/(v_0 k)$, where v_0 is the r.m.s. velocity (Kraichnan, 1959, Lesieur et al., 1971). This is different from the *E.D.Q.N.M.* where the time $\theta(k)$ can also be written as $(1/k v_k)$, v_k being the local velocity $[kE(k)]^{1/2}$. Therefore the *D.I.A.* inertial-range is

$$E(k) \sim (\epsilon v_0)^{1/2} k^{-3/2} \quad . \qquad (VII-5-5)$$

It has been shown by Kraichnan (1966) that the existence of a *D.I.A.* inertial-range exponent differing from $-5/3$ was related to the non-invariance of the model by random Galilean transformations: indeed this non-invariance implies that the large eddies have a direct action on the small eddies, which is in contradiction with the locality assumption implicitly contained in the Kolmogorov theory. So a necessary condition for obtaining the $k^{-5/3}$ law is to satisfy the random Galilean invariance principle. This is *not* a sufficient condition, as shown by the example of the *M.R.C.M.*, which is random Galilean invariant (at the level of the final spectral equation, not of the stochastic model itself, c.f. Lesieur, 1973) but does not admit the $-5/3$ inertial-range.

The case of the *Q.N.M.* is different: this theory seems to be quite far from the actual physics of turbulence, since only molecular viscosity damps the third-order moments. Nevertheless the markovianization guarantees the realizability, though the skewness factor (calculated from (VI-7-4')) reaches values far superior to the experimental values (c.f. Tatsumi et al., 1978, Frisch et al., 1980). Actually, the latter authors have shown that, in the limit of zero viscosity (with the *type S* initial conditions defined in Chapter VI), the skewness factor of the *Q.N.M.* blows up with the enstrophy, while for the *E.D.Q.N.M.* it remains bounded. The other interesting particularity of the *Q.N.M.* is to exhibit two inertial-ranges: indeed the time θ_{kpq} can then be obtained from eq (VII-3-7) by setting μ_{kpq} equal to zero. For a given time t, one can therefore introduce a wave number

$$k_t = \frac{1}{(\nu t)^{1/2}} \qquad (VII-5-6)$$

which is such that

$$\theta_{kpq} = \begin{cases} t, & \text{for } k << k_t \\ 1/\nu(k^2 + p^2 + q^2), & \text{for } k >> k_t \end{cases}$$

and eq. (VII-5-2) leads to

$$\Pi(k) \sim t k^4 E(k)^2; \; E(k) \sim (\frac{\epsilon}{t})^{1/2} k^{-2} \; (k << k_t) , \qquad (VII-5-7)$$

$$\Pi(k) \sim \frac{1}{\nu} k^2 E(k)^2; \; E(k) \sim (\epsilon\nu)^{1/2} k^{-1} \; (k >> k_t) . \qquad (VII-5-8)$$

These two inertial ranges separated by k_t are actually well recovered in the numerical resolution of the $Q.N.M.$ spectral equation done by Frisch et al. (1980), as is shown in Figure VII-2. The same inertial-ranges exist in the $Q.N.$ approximation, as recalled in Tatsumi (1980). In this reference, are also presented high Reynolds number $Q.N.M.$ calculations (independent of the above quoted calculations done by Frisch et al., 1980) which display without ambiguity both k^{-2} and k^{-1} ranges. Notice finally that the k^4 infrared ($k \to 0$) behaviour of the spectrum appearing in Figure VII-2 will be explained in section 10.

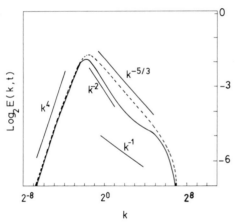

Figure VII-2: decaying energy spectra in: a) an $E.D.Q.N.M.$ calculation (dashed line); b) a Q.N.M. calculation (straight line). In the latter case, the two k^{-2} and k^{-1} inertial ranges are clearly displayed (from Frisch et al., 1980, courtesy J. Fluid Mech.)

6 - Numerical resolution of the closure equations

The closure equations of the markovian type can be solved numerically in the isotropic case, even at very high Reynolds numbers $R = v_0/\nu k_i(0)$ (v_0 and $k_i(0)$ are the initial values of the r.m.s. velocity and of the wave number where the energy spectrum peaks). The relative simplicity of the numerical resolution, as compared to the direct simulations of the Navier- Stokes equation itself, comes from the existence of one spatial variable (the wave number k), and from the fact that the energy spectrum varies smoothly with k. This allows us to take a logarithmic discretization for k

$$k_L = \delta k \ 2^{L/F} \qquad \text{(VII} - 6 - 1)$$

where δk is the minimum wave number, and F the number of wave numbers per octave. In the following results (due to André and Lesieur, 1977, and Lesieur and Schertzer 1978), the third-order velocity correlations relaxation rate μ_k is chosen according to eq. (VII-3-4), with $a_1 = 0.36$, which corresponds to a Kolmogorov constant of 1.4. These calculations take $F = 4$, and a maximum wave number k_c related to $k_i(0)$ by

$$\frac{k_c}{k_i(0)} = 8 \ R^{3/4} \qquad \text{(VII} - 6 - 2)$$

following the law (VI-6-2). The numerical factor 8 has been adjusted in the calculation, so as to take into account the dissipative range in the wave-number span. For instance, a calculation with 65 points taking $\delta k = 1/4$ and $k_i(0) = 1$ has a maximum wave number $k_c = 2^{14}$ and an initial Reynolds number of 26000 . The non-linear transfer of (VII-3-9) can be calculated with a numerical scheme developed by Leith (1971) which conserves exactly the quadratic invariants (kinetic energy, helicity, enstrophy in two-dimensions) when applied to the *E.D.Q.N.M.* spectral equations. The time-differencing in (VII-3-9) is approximated by a forward scheme, with a stability condition

$$\nu k_{max}^2 \ \delta t \leq 1 \quad . \qquad \text{(VII} - 6 - 3)$$

Due to the logarithmic discretization, a problem arises nevertheless when evaluating the kinetic energy transfer: indeed the "elongated" triads (k, p, q) whose ratio of the minimum to the maximum wave number is lower than $2^{1/F} - 1$ or $1 - 2^{-1/F}$ are not taken into account in the discretization (except for the isoscleles interactions $k = p$ or $k = q$). This may be at the origin of non negligible errors, especially for two-dimensional turbulence where, as will be seen in Chapter IX, the non local interactions play a major role in the enstrophy cascade. One possible way of getting rid of this difficulty is to discard all the remaining non local interactions in the transfer computed numerically, calculate analytically the transfers corresponding to the non local interactions, and reintroduce them in the calculation. This method has been developed by

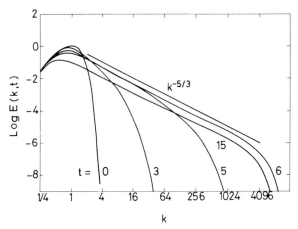

Figure VII-3: time evolution of a freely-evolving three-dimensional kinetic energy spectrum, within the $E.D.Q.N.M.$ theory (from André and Lesieur, 1977, courtesy J. Fluid Mech.)

Lesieur and Schertzer (1978), and also applies to two-dimensional turbulence (Pouquet et al., 1975, Basdevant et al., 1978). We will only give here the general approach of the method: to simplify, we assume that the two numbers $2^{1/F} - 1$ and $1 - 2^{-1/F}$ are equal (for $F = 4$, they are respectively 0.19 and 0.16) to a, and that a is small compared with 1. Therefore, as shown in Lesieur and Schertzer (1978), the non-local transfers $T_{NL}(k,t)$ due to triad-interactions in (VII-3-9) such that

$$inf(k,p,q)/sup(k,p,q) \leq a \qquad\qquad (VII-6-4)$$

can be written as

$$T_{NL}(k,t) = -\frac{\partial\Pi_{NL}(k,t)}{\partial k} \qquad\qquad (VII-6-5)$$

with

$$\Pi_{NL}(k,t) = \Pi^+_{NL}(k,t) - \Pi^-_{NL}(k,t) \qquad\qquad (VII-6-6)$$

$$\Pi^+_{NL}(k,t) = 2\int_0^{ak} dq \int_k^{k+q} dk' \int_{k'-q}^k S(k',p,q)dp \qquad (VII-6-7)$$

$$\Pi^-_{NL}(k,t) = 2\int_0^k dk' \int_{sup(k,k'/a)}^{\infty} dp \int_{p-k'}^p S(k',p,q)dq \ . \quad (VII-6-8)$$

These expressions are equivalent to eqs (VI-3-5,...-3-10) for the non-local interactions. $S(k,p,q)$ is the integrand in the r.h.s. of (VII-3-9), symmetrized with respect to p and q. The non-local flux $\Pi^+_{NL}(k,t)$ corresponds to a "semi-local" flux through k in the sense that the triads involved are such that $q << p \approx k < k'$. Then expansions with respect to the small parameters q/k' in (VII-6-7) and k'/p in (VII-6-8) can be performed (c.f. Kraichnan, 1966, 1971 a, Lesieur and Schertzer, 1978, and Métais, 1983). These expansions are somewhat tedious, but can be appreciably simplified, using the two following results derived in Lesieur and Schertzer (1978) and concerning the calculation of Π^+_{NL} and Π^-_{NL}:

-First result ($q << k$): if, for $q << k$, $(kq/p)S(k,p,q)$ is expanded in (q/k) as $S_1[E(k), E(q)] g_1(y)$, then

$$\Pi^+_{NL}(k,t) = 2 \int_0^{ak} q \, dq S_1[E(k), E(q)] \int_0^1 d\phi \int_\phi^1 g_1(y) dy$$

-Second result ($k << p$): if, for $k << p$, $(kp/q)S(k,p,q)$ is expanded in (k/p) as $S_2[E(p), E(k)] g_2(z)$, then

$$\Pi^-_{NL}(k,t) = 2 \int_0^k dk' \int_{sup(k,k'/a)}^\infty S_2[E(p), E(k')] dp \int_0^1 g_2(z) dz$$

Actually, these results are valid also for the non-local fluxes of other quantities such as the helicity or a passive scalar (see Chapter VIII). Notice also that generally the time θ_{kpq} is not expanded. In the three-dimensional isotropic case without helicity for instance, we have[9]

$$\Pi^+_{NL}(k,t) = \frac{2}{15} \int_0^{ak} \theta_{kkq} \, q^2 E(q) dq [kE(k) - k^2 \frac{\partial E}{\partial k}]$$

$$+ \frac{2}{15} \int_0^{ak} \theta_{kkq} \, q^4 dq \frac{E^2(k)}{k} \qquad (VII - 6 - 9)$$

$$\Pi^-_{NL}(k,t) = -\frac{2}{15} \int_0^k k'^2 E(k') dk' \int_{sup(k,k'/a)}^\infty \theta_{k'pp} [5E(p) + p\frac{\partial E}{\partial p}] dp$$

$$+ \frac{14}{15} \int_0^k k'^4 dk' \int_{sup(k,k'/a)}^\infty \theta_{k'pp} \frac{E(p)^2}{p^2} dp \qquad (VII - 6 - 10)$$

The physics of these terms will be interpreted later. The non-local transfer is then calculated from (VII-6-9) and (VII-6-10). Notice that with

---

[9] The non local fluxes given below have been carefully checked by Métais (1983), and correspond to the notations of the present book for the energy spectrum (density of $(1/2) < \vec{u}^2 >$). The reader is warned that an error of a factor 2 is contained in one of the terms of the analogous expressions given in Lesieur and Schertzer (1978).

this "flux form", these non-local transfers are energy conservative provided $\Pi_{NL}(0,t) = 0$.

The non-local energy transfers derived from (VII-6-9) and (VII-6-10) are then[10]

$$
\begin{aligned}
T_{NL}(k,t) = {} & \frac{2}{15}[4E(k) + 2k\frac{\partial E}{\partial k} + k^2\frac{\partial^2 E}{\partial k^2}]\int_0^{ak}\theta_{kkq}\,q^2E(q)dq \\
& -\frac{4}{15}[3E(k) + k\frac{\partial E}{\partial k}]\frac{E(k)}{k^2}\int_0^{ak}\theta_{kkq}\,q^4dq \\
& -\frac{2}{15}\theta_{k,k,ak}\,a^5k^3E(k)^2 \\
& -\frac{2}{15}\theta_{k,k,ak}\,a^3k^3E(ak)[E(k) - k\frac{\partial E}{\partial k}] \\
& -\frac{2}{15}k^2E(k)\int_{k/a}^{\infty}\theta_{kpp}[5E(p) + p\frac{\partial E}{\partial p}]dp \\
& +\frac{14}{15}k^4\int_{k/a}^{\infty}\theta_{kpp}\frac{E(p)^2}{p^2}dp \quad . \qquad (\text{VII} - 6 - 11)
\end{aligned}
$$

The numerical resolution of eq. (VII-3-9) consists then in solving the equation

$$
(\frac{\partial}{\partial t} + 2\nu k^2)E(k,t) = T_{NL}(k,t) + T_L(k,t) \qquad (\text{VII} - 6 - 12)
$$

where $T_L(k,t)$ is the local transfer calculated numerically with Leith's numerical scheme mentioned above. In fact, calculations which neglect the non-local transfers do not, for $F > 4$, depart very much from the complete calculation. The non-local expansions are anyhow extremely useful to understand the infrared dynamics, and also to provide subgrid-scale parameterizations, as will be seen in Chapter XII.

Figure VII-3 (taken from André and Lesieur, 1977) shows time-evolving energy spectra, initially peaked in the vicinity of $k = k_i(0)$: one can clearly see the spreading of the spectrum towards large wave numbers, due to non-linear interactions; a $k^{-5/3}$ spectrum establishes after a time of about $t_* = 5/v_0k_i(0)$. Since the initial energy spectrum is strongly peaked at $k_i(0)$, the time t_* is also of the order of $5\,D(0)^{-1/2}$

---

[10] These non-local energy transfers can also be obtained directly from non-local expansions of the transfer term $T(k,t)$ integrated on the domain satisfying the condition (VII-6-4). The calculation is equivalent, as noticed in Lesieur and Schertzer (1978), but is longer since a second order expansion in a is then needed. In fact, it has been verified by Métais (1983) that the two methods give the same result.

(it is actually of $5.6\,D(0)^{-1/2}$ in this calculation). For $t > t_*$, the spectrum decays self-similarly following what can be checked to be the law (VI-6-9). Figure VII-4 (also taken from André and Lesieur, 1977) shows the asymptotic tendency of the kinetic energy time-evolution: when the Reynolds number exceeds a few thousands, the kinetic energy reaches an asymptotic state where it is conserved for $t < t_*$ and then dissipated at a finite rate for $t > t_*$. Finally, Figure VII-5 shows, in the same calculation as that of Figure VII-4, the time-evolution of the velocity skewness factor (calculated from VI-7-4'), which also reaches at high Reynolds number an asymptotic evolution where it is conserved for $t > t_*$, as predicted in Chapter VI. The value of the skewness for $t > t_*$ is equal to 0.5, close to the 0.4 value found experimentally. This is a further argument in favor of the *E.D.Q.N.M.* theory which, once its adjustable constant a_1 arising in (VII-3-4) has been chosen to fit the Kolmogorov constant, predicts the skewness factor satisfactorily. It will be shown analytically below that the E.D.Q.N.M. skewness remains finite for $t < t_*$ when the viscosity goes to zero. The abrupt decrease of the skewness for $t > t_*$ may, from Chapter VI, be explained in the following manner: before t_*, still in the limit of zero viscosity, one may assume that the skewness obeys the Euler equation, with a kinetic-energy spectrum decreasing rapidly at infinity. At $t = t_*$, the "Euler skewness" would fall to 0, corresponding to the formation of a $k^{-5/3}$ spectrum extending to infinity (see Chapter VI). In fact, the skewness is now determined by the viscous balance (VI-7-6) and the shape of the spectrum in the dissipation range, and has no reason to be equal to the Euler value just before t_*. Hence there is a discontinuity between the inviscid value 0.75 before t_* and the viscous self-similar plateau 0.5 after t_*.

These numerical three-dimensional isotropic evolving *E.D.Q.N.M.* calculations show therefore after the critical time t_* a behaviour characterized by a finite dissipation of energy and a constancy of the skewness factor, agreeing well with conclusions in Chapter VI.

7 - The enstrophy divergence and energy catastrophe

The last consequence which can be drawn from Figure VII-4 is that at vanishing viscosity the enstrophy blows up at t_*: indeed, we recall from (VI-2-8) that in the freely-evolving turbulence (without external forces), the kinetic energy dissipation rate is equal to

$$\epsilon = -\frac{1}{2}\frac{d}{dt} < \vec{u}^2 > = 2\nu\,D(t) \qquad (VII-7-1)$$

Turbulence in fluids

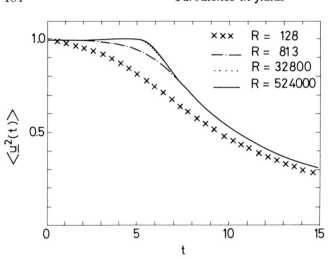

Figure VII-4: time evolution of the kinetic energy at different initial Reynolds numbers, in $E.D.Q.N.M.$ calculations performed by André and Lesieur (1977) (courtesy J. Fluid Mech.)

where $D(t)$ is the enstrophy. So the fact that the kinetic energy is conserved at vanishing viscosity for $t < t_*$, and dissipated at a finite rate afterwards, implies that the enstrophy diverges at t_* in the limit of zero viscosity: this is not surprising, since we have already seen that a Kolmogorov $k^{-5/3}$ spectrum appears at $t = t_*$; for vanishing viscosity the Kolmogorov wave number k_d defined in (VI-4-2) goes to infinity, and the total enstrophy diverges with $\int^{k_d} k^{1/3} dk$. The divergence of the enstrophy at t_* cannot be proved exactly analytically in the frame of the $E.D.Q.N.M.$, but an exact result can be derived with the $M.R.C.M.$, as demonstrated in Lesieur (1973) and André and Lesieur (1977): by multiplication of eq. (VII-3-9) by k^2, integration over k from zero to infinity, and the exchange of the (k,p) variables in the $E(q)E(k)$ term, one obtains

$$\frac{dD}{dt} = \int \frac{k^2}{q}(k^2 - p^2)(xy + z^3)\theta_{kpq}E(p)E(q)dp\,dq\,dk - 2\nu \int_0^\infty k^4 E(k)dk .$$

$$(VII - 7 - 2)$$

We start initially with *type S* conditions (see Chapter VI) in the Euler case. A symmetrization with respect to p and q, and the change of variable $k \to x, dk = -(pq/k)dx$, yields

$$\frac{dD}{dt} = \frac{1}{2}\int_0^{+\infty}\int_0^{+\infty} p^2 q^2 E(p)E(q) \int_{-1}^{+1} \theta_{kpq}(1 - x^2)dx\,dp\,dq$$

— {a term which would be zero if θ_{kpq} were an even function of

x for p and q fixed} .

$$(VII - 7 - 3)$$

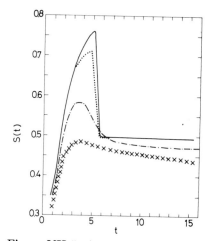

Figure VII-5: time evolution of the skewness factor at different Reynolds numbers (same conditions as in Figure VII-4, courtesy J. Fluid Mech.)

Thus, if $\theta_{kpq} \equiv \theta_0$, (VII-7-3) becomes

$$\frac{dD}{dt} = \frac{2}{3}\theta_0 D^2 \qquad\qquad (VII-7-4)$$

whose solution is

$$D(t) = \frac{3}{2\theta_0} \frac{1}{t_* - t} \qquad\qquad (VII-7-5)$$

$$t_* = \frac{3}{2D(0)\theta_0} \quad . \qquad\qquad (VII-7-6)$$

This result shows analytically that whithin the $M.R.C.M.$ the enstrophy also diverges at a critical time t_* (of course different from the critical time found numerically in the $E.D.Q.N.M.$). Note that the replacement of θ_{kpq} by θ_{00q} in (VII-7-2) (which is extremely arbitrary here, but will be shown to be physically plausible when studying the passive scalar diffusion) will give an enstrophy evolution equation of the type

$$\frac{dD}{dt} \sim D^{3/2} - 2\nu P(t) \qquad\qquad (VII-7-7)$$

which will blow up at a time t_* proportional to $D(0)^{-1/2}$, when $\nu \to$ 0. Notice also, as stressed by Orszag (1977), that a calculation similar to the analysis leading to (VII-7-4) was performed by Proudman and Reid (1954) in the Quasi-Normal case. The resulting equation for the enstrophy is

$$\frac{d^2 D}{dt^2} = \frac{2}{3} D^2 \quad ,$$

as the reader may easily check using eq. (VII-2-9).

All these results, allied with the phenomenology of Chapter VI, go in favour of the same singular behaviour for the decaying Navier-Stokes equation in the limit of zero viscosity: this divergence of the enstrophy, due to the stretching of vortex filaments by turbulence, would be accompanied by the formation of the Kolmogorov $k^{-5/3}$ spectrum. The further evolution would be characterized by a finite dissipation of kinetic energy and a constancy of the skewness factor. The rigourous derivation of these conjectures, if correct, would certainly constitute a major breakthrough in the theory of isotropic turbulence. Notice finally that the value of 5.6 $D(0)^{-1/2}$ predicted for t_* by the E.D.Q.N.M. calculation is very close to the value predicted in (VI-7-9) assuming a constant skewness factor of 0.4.

A last remark concerning these enstrophy equations refers to the Q.N.M. approximation and is developed in Frisch et al. (1980): from (VII-7-4), the enstrophy equation of this latter theory is (when $\nu \to 0$)

$$\frac{dD}{dt} = \frac{2}{3}t \, D^2 \qquad\qquad (VII - 7 - 8)$$

since the Q.N.M. θ_{kpq} time is then equal to t (for a fixed t, the wave number k_t introduced in (5-6) goes to infinity at vanishing viscosity). Therefore the enstrophy blows up at $t_* = [3/D(0)]^{1/2}$. Since in the early stage of evolution (that is for $t < t_*$) the skewness factor is, from (VI-7-7), proportional to $D^{-3/2}(dD/dt)$, it will grow like $tD^{1/2}$ and become infinite with the enstrophy. In the E.D.Q.N.M. on the contrary, it can be shown (see André and Lesieur, 1977) that $D^{-3/2}(dD/dt) \leq 1.51$: hence the skewness will be upper bounded by 1.77 before t_*.

8 - The Burgers-$M.R.C.M.$ model

The Burgers equation

$$\frac{\partial}{\partial t}u(x,t) + u\frac{\partial u}{\partial x} = \nu\frac{\partial^2 u}{\partial x^2} \qquad\qquad (VII - 8 - 1)$$

has been widely studied as a unidimensional model of turbulence (see e.g. Tatsumi, 1980), Fournier and Frisch, 1983 a). Actually it has been shown to develop randomly distributed shocks (*sawtooth profile* in the inviscid case and starting with random initial conditions. These shocks correspond to a k^{-2} energy spectrum extending to infinity. When the viscosity is finite, but small, the k^{-2} inertial range is terminated at high wave numbers by a dissipation range.

employed the term "linear" for the cascade, because the same behaviour holds for a passive scalar, as will be seen in Chapter VIII. These energy and helicity spectra are shown in Figure VII-7.

• iii) the *relative helicity* $H(k)/kE(k)$ is, in the inertial-range, proportional to k^{-1}, and decreases rapidly with k. Thus the helicity has no real influence on the energy flux, expressed in terms of the energy dissipation rate ϵ. It follows that the Kolmogorov constant in the energy cascade is not modified by the presence of helicity.

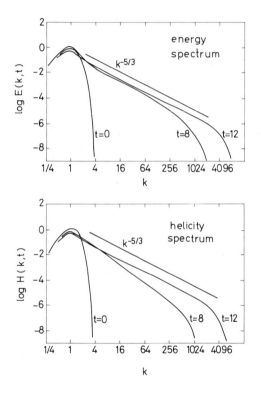

Figure VII-7: time evolution of the energy and helicity spectra (the unit of time is $1/v_0 k_i(0)$), in the same calculation as in Figure VII-6. Now the inertial-ranges establish at about $t_* = 9$ (from André and Lesieur, 1977, courtesy J. Fluid Mech.)

By analogy with two-dimensional turbulence (see Chapter IX), Brissaud et al. (1973 b) have conjectured the possibility of a pure helicity cascade towards large wave numbers, with a zero energy flux, together with an inverse energy cascade towards low wave numbers, with no helicity flux. The latter cascades were determined by phenomenological arguments, and such that

$$E(k) \sim \epsilon^{2/3} k^{-5/3}; \ H(k) \sim \epsilon^{2/3} k^{-2/3} \ . \qquad (\text{VII} - 9 - 8)$$

These cascades did not appear in the calculations of André and Lesieur (1977) which, however, were unforced calculations. Nevertheless, it was checked in the same reference that the *E.D.Q.N.M.* energy flux through stationary energy and helicity spectra given by (VII-9-8) was positive, which eliminates the possibility of such inverse cascading spectra. This result was corroborated by the study of the absolute equilibrium ensemble solutions of the truncated Euler equations (see Chapter X) with helicity, made by Kraichnan (1973), which did not show any inverse energy transfer tendency. So the possibility of strong inverse energy transfers in the presence of maximal helicity seems to be ruled out.

New ideas were more recently proposed by Tsinober and Levich (1983) and Moffatt (1985), where there would exist in a flow with zero mean helicity local regions (in the $\vec{x}$ space) with non-zero helicity (positive or negative) where the kinetic energy dissipation would be less active than in the non-helical regions, because of the preceding results concerning the inhibition of kinetic energy dissipation by helicity. The flow would then evolve towards a set of "coherent" helical structures separated by non-helical dissipative structures (maybe fractal). The coherent structures of same sign could possibly pair, leading to the inverse transfer sought for in vain above. Up to now this is nothing more than a conjecture. This tendency has been found in direct numerical simulations of the Taylor-Green vortex (Shtilman et al., 1985). However, other simulations of isotropic, homogeneous or inhomogeneous turbulence done by Rogers and Moin (1987 b) do not exhibit a correlation between the local relative helicity and the kinetic energy dissipation. Remark finally that an investigation of the transfers within the complex helical waves decomposition introduced in IV-5 has not shown any tendency to significant positive transfer between helical waves of same polarization (Lesieur, 1972, Jallade, 1986).

10 - The decay of kinetic energy

Section 6 has shown that after a critical time the kinetic energy of freely-evolving (we will also say "decaying") three-dimensional isotropic turbulence would decay at a finite rate. The asymptotic laws of decay pose an interesting question, and such information can be very useful for the one-point closure modelling of turbulence for instance, or problems such as the action of rotation or stratification on initially isotropic turbulence, already mentioned in Chapter III. Experimentally these questions are studied in grid turbulence facilities, where the turbulence observed at a distance x downstream of the grid has decayed during a time $t = x/U$ since it was formed behind the grid (U is the mean velocity of the flow in the apparatus): for instance Comte-Bellot and Corrsin (1966) found a decay exponent of the kinetic energy equal to -1.26

$$\frac{1}{2} < \vec{u}^2 > \propto t^{-1.26}$$

while Warhaft and Lumley (1978) found -1.34. The latter law was valid up to about 60 initial large-eddy-turnover times. We will see in this section that the *E.D.Q.N.M.* closures give valuable information about the possible decay laws, according to the shape of the initial energy spectrum. It must also be stressed that if it could be shown experimentally that this finite decay of kinetic energy occurs at a finite time when the viscosity goes to zero, it would be a further argument in favour of an inviscid enstrophy blow up at a finite time.

Let us first return to the concept of non-local interactions introduced in section 6, and utilized here to calculate the energy transfer when $k \to 0$ [$k << k_i$ and $E(k) < E(k_i)$] : the "non-local parameter" a will be taken equal here to k/k_i. The predominant terms in (VII-6-11) are the last two terms, which correspond to non-local interactions $k << p \approx q \approx k_i$. We have to the lowest order in k/k_i;

$$T(k,t) = -\frac{2}{15} k^2 E(k) \int_{k_i}^{\infty} \theta_{0pp} [5E(p) + p\frac{\partial E}{\partial p}] dp$$
$$+ \frac{14}{15} k^4 \int_{k_i}^{\infty} \theta_{0pp} \frac{E(p)^2}{p^2} dp + O[kE(k)]^{3/2} \qquad (VII-10-1)$$

where the $O[kE(k)]^{3/2}$ term corresponds to the local interactions, which we will neglect in this spectral region. The first term in the r.h.s. of (VII-10-1) is an "eddy-viscous" term of the form

$$-2\nu_t k^2 E(k)$$

with

$$\nu_t = \frac{1}{15} \int_{k_i}^{\infty} \theta_{0pp} [5E(p) + p\frac{\partial E}{\partial p}] dp \qquad (VII-10-2)$$

which represents the damping action of the turbulence on the low frequency modes. Throughout this book we will widely discuss this eddy-viscosity concept in spectral space, introduced by Kraichnan (1976). The second term is positive and injects a k^4 transfer in low wave numbers, through some kind of resonant interaction between two modes $\approx k_i$. It is this term which is responsible for the sudden appearance of a k^4

spectrum at $k < k_i$ when the initial energy spectrum is sharply peaked at k_i (or simply $\propto k^s$ with $s > 4$). Eq. (VII-10-1), derived from the *E.D.Q.N.M.*, is important, for it contains the two leading terms which govern the dynamics of three-dimensional isotropic turbulence in the low wave numbers: if turbulence is stationary and sustained by a forcing spectrum concentrated at k_i (k_i fixed), the balance between these two terms yields a k^2 energy spectrum for $k \to 0$. It is called an energy equipartition spectrum, for it corresponds to the same amount of energy at each wave vector $\vec{k}$. If turbulence is unforced and decays freely, the behaviour of the "infrared" energy spectrum (that is at low wave numbers) depends on the infrared spectral exponent s of the initial conditions, as discussed in Lesieur and Schertzer (1978). This is due to the k^4 nonlocal transfer: if

$$E(k,0) \propto k^s, \quad 1 < s < 4 \qquad (VII-10-3)$$

then the k^4 and eddy-viscous non-local transfers will modify negligibly the initial spectrum, which will keep its original shape in the low k, that is

$$E(k,t) = C_s k^s \qquad (VII-10-4)$$

C_s being a constant (the lower bound 1 in (VII-10-3) corresponds to a local transfer of the order of the eddy-viscous transfer). Eq. (VII-10-4) is a situation of perfect "permanence of big eddies". This is no longer the case for

$$E(k,0) \propto k^s, \quad s \geq 4 \qquad (VII-10-5)$$

where the spectrum will immediately pick up the k^4 component of the non-local transfer (unless it already possesses it) and evolve as

$$E(k,t) = C_4(t) \, k^4 \qquad (VII-10-6)$$

with, from (VII-10-1)

$$\frac{dC_4}{dt} = \frac{14}{15} \int_{k_i}^{\infty} \theta_{0pp} \frac{E(p)^2}{p^2} dp \quad . \qquad (VII-10-7)$$

As mentioned in Lesieur and Schertzer (1978), this expresses the non-invariance with time of the Loitzianskii integral (see Orszag, 1977). It must nevertheless be stressed that the *E.D.Q.N.M.* calculations show that C_4 increases very slowly and that an assumption like the invariance with time of the Loitzianskii integral (when $s = 4$) would not greatly alter the turbulence decay.

In order to now solve the kinetic energy decay problem, it suffices to simply assume a self-similar decaying energy spectrum already considered in (VI-6-9)

$$E(k,t) = v^2 l \, F(kl), \quad l = \frac{v^3}{\epsilon}, \quad \epsilon = -\frac{1}{2}\frac{dv^2}{dt} \qquad (VII-10-8)$$

where F is a dimensionless function. This form is not exact in the dissipative range, but the error thus introduced in the kinetic energy decay exponent is very small, as the closures will show[13]. Eq. (VII-10-8) developed for $k \to 0$ allows us to obtain

$$v^2 l^{s+1} = \text{constant}, \quad s < 4 \qquad (\text{VII} - 10 - 9)$$

$$v^2 l^{s+1} \propto t^\gamma, \quad s = 4; \quad \gamma = \frac{1}{C_4} \frac{dC_4}{d\ln t} \qquad (\text{VII} - 10 - 10)$$

where dC_4/dt is given by (VII-10-7) if one trusts the closures. It is now very easy to solve the problem (see Lesieur and Schertzer, 1978). Assuming that the kinetic energy and the integral scale follow time power laws, one finds

$$\frac{1}{2}v^2(t) \propto t^{-2(s+1-\gamma)/(s+3)} \qquad (\text{VII} - 10 - 11)$$

$$l(t) \propto t^{(2+\gamma)/(s+3)} \qquad (\text{VII} - 10 - 12)$$

where γ is zero for $s < 4$ and found (through an *E.D.Q.N.M.* calculation) to be equal to 0.16 when $s = 4$. We recall that s cannot exceed 4. For $s = 2$, the energy follows Saffman's (1967) $t^{-6/5}$ law. For $s = 4$, it follows a $t^{-1.38}$ law instead of the $t^{-10/7}$ (i.e. $t^{-1.43}$) Kolmogorov's (1941 b) law obtained in this case when discarding γ. In any case $v/l = \epsilon/v^2$ decays as t^{-1}, as mentioned in Chapter III, and the integral scale grows as $t^{2/5}$ for $s = 2$ and $t^{0.31}$ for $s = 4$. The large-eddy turn-over time l/v grows like t, and the Reynolds number vl/ν decays like $t^{(2\gamma+1-s)/(s+3)}$. To confirm the existence of self-similar evolving energy spectra, at least in the frame of the *E.D.Q.N.M.* closure, Figure VII-8 (taken from Lesieur and Schertzer (1978)) shows the function $F(kl)$ introduced in (VII-10-8) at various times of the evolution, for two different initial conditions corresponding to $s = 2$ and $s = 4$: it indicates a perfect self similarity in the energy-containing and inertial-ranges, and only a very slight departure from self-similarity in the dissipation range.

The laws (VII-10-11) and (VII-10-12) can also be derived, as in Comte-Bellot and Corrsin (1966), by assuming a crude energy spectrum model of the form

$$E(k,t) = C_s k^s, \quad k < k_i(t)$$

$$E(k,t) = C_K \epsilon^{2/3} k^{-5/3}, \quad k > k_i(t)$$

---

[13] which proves that the decay of kinetic energy at high Reynolds numbers is inviscid, and can in principle be handled within Euler equation

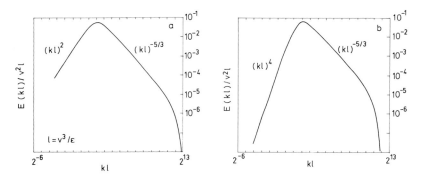

Figure VII-8: *E.D.Q.N.M.* calculation of the normalized kinetic energy spectra $E(kl)/v^2 l$ at times 125, 150, 175 and 200, for two initial conditions corresponding to $s = 2$ (a) and $s = 4$ (b). The figure, where the spectra have been superposed, shows that they evolve self-similarly, according to (VII-10-8) (from Lesieur and Schertzer, courtesy Journal de Mécanique).

with $k_i(t) \approx l(t)^{-1}$. Notice finally that (VII-10-11) and (VII-10-12) lead to a Richardson-type law (VI-5-1)

$$\frac{1}{2}\frac{dl^2}{dt} \sim \epsilon^{1/3} l^{4/3} \quad . \tag{VII-10-13}$$

For $s < 4$, these decay results are not dependent on the closure used, provided the latter gives a non local transfer of the form (VII-10-1) and a self-similar evolving spectrum (VII-10-8). For $s = 4$, the results obtained by Tatsumi et al. (1978) with the aid of the $Q.N.M.$ theory are very close to the $E.D.Q.N.M.$ ones. It is more difficult to interpret the experimental grid turbulence results: indeed, a grid of mesh M (distance between two bars) will produce immediately downstream a turbulence of integral scale M, turbulence due to the interaction of the wakes of neighbouring bars. Such a turbulence will not have, at $t = 0$, much energy at wave numbers different from M^{-1}. Therefore one might conjecture an energy spectrum that adjusts quickly on k^4 for $k \to 0$. Hence the kinetic energy would decay as $t^{-1.38}$. This is close to Warhaft and Lumley's (1978) $t^{-1.34}$ result, but at variance with Comte-Bellot and Corrsin's (1966) $t^{-1.26}$ measurements. The reason for the discrepancy is not obvious, and since it does not seem possible to measure experimentally the $k \to 0$ part of the energy spectrum, we can only propose various possible reasons such as the relatively low Reynolds number of the experiments, the lack of isotropy in the large scales, the problem of the determination of the origin of time, the insufficient length of the apparatus, etc.

Numerically, the unforced direct or Large-Eddy numerical simulations having enough resolution in the small k and taking an initial kinetic energy spectrum $\propto k^s$ with $s > 4$, do show the formation of the infrared k^4 energy spectrum: this is obvious on Figure VII-9, taken from Lesieur and Rogallo (1989), where $s = 8$ initially. These simulations (see also Lesieur et al., 1989) give a kinetic-energy decay exponent of the order of 1.4, in good agreement with the above theory. It is however difficult in these simulations to have reliable data on the infrared spectrum, due both to the shift of $k_i(t)$ towards the lower modes, and to the lack of statistical reliability of the spectra in these modes[14].

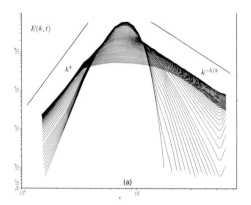

Figure VII-9: time-evolution of the kinetic-energy spectrum in a pseudo-spectral large-eddy simulation done by Lesieur and Rogallo (1989), starting initially with a k^8 spectrum at low k. The resolution is 128^3 modes. See Chapter XII for details (courtesy The Physics of Fluids).

11 - The Renormalization-Group techniques

The Renormalization-Group techniques ($R.N.G.$) were first developed with great success in the Physics of *critical phenomena*, in for instance, studies relating to non-linear spin dynamics in ferromagnetic systems. These analysis consider generally the dimension of space d as a variable parameter, and it turns out that the problem can be solved for $d = 4$. The solutions for $d = 4 - \epsilon$, where ϵ is a small parameter, are then expanded in powers of ϵ, and ϵ is taken equal to 1 in order to recover the solution for $d = 3$.

Noticing certain similarities between these non-linear spin dynamics and Navier-Stokes equations with random forcing, Forster, Nelson

---

[14] In these isotropic calculations, the spectra are calculated by an average on a sphere of radius k, and very few modes will contribute in the statistics at low k.

and Stephen (1977) applied the *R.N.G.* to the latter equations. In this analysis, the dimension of space was still a variable parameter. The formalism has been adapted by Fournier (1977) to Navier-Stokes equations for a fixed dimension d, the variable parameter becoming the exponent of the forcing spectrum in Fourier space. It is this analysis (taken from Frisch and Fournier, 1978), which will be summarized now:

11.1 The *R.N.G.* algebra

One considers Navier-Stokes equation with a gaussian random forcing of zero mean, Fourier expanded both with respect to space and time[15]. The velocity field

$$V_i(\vec{k}, \omega) = \frac{1}{2\pi} \int \hat{u}_i(\vec{k}, t) \, [\exp i\omega t] \, dt$$

satisfies

$$(-i\omega + \nu k^2) \, V_i(\vec{k}, \omega) = f_i(\vec{k}, \omega)$$
$$- \frac{i}{2} \lambda P_{ijm}(\vec{k}) \int_{\vec{p}+\vec{q}=\vec{k}} \int d\vec{p} \, d\Omega \, V_j(\vec{p}, \Omega) V_m(\vec{p}, \omega - \Omega) \ .$$

$$(VII - 11 - 1)$$

λ is a parameter equal to 1 for Navier-Stokes equations, but which will be considered as small at certain stages of the *R.N.G.* theory[16]. The forcing spectrum, statistically stationary, is still defined by (VII-4-6), hence

$$< f_i(\vec{k}', \omega) f_n(\vec{k}, \omega') > = (\frac{1}{2\pi})^2 \iint < f_i(\vec{k}', \tau) f_n(\vec{k}, \tau') >$$
$$\exp i(\omega\tau + \omega'\tau') d\tau d\tau' = \frac{1}{2\pi} F_{in}(\vec{k}) \, \delta(\vec{k} + \vec{k}') \, \delta(\omega + \omega') \ .$$

$$(VII - 11 - 2)$$

The following analysis will be carried out for three-dimensional isotropic turbulence, where it is assumed that

$$F_{in}(\vec{k}) = 2DP_{ij}(\vec{k}) \frac{F(k)}{4\pi k^2} \ , \qquad\qquad (VII - 11 - 2')$$

(the forces have no helicity), with

$$F(k) = k^{-r} \ . \qquad\qquad (VII - 11 - 3)$$

---

[15] which requires that the velocity field should be statistically stationary. This is one of the limitations of the theory, which cannot handle freely-decaying turbulence.

[16] In fact, the *R.N.G.* resembles the *D.I.A.* in this respect (see Leslie, 1973), and both theories can be considered in this sense as weakly nonlinear.

D is a dimensional parameter which allows to vary the intensity of the forcing.

If $\lambda = 0$ in eq. (VII-11-1), the solution of this Langevin equation is

$$V_i^{(0)}(\vec{k},\omega) = G_0(\vec{k},\omega)\, f_i(\vec{k},\omega)$$

$$G_0 = \frac{1}{-i\omega + \nu k^2} \quad . \qquad \qquad (VII-11-4)$$

Eq. (VII-11-1) (with $\lambda \neq 0$) may be written as

$$V = G_0 f + \lambda G_0 \gamma(V,V) \quad . \qquad \qquad (VII-11-5)$$

Now, one looks for solutions where V is expanded in series of the small parameter λ:

$$V = V^{(0)} + \lambda V^{(1)} + \lambda^2 V^{(2)} + ...$$

An easy way to generate this expansion is to substitute V given by eq. (VII-11-5) into (VII-11-5) itself. It is found

$$V = G_0 f + \lambda G_0 \gamma(G_0 f, G_0 f) + 2\lambda^2 G_0 \gamma[\gamma(G_0 f, G_0 f), G_0 f] + O(\lambda^3) \ .$$

In fact, the solution may be written formally up to any order, using diagrams built with the *propagator* $G_0(k,\omega)$ and the *vertex* $\lambda\gamma(V,V)$.

The above technique is used in the following way: one considers now eq. (VII-11-1) on a wave-number span extending from 0 to Λ (*Problem A*). Let Λe^{-l} be a "cutoff wave number" ($0 < l << 1$), $V^<$ and $V^>$ the velocity corresponding to the modes below and above the cutoff. Eq. (VII-11-5) written for $V^>$ and $V^<$ yield

$$V^> = G_0^> f^> + \lambda G_0^> \gamma(V^>,V^>) + \lambda G_0^> \gamma(V^<,V^<)$$
$$+ 2\lambda G_0^> \gamma(V^<,V^>) \ , \qquad (VII-11-6)$$

$$V^< = G_0^< f^< + \lambda G_0^< \gamma(V^<,V^<) + \lambda G_0^< \gamma(V^>,V^>)$$
$$+ 2\lambda G_0^< \gamma(V^<,V^>) \ , \qquad (VII-11-7)$$

where the symbols $<$ and $>$ refer to modes (or interaction with modes) below or above the cutoff. Afterwards, eq. (VII-11-6) is solved up to order 2 by the same perturbation techniques as above for the Langevin equation, and the result is substituted into eq.(VII-11-7) for $V^<$. A last step is, in this equation, to average on the terms $f^>$, assuming that they vary on a much smaller time scale than $V^<$ (which is in fact an assumption of separation of spatial scales at the cutoff). The resulting equation for $V^<$ is

$$[-i\omega + \nu(l)k^2]\, V_i^<(\vec{k},\omega) = \hat{f}_i(\vec{k},\omega) - \frac{i}{2}\lambda P_{ijm}(\vec{k})$$

$$\int_{\vec{p}+\vec{q}=\vec{k}} \int d\vec{p}\, d\Omega\, V_j^<(\vec{p},\Omega)\, V_m^<(\vec{p},\omega - \Omega) \ , \qquad (VII-11-8)$$

with

$$\nu(l) = \nu[1 + \bar{\lambda}^2 H(r)\frac{1 - \exp-(3+r)l}{3+r}]$$

$$\bar{\lambda} = \lambda \ D^{1/2} \ \nu^{-3/2}\Lambda^{-(r+3)/2} \qquad\qquad (VII-11-9)$$

$$H(r) = \frac{1}{60\pi^2}(3-r) \quad .$$

The forcing has not been modified, which may be shown to be valid for $r \geq -4$. Notice that the non-dimensional parameter $\bar{\lambda}$ (called by Fournier and Frisch (1983 b) *reduced coupling constant* characterizes the relative importance of non-linear and viscous terms (equivalent Reynolds number)[17]. Thus, the wave-number range $[\Lambda e^{-l}, \lambda]$ has been eliminated. This operation is called a *decimation*. Notice that, in eq. (VII-11-8), terms called *non pertinent* have been discarded, not that they are negligible at this stage, but because they vanish after an infinite number of decimations, when the cutoff wave number has tended to 0. Hence, eq. (VII-11-8) is questionable for subgrid-scale modelling purposes if the cutoff wave number Λe^{-l} remains fixed.

Afterwards, expansions are done with respect to the small parameter $\epsilon = 3 + r$. Eq. (VII-11-9) writes

$$\frac{d\nu}{dl} = \nu\bar{\lambda}^2 \ H \quad , \qquad\qquad (VII-11-10)$$

equivalent to

$$\frac{\delta\nu}{\nu} = \bar{\lambda}^2 \ H \ \frac{\delta\Lambda}{\Lambda} \quad , \qquad\qquad (VII-11-11)$$

and showing the increase of viscosity on modes $[0, \Lambda - \delta\Lambda]$ due to modes $[\Lambda - \delta\Lambda, \Lambda]$. Now, the following changes of variable are done in eq. (VII-11-8):

$$\tilde{k} = e^l \ k \ , \ \tilde{\omega} = e^l Z\omega \ , \ \tilde{V}(\tilde{k}, \tilde{\omega}) = e^{-lX} \ V^<(k, \omega) \quad .$$

The wave-number span is again $[0, \Lambda]$, and the coefficients $\nu(l), D, \lambda$ are changed into:

$$\nu_N(l) = \nu(l) \ \exp(Z-2)l \ ; \ \lambda_N(l) = \lambda \ \exp(X-4) \ ,$$

---

[17] Indeed, let eq. (VII-11-1) be written in physical space: the velocity will be evaluated by assuming a balance between the forcing and the viscous damping, which yields $V \sim f/\nu\Lambda^2$. Therefore, the equivalent Reynolds number $R = \lambda V/\nu\Lambda$ is equal to $\lambda f/\nu^2\Lambda^3$. In order to evaluate f , one writes from eqs (VII-11-2') and (VII-11-3): $f^2 \sim DT_\nu \int_0^\Lambda k^{-r}dk$, where $T_\nu = 1/\nu\Lambda^2$ is the shortest time correlation time permitted by the viscous forces. This yields $R = \bar{\lambda}$.

$$D_N(l) = D \ \exp(3Z - 2X + r + 5)l \quad .$$

Supposing that $\nu_N(l)$ and D_N will remain unchanged (equal respectively to ν and D), it is found:

$$Z - 2 + \bar{\lambda}^2 H = 0 \quad ; \quad 3Z - 2X + r + 5 = 0 \quad ,$$

$$\lambda(l) = \lambda \ \exp\frac{1}{2}(r + 3 - 3H\bar{\lambda}^2) \quad .$$

Hence we are back to the initial problem (Problem A), with the same ν, D and Λ but with a new parameter $\lambda(l)$ which satisfies

$$\frac{1}{\lambda}\frac{d\lambda}{dl} = \frac{1}{\bar{\lambda}}\frac{d\bar{\lambda}}{dl} = \frac{1}{2}[r + 3 - 3H(r)\lambda^2] \quad . \qquad (VII - 11 - 12)$$

The process can now be iterated, letting k go to zero. Then, two cases have to be considered: if $\epsilon = r + 3$ is negative[18], $\bar{\lambda}(l)$ will tend to zero when $l \to \infty$, and the eventual solution for $V^<$ will be given by the Langevin equation (VII-11-4): the stationary spatial simultaneous spectral tensor may thus be easily determined as

$$<\hat{U}_{ij}(\vec{k}, t)> = \frac{F_{ij}(\vec{k})}{2\nu k^2} \quad ,$$

and the kinetic-energy spectrum will thus be equal to, from eq. (VII-11-2')

$$E(k) = 2DF(k)/2\nu k^2 \propto k^{-r-2} \quad . \qquad (VII - 11 - 13)$$

If $\epsilon > 0$, there will be a fixed point given by

$$\bar{\lambda}_* = \sqrt{\frac{\epsilon}{3H}} \quad . \qquad (VII - 11 - 14)$$

At the fixed point, the kinetic energy spectrum $E(k) \propto k^{-m}$ does not vary during one iteration. This allows us to show, taking into account the changes of variable done, that

$$m = Z + r = 2 - \bar{\lambda}_*^2 H + r = 1 + \frac{2r}{3} \quad .$$

In this case eq. (VII-11-11) may be easily integrated, yielding an exact expression for the renormalized viscosity, as shown by Fournier and Frisch (1983):

$$\nu(\Lambda) = (3H)^{1/3} \ D^{1/3} \ \epsilon^{-1/3} \ \Lambda^{-\epsilon/3} \quad . \qquad (VII - 11 - 15)$$

---

[18] the parameter ϵ employed here has of course nothing to do with the kinetic-energy dissipation rate considered above

The corresponding asymptotic kinetic energy spectrum may be calculated by perturbative methods from the above ($\bar{\lambda} = 0$) Langevin equation, since the fixed-point value for $\bar{\lambda}$ is $O(\epsilon)$, and we get (Fournier and Frisch, 1983):

$$E(k) = \frac{2DF(k)}{2\nu(k)k^2} = D^{2/3} \, (3H)^{-1/3} \, \epsilon^{1/3} \, k^{1-2\epsilon/3} \quad . \qquad (VII-11-16)$$

Eq. (VII-11-15) is written:

$$\nu(\Lambda) = (3H)^{1/2} \, \epsilon^{-1/2} \, \sqrt{\frac{E(\Lambda)}{\Lambda}} \quad , \qquad (VII-11-17)$$

which is independent of the forcing parameter D. This yields in particular $E(k) \propto k^{-5/3}$ for $F(k) \propto k^{-1}$, although the convergence of the method may be questioned, since $\epsilon = r + 3 = 4$, which by no means may be a small parameter. This last result has been used by Yakhot and Orszag(1986) in order to implement the $R.N.G.$ techniques for subgrid-scale modelling purposes (see Chapter XII). Eqs (VII-11-16) and (VII-11-17) yield in this case (with, from (VII-11-9), $H = 1/30\pi^2$):

$$E(k) = 7.336 \, D^{2/3} \, k^{-5/3}$$
$$\nu(\Lambda) = 0.050 \, \sqrt{\frac{E(\Lambda)}{\Lambda}} \quad . \qquad (VII-11-18)$$

11.2 Two-point closure and $R.N.G.$ techniques

11.2.1 The $k^{-5/3}$ range

We will see in Chapter XII that an eddy-viscosity equivalent to (VII-11-18), but with a higher value of the numerical constant (0.267 instead of 0.050) may be derived from the non-local interactions theory within the $E.D.Q.N.M.$ approximation. Here the small parameters remain smaller than 1, but there is an adjustable constant a_1 which has to be tuned on the Kolmogorov constant.

In the $R.N.G.$ analysis as developed by Yakhot and Orszag (1986) (see also Dannevik et al., 1987), certain expansions about the fixed point show that the kinetic energy spectrum must annul a transfer whose form is similar to the $E.D.Q.N.M.$ transfer given by eqs (VII-3-9) and (VII-3-7) (μ_k being set equal to $\nu(k) \, k^2$, where $\nu(k)$ is the eddy-viscosity of the $R.N.G.$ theory). This yields a Kolmogorov spectrum, with

$$\nu(k) = 0.388 \, \sqrt{\frac{E(k)}{k}} \quad , \qquad (VII-11-19)$$

result which is strongly at variance with the prediction of the original *R.N.G.* theory given in eq. (VII-11-18). Therefore there seems to be some inconsistency in using an *E.D.Q.N.M.*-type spectral equation within the *R.N.G.* formalism. Afterwards, the parameter D characterizing the forcing is adjusted in such a way that the energy flux is, for the Kolmogorov spectrum, constant and equal to $\bar{\epsilon}$. This yields a constant value for $D/\bar{\epsilon}$, and a Kolmogorov constant equal to 1.617. It is however questionable to use D, fixing the intensity of the forcing, as an adjustable parameter: indeed, D should satisfy, from the original equations,

$$\bar{\epsilon} = 2D \int_0^\Lambda k^{-r} \, dk \quad ,$$

where $\bar{\epsilon}$ is the kinetic-energy injection rate. When $r \geq 1$, a lower cutoff wave number $k_{\min}$ is required in order to prevent the divergence of $\bar{\epsilon}$. For $r = 1$, $\bar{\epsilon}/D = 2\ln(\Lambda/k_{\min})$ depends logarithmically on Λ and $k_{\min}$, and this is certainly not a constant.

11.2.2 The infrared spectrum

Returning to the case of the infrared spectrum, it may be of interest to compare the *R.N.G.* results obtained in the above case (turbulence forced up to a cutoff Λ by a forcing k^{-r} with $r \approx -3$, where the theory has good chances to converge) with the two-point closure predictions. We use (VII-10-1) to write in the stationary case

$$2(\nu + \nu_t)k^2 E(k) = Ak^4 + F(k) + O[kE(k)]^{\frac{3}{2}} \qquad (\text{VII} - 11 - 20)$$

where A is the coefficient of the k^4 transfer in (VII-10-1). As already noticed, a k^1 infrared kinetic-energy spectrum gives a "crossover" exponent over which non-local transfers dominate, and under which the local transfers are preponderant. Contrary to the *R.N.G.*, two-point closures may consider wide variations of r. They yield:
 • i) for $r \leq -4$, $E(k) = (A/\nu_t)k^2$ (which justifies discarding the local transfer), result already mentioned above.
 • ii) for $-4 < r \leq -3$, the non-local eddy-viscous term is still greater than the local transfer, and is balanced by the forcing: one has $E(k) \propto k^{-r-2}$.
 • iii) for $r > -3$, the local term balances the forcing, which yields $E(k) \propto k^{-1-(2r/3)}$. These two results are in agreement with the *R.N.G.* predictions.

It turns out that the *R.N.G.* method seems to handle correctly the isotropic forced turbulence in the infrared limit $k \to 0$, with a forcing exponent $r \geq -4$. However, there are severe convergence problems if one wants to describe the dynamics of the Kolmogorov inertial range.

Furthermore, the method is unable to capture both the infrared and ultraviolet dynamics, and cannot be implemented if there is no forcing, unless an $E.D.Q.N.M.$-type equation is associated to the theory. Remarking also that it seems difficult to apply it to inhomogeneous problems, we will conclude that two-point closures are tools which are much more controllable, and offer a much wider range of possibilities than the $R.N.G.$, even for subgrid-scale modelling purposes.

Chapter VIII

DIFFUSION OF PASSIVE SCALARS

1 - Introduction

We have already seen that under certain approximations consisting in neglecting the buoyancy in the Boussinesq approximation derived from the Navier-Stokes equations, the temperature $T(\vec{x}, t)$ satisfied a passive scalar type diffusion equation

$$\frac{\partial T}{\partial t} + \vec{u}.\vec{\nabla}T = \kappa \nabla^2 T \qquad (VIII - 1 - 1)$$

and was simply transported by the fluid particle (and diffused by molecular effects) without any action on the flow dynamics. More generally, one can consider any passive quantity which diffuses according to eq. (VIII-1-1), such as a dye which marks the flow. The Schmidt number of the passive scalar is ν/κ, where κ is the molecular diffusivity of the scalar. It corresponds to the Prandtl number when the passive scalar is the temperature. Since we will consider only one diffusing quantity here, we will associate it with the temperature, and speak of the Prandtl number of the passive scalar.

When a passive scalar diffuses in homogeneous turbulence, one is interested mainly in two problems: the first concerns the small-scale statistics of the scalar, and will be studied by assuming that the scalar fluctuations are also homogeneous. This is, for instance, the case of a grid turbulence, where a slight statistically homogeneous heating in the fluid close to the grid will produce a random temperature fluctuation field whose intensity will decay downstream of the grid, due to the molecular-conductive effects that will tend to homogenize the temperature within the fluid. The second problem concerns the dispersion of a localized scalar cloud (or a heated spot) by the turbulence: this is an inhomogeneous problem as far as the temperature is concerned, which requires to look both at the absolute diffusion of the cloud gravity centre, and the particle pair relative dispersion: the latter question gives information on the spreading rate of the cloud, and can be studied within a homogeneous formalism. The present chapter will only deal with three-dimensional isotropic turbulence. It will be organized as follows: the second section will recall the phenomenology of the passive scalar turbulent diffusion problem, as described for instance by Tennekes and Lumley (1972) and Leslie (1973). The third section will show how to apply the *E.D.Q.N.M.* closure to the statistically homogeneous scalar, and will focus on singular behaviour accompanying the catastrophic stretching of vortex filaments described in Chapter VII. The fourth section will be devoted to the homogeneous scalar decay, and the fifth section to the particle pair dispersion problem, which appears to be a special case of scalar decay. Finally, the sixth section will present an attempt to apply the *E.D.Q.N.M.* approximation to the diffusion of an inhomogeneous scalar in isotropic turbulence.

2 - Phenomenology of the homogeneous passive scalar diffusion

Assuming that the temperature $T(\vec{x}, t)$ is statistically homogeneous and isotropic, of zero mean, we first introduce the conductive wave number, in a similar way we have earlier introduced the Kolmogorov dissipative wave number: let $r_c = k_c^{-1}$ be the scale at which the molecular diffusive effects in (VIII-1-1) are of the same order as the convective term $\vec{u}.\vec{\nabla}T$, that is such that the local Peclet number

$$P_e(r_c) = \frac{r_c v_{r_c}}{\kappa} \qquad (VIII - 2 - 1)$$

should be about one. Then two cases have to be considered: if r_c^{-1} lies in the $k^{-5/3}$ Kolmogorov energy inertial range, we have from (VI-4-4) and (VIII-2-1):

$$v_{r_c} \sim (\epsilon r_c)^{1/3} \qquad (VIII - 2 - 2)$$

which yields

$$k_c \sim (\frac{\epsilon}{\kappa^3})^{1/4} \sim (\frac{\nu}{\kappa})^{3/4} k_d \quad . \qquad (VIII - 2 - 3)$$

But (VIII-2-3) is only valid when the Prandtl number ν/κ is smaller than one, since this analysis has been done assuming that r_c^{-1} lies in the kinetic energy inertial-range, or equivalently that $k_c < k_d$.

If the Prandtl number is greater than one, k_c is greater than k_d and is in the dissipative range (otherwise the preceding analysis should still hold, and (VIII-2-3) would yield $k_c > k_d$, in contradiction with the hypothesis): one can imagine that a small blob of temperature of diameter r_c is submitted to a velocity shear of characteristic scale and velocity k_d^{-1} and v_d corresponding to a local Reynolds number equal to one and such that

$$v_d k_d^{-1} = \nu \quad . \qquad (VIII-2-4)$$

This scalar blob will therefore be elongated in the direction of the shear, and develop smaller transverse scales satisfying $P_e(r_c) = 1$, $P_e(r_c)$ being defined according to (VIII-2-1), with v_{r_c} equal to v_d . It yields in this case

$$k_c \sim \frac{\nu}{\kappa} k_d \quad . \qquad (VIII-2-5)$$

Thus (VIII-2-3) and (VIII-2-5) allow us to determine the conductive wave number, according to the respective value of the Prandtl number compared to one. When the Prandtl number goes to infinity, the scalar transported by turbulence is thus expected to develop infinitely small structures.

2.1 The inertial-convective range

We recall that the temperature spectrum, defined in (V-5-16), satisfies

$$\frac{1}{2} < T(\vec{x}, t)^2 > = \int_0^{+\infty} E_T(k, t) dk \quad . \qquad (VIII-2-6)$$

Let ϵ_T be the scalar dissipative rate

$$\epsilon_T = -\frac{1}{2} \frac{d}{dt} < T(\vec{x}, t)^2 > \quad . \qquad (VIII-2-7)$$

It was shown in (VI-2-20) that

$$\epsilon_T = 2\kappa \int_0^{+\infty} k^2 E_T(k, t) dk \quad . \qquad (VIII-2-8)$$

This allows one to define the temperature enstrophy

$$D_T(t) = \frac{1}{2} < \vec{\nabla} T^2 > = \int_0^{+\infty} k^2 E_T(k, t) dk \quad , \qquad (VIII-2-9)$$

characteristic of the mean temperature gradients. Let k_i and k_i^T be the wave numbers characteristic of the peaks of respectively the energy and the temperature spectra: these wave numbers could be imposed by an external stationary forcing of kinetic energy and temperature at rates

ϵ and ϵ_T, or correspond to freely-evolving situations, where we will see that $k_i^T(t)$ decreases with time, like $k_i(t)$. We assume that

$$sup(k_i, k_i^T) << inf(k_d, k_c) \quad . \qquad (VIII-2-10)$$

Oboukhov (1949) and Corrsin (1951) have independently proposed that for k lying in the range defined by (VIII-2-10), the temperature spectrum should be proportional to $(\epsilon_T/\epsilon)E(k)$. Such an hypothesis is due to the linear character of the diffusion equation (VIII-1-1). It leads to

$$E_T(k) \sim \epsilon_T \epsilon^{-1/3} k^{-5/3} \quad . \qquad (VIII-2-11)$$

In fact, the easiest way to obtain (VIII-2-11) is to apply an Oboukhov-type theory, already used in Chapter VII to study the helicity spectrum, that is

$$\epsilon_T \sim k \, E_T(k)\mu_k \quad , \qquad (VIII-2-12)$$

where μ_k is the triple-correlation relaxation rate introduced in Chapter VII, proportional to $\epsilon^{1/3}k^{2/3}$ when k lies in the Kolmogorov inertial range. The linear-cascade assumption actually assumes that the rate at which the scalar cascade proceeds is governed by the local velocity gradients at k. An alternative way of obtaining this result is to write that a typical temperature fluctuation δT_r at a scale $r \approx k^{-1}$ is such that

$$\epsilon_T \sim \frac{\delta T_r^2}{r/v_r} \quad . \qquad (VIII-2-13)$$

Remembering that $v_r \sim (\epsilon r)^{1/3}$, this yields

$$\delta T_r^2 \approx <[T(\vec{x} + \vec{r}, t) - T(\vec{x}, t)]^2> \sim \epsilon_T \epsilon^{-1/3} r^{2/3} \quad , \qquad (VIII-2-14)$$

which is the equivalent Kolmogorov law for the second-order structure function of the temperature. Eqs (VIII-2-11) and (VIII-2-14) character-ize the inertial-convective range, where the velocity is *inertial* (no influence of viscosity) and the scalar is *convective* (that is, simply transported by the velocity field). A very impressive experimental confirmation of the inertial-convective range was provided by Gagne (1987).

2.2 The inertial-conductive range

We assume that the Prandtl number is smaller than one, so that the conductive wave number k_c is smaller than k_d and given by (VIII-2-3). For $k < k_c$, the temperature spectrum displays an inertial-convective range described above. For $k_c < k < k_d$, we are still in the Kolmogorov energy cascade (that is, "inertial"), but the molecular-conductive effects are predominant for the scalar (see Figure VIII-1). This allows us in eq.

(VIII-1-1) to neglect the time-derivative term $\partial T/\partial t$. It has been pro-posed by Batchelor et al. (1959), that the Quasi-Normal theory should be valid in this case. This makes it easy to calculate the temperature spectrum. This calculation, already developed by Leslie (1973), is re-called here: let $\hat{T}(\vec{k}, t)$ be the Fourier transform of the temperature. Eq. (VIII-1-1) is then written for two wave vectors $\vec{k}$ and $\vec{k'}$:

$$-\kappa\, k^2 \hat{T}(\vec{k}, t) = i \int \hat{u}_j(\vec{p}, t) q_j \hat{T}(\vec{q}, t) \delta(\vec{k} - \vec{p} - \vec{q}) d\vec{p} d\vec{q} \ ;$$

$$-\kappa k'^2 \, \hat{T}(\vec{k'}, t) = i \int \hat{u}_l(\vec{p'}, t) q'_l \hat{T}(\vec{q'}, t) \delta(\vec{k'} - \vec{p'} - \vec{q'}) d\vec{p'} d\vec{q'} \ ,$$

and hence, after multiplication of both equations and ensemble averaging

$$\kappa^2 k^2 k'^2 < \hat{T}(\vec{k'}, t) \hat{T}(\vec{k}, t) >= - \int < \hat{u}_j(\vec{p}) \hat{u}_l(\vec{p'}) \hat{T}(\vec{q}) \hat{T}(\vec{q'}) > q_j q'_l$$

$$\delta(\vec{k} - \vec{p} - \vec{q}) \delta(\vec{k'} - \vec{p'} - \vec{q'}) d\vec{p} \, d\vec{q} \, d\vec{p'} \, d\vec{q'} \ . \qquad \text{(VIII} - 2 - 15)$$

The Quasi-Normal approximation gives

$$< \hat{u}_j(\vec{p}) \hat{u}_l(\vec{p'}) \hat{T}(\vec{q}) \hat{T}(\vec{q'}) >=< \hat{u}_j(\vec{p}) \hat{u}_l(\vec{p'}) ><\hat{T}(\vec{q}) \hat{T}(\vec{q'}) >$$
$$\text{(VIII} - 2 - 16)$$

since all the other terms $< \hat{u}\hat{T} ><\hat{u}\hat{T} >$ are zero: indeed we recall that in three-dimensional isotropic turbulence, and if the isotropy assump-tions are done both on the velocity and the scalar, the scalar-velocity correlations are zero. Using the relations (V-4-2) and (V-5-16) yields for (VIII-2-15), after integration on $\vec{p'}$ and $\vec{q'}$

$$\kappa^2 k^2 k'^2 < \hat{T}(\vec{k'}) \hat{T}(\vec{k}) > = \int \hat{U}_{jl}(\vec{p}) \frac{E_T(q)}{2\pi q^2} q_j q_l$$

$$\delta(\vec{k} - \vec{p} - \vec{q}) \, \delta(\vec{k'} + \vec{p} + \vec{q}) d\vec{p} \, d\vec{q} \ .$$

Further integrations on $\vec{p}$ and $\vec{k'}$ lead to

$$\kappa^2 k^4 \frac{E_T(k)}{2\pi k^2} = \int \hat{U}_{jl}(\vec{k} - \vec{q}) \frac{E_T(q)}{2\pi q^2} q_j q_l \, d\vec{q} \ . \qquad \text{(VIII} - 2 - 17)$$

As will be shown soon, the temperature spectrum decreases very rapidly to infinity, and the essential part of the integral arising in (VIII-2-17) comes from the triads such that $q << k$. Once this non local approx-imation has been made, and expressing $\hat{U}_{jl}(\vec{k})$ with the aid of (V-5-9), eq. (VIII-2-17) reduces to

$$\kappa^2 k^4 \frac{E_T(k)}{2\pi k^2} = \frac{E(k)}{4\pi k^2} \int \frac{E_T(q)}{2\pi q^2} q^2 \sin^2 \beta d\vec{q}$$

where $q^2 \sin^2 \beta$ stands for $q_j q_l P_{jl}(\vec{k})$, and the angle β is the interior angle opposite to the side p in the triangle (k, p, q). The integration on $\vec{q}$ is then carried out using the polar coordinates β, q, and ϕ, where ϕ is the angle defining the rotation about the vector $\vec{k}$. One obtains

$$\kappa^2 k^2 E_T(k) = \frac{E(k)}{4\pi k^2} \, 2\pi \int_0^{+\infty} q^2 E_T(q) dq \int_0^\pi \sin^3 \beta d\beta \ , \quad \text{(VIII} - 2 - 18)$$

and finally

$$E_T(k) = \frac{2}{3} \kappa^{-2} \int_0^{+\infty} q^2 E_T(q) dq \ k^{-4} E(k) \qquad \text{(VIII} - 2 - 19)$$

or, using (VIII-2-8)

$$E_T(k) = \frac{1}{3} \, \epsilon_T \, \kappa^{-3} k^{-4} E(k) \quad . \qquad \text{(VIII} - 2 - 20)$$

Then the assumption of a $k^{-5/3}$ kinetic energy spectrum yields

$$E_T(k) \sim \epsilon_T \, \kappa^{-3} \, \epsilon^{2/3} \, k^{-17/3} \quad , \qquad \text{(VIII} - 2 - 21)$$

which is the inertial-conductive range predicted by Batchelor et al. (1959). Both kinetic energy and temperature spectra are schematically shown on Figure VIII-1: the inertial-convective range extending up to k_c is then followed by the $k^{-17/3}$ inertial-conductive range extending from k_c to k_d. Experimentally, such a range could be expected to exist for turbulence in liquid metals, but the $-17/3$ slope is too steep to allow such a verification. A numerical simulation of a passive scalar convected by a frozen velocity field (that is, with an infinite correlation time) done by Chasnov et al. (1988) confirms the $-17/3$ exponent. We also anticipate that in two-dimensional turbulence the preceding calculation leading to (VIII-2-20) is still valid, except for the factor 2π of the integration on the angle ϕ (see Lesieur et al., 1981).

2.3 The viscous-convective range

We will now consider the case of a fluid with a Prandtl number larger than one. The conductive wave number k_c is given by (VIII-2-5). For $k_d < k < k_c$, we have seen that the scalar is strained by a uniform shear of vorticity $v_d k_d$. This vorticity can easily be shown from Chapter VI to be of the order of $(\epsilon/\nu)^{1/2}$, that is the square root of the enstrophy. Therefore, we assume that there is a scalar cascade of constant rate ϵ_T

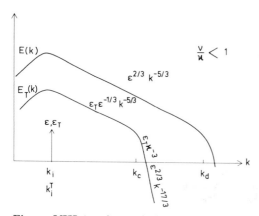

Figure VIII-1: schematic inertial-convective and inertial-conductive ranges of the temperature spectrum in the Kolmogorov inertial range of the kinetic energy spectrum (low Prandtl number).

(recalling that the scalar is "convective" and not affected by molecular diffusive effects) such that

$$\epsilon_T \sim k \, E_T(k) \, (\frac{\epsilon}{\nu})^{1/2} \qquad\qquad (VIII - 2 - 22)$$

which yields

$$E_T(k) \sim \epsilon_T (\frac{\nu}{\epsilon})^{1/2} k^{-1} \quad . \qquad\qquad (VIII - 2 - 23)$$

This is the k^{-1} viscous-convective range proposed by Batchelor (1959). Actually this range cannot, as stressed in this reference (see also Leslie, 1973), extend up to k_c: indeed, if k_{vc} is the maximum wave number for which the k^{-1} range is valid, we must have from (VIII-2-23)

$$\epsilon_T \sim 2\kappa \int_0^{k_{vc}} \epsilon_T (\frac{\nu}{\epsilon})^{1/2} k \, dk \quad ,$$

which gives

$$\frac{k_{vc}}{k_d} = (\frac{\nu}{\kappa})^{1/2} \qquad\qquad (VIII - 2 - 24) \quad ,$$

to be compared with (VIII-2-5). Thus, this wave number k_{vc} is smaller than k_c. It has been shown by Batchelor (1959) that the temperature spectrum decreases exponentially for $k_{vc} < k < k_c$. Figure VIII-2 shows the kinetic energy and temperature spectra in the case of the viscous-convective range.

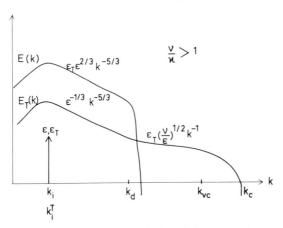

Figure VIII-2: schematic inertial-convective and viscous-convective ranges of the temperature spectrum at high Prandtl number.

3 - The *E.D.Q.N.M.* isotropic passive scalar

The *E.D.Q.N.M.* approximation can easily be derived for the passive scalar (Herring et al., 1982, Larchevêque et al., 1980, Larchevêque, 1981, Larchevêque and Lesieur, 1981). With the same symbolic notation as in VII-1, one can write:

$$\frac{\partial \hat{T}(\vec{k})}{\partial t} = \hat{T}\hat{u} - \kappa k^2 \hat{T}(\vec{k}) \; ;$$

$$\frac{\partial}{\partial t} <\hat{T}(\vec{k},t)\hat{T}(\vec{k}',t)> = <\hat{T}\hat{T}\hat{u}> - \kappa(k^2 + k'^2) <\hat{T}(\vec{k},t)\hat{T}(\vec{k}',t)> \; ;$$

$$\frac{\partial}{\partial t} <\hat{T}(\vec{k})\hat{T}(\vec{p})\hat{u}(\vec{q})> = <\hat{T}\hat{T}\hat{u}\hat{u}>$$
$$- [\kappa(k^2 + p^2) + \nu q^2] <\hat{T}(\vec{k})\hat{T}(\vec{p})\hat{u}(\vec{q})> \; ,$$

and the same Quasi-Normal type approximations as for the velocity moments hierarchy can be made. Since the calculation is simpler than for the kinetic energy spectrum equation, we will derive it in detail: let us write the temperature equation in Fourier space for two wave vectors $\vec{k}$ and $\vec{k}'$

$$(\frac{\partial}{\partial t} + \kappa k^2)\hat{T}(\vec{k},t) = -i \int \hat{u}_j(\vec{p},t)q_j\hat{T}(\vec{q},t)\delta(\vec{k} - \vec{p} - \vec{q})d\vec{p} \, d\vec{q} \; ,$$

$$(VIII - 3 - 1)$$

$$(\frac{\partial}{\partial t} + \kappa k'^2)\hat{T}(\vec{k}',t) = -i \int \hat{u}_l(\vec{p}',t)q'_l\hat{T}(\vec{q}',t)\delta(\vec{k}'-\vec{p}'-\vec{q}')d\vec{p}' d\vec{q}' \ .$$

$$(VIII - 3 - 2)$$

Multiplying (VIII-3-1) by $\hat{T}(\vec{k}',t)$ and (VIII-3-2) by $\hat{T}(\vec{k},t)$, adding and averaging yields

$$[\frac{\partial}{\partial t} + \kappa(k^2 + k'^2)] < \hat{T}(\vec{k}',t)\hat{T}(\vec{k},t) >=$$

$$- i \int q_j < \hat{T}(\vec{k}')\hat{T}(\vec{q})\hat{u}_j(\ p) > \delta(\vec{k} - \vec{p} - \vec{q})d\vec{p} \ d\vec{q}$$

$$- i \int q'_l < \hat{T}(\vec{k})\hat{T}(\vec{q}')\hat{u}_l(\vec{p}') > \delta(\vec{k}' - \vec{p}' - \vec{q}')d\vec{p}' d\vec{q}' (VIII - 3 - 3)$$

We also need the time evolution equation for $\vec{\hat{u}}(\vec{k},t)$:

$$(\frac{\partial}{\partial t} + \nu k''^2)\hat{u}_m(\vec{k}'',t) = -ik_n'' P_{ma}(\vec{k}'')$$

$$\int \hat{u}_n(\vec{p}'',t)\hat{u}_a(\vec{q}'',t)\delta(\vec{k}'' - \vec{p}'' - \vec{q}'')d\vec{p}'' d\vec{q}'' \ . \ (VIII - 3 - 4)$$

From eqs (VIII-3-1), (VIII-3-2) and (VIII-3-4), one can form the equation for the triple correlation $< \hat{T}\hat{T}\hat{u} >$:

$$[\frac{\partial}{\partial t} + \kappa(k^2 + k'^2) + \nu k''^2] < \hat{T}(\vec{k}',t)\hat{T}(\vec{k},t)\hat{u}_m(\vec{k}'',t) >=$$

$$- i \int q_b < \hat{T}(\vec{k}')\hat{T}(\vec{q})\hat{u}_b(\vec{p})\hat{u}_m(\vec{k}'') > \delta(\vec{k} - \vec{p} - \vec{q})d\vec{p} \ d\vec{q}$$

$$- i \int q'_c < \hat{T}(\vec{k})\hat{T}(\vec{q}')\hat{u}_c(\vec{p}')\hat{u}_m(\vec{k}'') > \delta(\vec{k}' - \vec{p}' - \vec{q}')d\vec{p}' d\vec{q}'$$

$$- ik_n'' P_{ma}(\vec{k}'') \int < \hat{T}(\vec{k})\hat{T}(\vec{k}')\hat{u}_n(\vec{p}'')\hat{u}_a(\vec{q}'') > \delta(\vec{k}''-\vec{p}''-\vec{q}'')d\vec{p}'' d\vec{q}'' \ .$$

$$(VIII - 3 - 5)$$

We are now in a situation to apply the Quasi-Normal approximation. As already emphasized, the temperature-velocity correlations are zero, and the Quasi-Normal expression of (VIII-3-5) becomes

$$[\frac{\partial}{\partial t}+\kappa(k^2 + k'^2) + \nu k''^2] < \hat{T}(\vec{k}',t)\hat{T}(\vec{k},t)\hat{u}_m(\vec{k}'',t) >=$$

$$i \ k'_b \frac{E_T(k',t)}{2\pi k'^2}\hat{U}_{bm}(\vec{k}'',t)\delta(\vec{k} + \vec{k}' + \vec{k}'')$$

$$+i \ k_c \frac{E_T(k,t)}{2\pi k^2}\hat{U}_{cm}(\vec{k}'',t)\delta(\vec{k} + \vec{k}' + \vec{k}'') \ . \qquad (VIII - 3 - 6)$$

We solve eq. (VIII-3-6) for the triple correlation $< \hat{T}\hat{T}\hat{u} >$, substitute in

(VIII-3-3), integrate on $\vec{k}'$, and obtain finally

$$(\frac{\partial}{\partial t} + 2\kappa k^2)\hat{\Phi}(k,t) = 2\int_{\vec{p}+\vec{q}=\vec{k}} d\vec{p} \int_0^t d\tau \; k_b k_j$$

$$\hat{U}_{bj}(\vec{q},\tau)[\hat{\Phi}(\vec{p},\tau) - \hat{\Phi}(k,\tau)] \exp -[\kappa(k^2 + p^2) + \nu q^2](t - \tau)$$

$$(\text{VIII} - 3 - 7)$$

where $\hat{\Phi}(k,t) = E_T(k,t)/2\pi k^2$ is the equivalent of the spectral tensor for the temperature (cf V-5-16). $\hat{\Phi}(\vec{k},t)$ is the spatial Fourier transform[1] of the temperature spatial correlation

$$\Phi(\vec{r},t) = <T(\vec{x},t)T(\vec{x}+\vec{r},t)> .$$

If linear-damping rates had been introduced to model the action of discarded fourth-order cumulants, the *E.D.Q.N.* approximation for the temperature equation would be obtained from (VIII-3-7) by replacing the exponential term by

$$\exp -[\kappa(k^2 + p^2) + \mu'(k) + \mu'(p) + \nu q^2 + \mu"(q)](t - \tau) \; , \quad (\text{VIII} - 3 - 8)$$

where $\mu'(k)$ and $\mu"(k)$ are two functions of same structure as the triple-velocity correlation relaxation rate μ_k introduced in eq. (VII-3-4), but with possibly different constants, that is

$$\mu_k = a_1(\int_0^k p^2 E(p,t)dp)^{1/2} \qquad (\text{VIII} - 3 - 9)$$

$$\mu'(k) = a_2(\int_0^k p^2 E(p,t)dp)^{1/2} \qquad (\text{VIII} - 3 - 10)$$

$$\mu"(k) = a_3(\int_0^k p^2 E(p,t)dp)^{1/2} \; . \qquad (\text{VIII} - 3 - 11)$$

Then the markovianization yields the *E.D.Q.N.M.* equation for $\hat{\Phi}(k,t)$:

$$(\frac{\partial}{\partial t} + 2\kappa k^2)\hat{\Phi}(k,t) = 2\int_{\vec{p}+\vec{q}=\vec{k}} d\vec{p} \; k_b k_j \; \theta_{kpq}^T$$

$$\hat{U}_{bj}(\vec{q},t)[\hat{\Phi}(\vec{p},t) - \hat{\Phi}(\vec{k},t)] \quad , \quad (\text{VIII} - 3 - 12)$$

with

$$\theta_{kpq}^T = \frac{1 - \exp -[\kappa(k^2 + p^2) + \mu'(k) + \mu'(p) + \nu q^2 + \mu"(q)]t}{\kappa(k^2 + p^2) + \mu'(k) + \mu'(p) + \nu q^2 + \mu"(q)} .$$

$$(\text{VIII} - 3 - 13)$$

---

[1] because of the isotropy assumption, $\hat{\Phi}(\vec{k},t)$ does not depend on the orientation of $\vec{k}$.

The choice of the new adjustable constants a_2 and a_3 in (VIII-3-10) and (VIII-3-11) is not simple. We recall that $a_1 = 0.36$ has been taken in order to recover a value of 1.4 for the Kolmogorov constant of the inertial-range kinetic energy spectrum. It has been shown by Herring et al. (1982) that a given value of the "Corrsin-Oboukhov constant" arising in the r.h.s. of (VIII-2-11) (here taken equal to 0.67 from experimental measurements of Champagne et al., 1977) imposes a certain one to one correspondence between a_2 and a_3. The last condition, allowing to determine a_2 and a_3, comes from considerations on the "turbulent Prandtl number", defined in the following way: it will be shown soon that an eddy-conductivity κ_t of the same genre as the eddy-viscosity ν_t defined in (VII-10-2), can be introduced. For certain kinetic energy spectra, the turbulent Prandtl number ν_t/κ_t is equal to $(a_2 + a_3)/6a_1$, as shown in Larchevêque and Lesieur (1981, see also Herring et al. ,1982, Chollet, 1983, and Section 4 of the present chapter). It is then possible to express this number in function of a_2/a_3 only, in such a way that the Corrsin-Oboukhov constant should be fixed to the value given above. This leads to a turbulent Prandtl number decreasing continuously from 0.6 to 0.325 for a_2/a_3 going from zero to infinity. Since the values of turbulent Prandtl numbers found experimentally in the boundary layer are in the range $0.6 \approx 0.8$ (see e.g. Fulachier and Dumas, 1976), this could lead to the choice $a_2 = 0$ (and hence $a_3 = 1.3$, from Herring et al., 1982). One could object that the analogy between both theoretical and experimental turbulent Prandtl number is not obvious. Nevertheless the choice $a_2 = 0$ has the further advantage of allowing analytical resolutions of the *E.D.Q.N.M.* temperature spectral equation. It has to be stressed that the simpler choice $a_2 = a_3 = a_1$ gives the same Corrsin-Oboukhov constant and a turbulent Prandtl number of 0.35. As shown by Herring et al. (1982), both choices support a very good comparison with atmospheric kinetic energy and temperature spectra reported in Champagne et al. (1977). Eq. (VIII-3-12) can immediately be written for the temperature spectrum, if one remarks that

$$\int d\vec{p} = \iint_{\Delta_k} dp \, dq \, 2\pi \, \frac{pq}{k}$$

and, from (V-5-9)[2]

$$k_b k_j \hat{U}_{bj}(\vec{q}, t) = k^2 (1 - y^2) \frac{E(q, t)}{4\pi q^2}$$

---

[2] This shows that the helicity has no direct influence on the scalar spectrum.

and the resulting equation is

$$\frac{\partial}{\partial t} E_T(k,t) = \iint_{\Delta_k} dp \ dq \ \theta_{kpq}^T \frac{k}{pq} (1-y^2) E(q,t)[k^2 E_T(p) - p^2 E_T(k)]$$
$$- 2\kappa \ k^2 E_T(k,t) \quad . \qquad\qquad\qquad (\text{VIII} - 3 - 14)$$

The generalization of this equation to a Test-Field Model study of the passive scalar may be found in Newman and Herring (1979).

3.1 A simplified *E.D.Q.N.M.* model

With the choice $a_2 = 0, \theta^T$ is equal to a function θ_q^T, and (VIII-3-12) can be easily written in the physical space with the aid of an inverse Fourier transform (Larchevêque and Lesieur, 1981): indeed, the inverse Fourier transform of the convolution

$$\int \theta_q^T \hat{U}_{bj}(\vec{q}) \ \hat{\Phi}(\vec{p}) d\vec{p}$$

is

$$\int \theta_q^T \hat{U}_{bj}(\vec{q})[\exp i\vec{q}.\vec{r}] \ d\vec{q} \ \Phi(\vec{r}),$$

and the inverse Fourier transform of

$$\int \theta_q^T \hat{U}_{bj}(\vec{q}) \hat{\Phi}(\vec{k}) d\vec{p} = \int \theta_q^T \hat{U}_{bj}(\vec{q}) d\vec{q} \ \hat{\Phi}(\vec{k})$$

is

$$\int \theta_q^T \hat{U}_{bj}(\vec{q}) d\vec{q} \ \Phi(\vec{r}).$$

Then, by introducing the generalized turbulent diffusion tensor

$$K_{ij}(\vec{r},t) = 2 \int \theta_q^T \hat{U}_{ij}(\vec{q},t)(1 - \exp i\vec{q}.\vec{r}) d\vec{q} \ , \qquad (\text{VIII} - 3 - 15)$$

it is possible to write eq. (VIII-3-12) as a diffusion equation for the spatial temperature covariance $\Phi(r,t)$:

$$\frac{\partial}{\partial t} \Phi(r,t) = \frac{\partial^2}{\partial r_i \partial r_j} [K_{ij}(\vec{r},t \ )\Phi(r,t)] + 2\kappa \nabla^2 \Phi(r,t) \quad . \quad (\text{VIII} - 3 - 16)$$

This equation has close analogies with an equation obtained by Kraichnan (1966 b) using the Lagrangian History Direct Interation techniques, and widely discussed by Leslie (1973). It might also be of interest for the determination of the p.d.f. in turbulent reacting flows (Eswaran and O'Brien, 1989).

Then it is shown in Larchevêque and Lesieur (1981), following Kraichnan (1966 b), that the isotropy and zero-divergence of $K_{ij}(\vec{r},t)$ allows
us to introduce a scalar $K_{//}(r,t)$ such that (VIII-3-16) reduces to

$$\frac{\partial}{\partial t}\Phi(r,t) = r^{-2}\frac{\partial}{\partial r}[r^2 K_{//}(r,t)\frac{\partial}{\partial r}\Phi(r,t)] + 2\kappa\nabla^2\Phi(r,t) .$$
$$(VIII-3-17)$$

The turbulent diffusion coefficient $K_{//}(r,t)$, which is a function of θ_q^T
and $E(q,t)$, is equal to

$$K_{//}(r,t) = 0.696\, a_3^{-1} C_K^{1/2}\epsilon^{1/3}r^{4/3} \qquad (VIII-3-18)$$

if one assumes that the wave number r^{-1} lies in an extended $k^{-5/3}$
kinetic energy inertial-range, with $\theta_q^T = \mu"(q)^{-1}$. The constant C_K
arising in (VIII-3-18) is the Kolmogorov constant.

Eqs (VIII-3-17) and (VIII-3-18) can then be solved, using self-similarity arguments, giving in particular information on the decay law of
temperature variance (see Eswaran and O'Brien, 1989), a problem which
will be considered in section 4 mainly in the Fourier space. We will
however come back to this physical space point of view in section 5,
concerning the dispersion of pairs of Lagrangian tracers.

3.2 *E.D.Q.N.M.* scalar-enstrophy blow up

We end this section with a study paralleling the enstrophy divergence
study done in Chapter VII, and showing how the *E.D.Q.N.M.* closure
predicts that this enstrophy divergence at a finite time t_c (in the limit
of zero viscosity) will imply a blow up of the scalar enstrophy (defined
in (VIII-2-9)): from eq. (VIII-3-14), after multiplication by k^2 and integration from $k = 0$ to ∞, one obtains, using the same techniques as in
VIII-7

$$\frac{d}{dt}D_T(t) = \int_0^{+\infty}\int_0^{+\infty}\theta_q^T p^2 q^2 E(q)E_T(p)dpdq\int_{-1}^{+1}(1-x^2)dx$$
$$- 2\kappa\int_0^{+\infty}k^4 E_T(k,t)dk , \qquad (VIII-3-19)$$

where the time θ_{kpq}^T has again been approximated by θ_q^T , itself taken
equal to $\mu"(q)^{-1}$. This yields

$$\frac{d}{dt}D_T(t) = \frac{8}{3}a_3 D_T(t)D(t)^{1/2} - 2\kappa\int_0^{+\infty}k^4 E_T(k,t)dk . \quad (VIII-3-20)$$

We assume first a perfect flow with no molecular diffusive effects (and
hence $\nu = \kappa = 0$) and suppose that, initially, the energy spectrum and

the temperature spectrum decrease rapidly (for instance exponentially) for $k > k_i \approx k_i^T$. Therefore the temperature enstrophy will diverge together with the enstrophy and at the same time t_c. The $M.R.C.M.$ equation equivalent to (VIII-3-20) is

$$\frac{d}{dt}D_T(t) = \frac{4}{3}\theta_0 D_T(t)D(t) \quad , \qquad \qquad (VIII-3-21)$$

and gives qualitatively the same result. Physically, one can say that the catastrophic stretching of vortex filaments by turbulence will at the same time steepen the temperature gradients in the fluid, leading to singularities which occur at the same time t_* at which the velocity gradients become singular[3]. More precisely, the $M.R.C.M.$ enstrophies are, from eqs (VII-7-5) and (VIII-3-21):

$$D(t) \propto (t_c - t)^{-1} \quad ; \quad D_T(t) \propto (t_c - t)^{-2} \quad , \qquad (VIII-3-22)$$

which shows that the scalar enstrophy diverges at t_c faster than the velocity enstrophy. The same tendency may be obtained with the $E.D.Q.$-$N.M.$ (non-diffusive) scalar equation (VIII-3-20), where the constant-skewness model prediction (VI-7-8) has been used for the velocity. It is easily found

$$D_T(t) \propto (t_c - t)^{-(8/3)a_3 t_c \sqrt{D(0)}} \quad , \qquad \qquad (VIII-3-23)$$

and the critical exponent of the divergence is of the order of 19 (with $a_3 = 1.3$ and $t_c = 5.6\ D(0)^{-1/2}$), which is huge compared with the $(t_c - t)^{-2}$ divergence of the velocity enstrophy. Therefore, it is expected that the temperature will cascade faster than the velocity towards small scales, due certainly to the fact that there is no pressure term in the scalar equation, and hence the scalar will directly react to the velocity gradients. The velocity, on the contrary will have to satisfy the incompressibility condition, and hence redistribute among its three components and throughout the fluid, with the aid of the pressure.

A numerical resolution of the $E.D.Q.N.M.$ temperature spectral equation (including viscous and conductive effects) is possible, using the same methods as for the kinetic energy spectrum. Again the problem arises of the non-local interactions modelling, which will be developed in the next section. Figure VIII-3a, taken from Lesieur et al. (1987), shows the evolution in time of both enstrophies (velocity and temperature) for

---

[3] A similar result has been predicted by Fal'kovich and Shafarenko (1988) in a study of weakly-interacting acoustic waves.

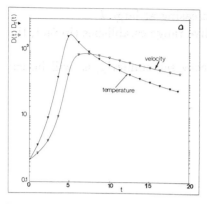 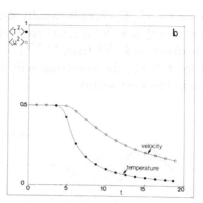

Figure VIII-3: a) time-evolution (at an initial Reynolds number of $R_l = 40000$ and a Prandtl number of one) of the velocity and temperature enstrophies in an *E.D.Q.N.M.* initial-value problem where the initial temperature and velocity fluctuations are confined in the same large scales (from Lesieur et al., 1987); b) time-evolution of the velocity and temperature variances in the same calculation. (from Lesieur et al., 1987, courtesy The Physics of Fluids).

an unforced calculation done in the same conditions as in Figures VII-3, VII-4, for an initial Reynolds number $R_l = v(0)/\nu k_i(0) = 40000$, a Prandtl number of 1, and $a_2 = 0$. One can see both the overshoot at $t_* \approx 5/v_0 k_i(0)$, with a faster divergence of the temperature enstrophy, in agreement with the above Euler flow considerations. Since ν and κ are small but not zero, D and D_T will saturate at values of the order of respectively ϵ/ν and ϵ_T/κ. The velocity and temperature variance decay is shown on Figure VIII-3-b, which indicates that the passive scalar follows approximately the asymptotic tendency of the kinetic energy, that is no dissipation before t_* and a finite dissipation ϵ_T after t_*. We recall from (VIII-2-8) that $\epsilon_T = 2\kappa D_T(t)$, and hence the temperature enstrophy is finite (for $\kappa \to 0$) for $t < t_*$ and infinite for $t > t_*$, in good agreement with the analytical calculations already made. Notice however that the temperature-variance starts being dissipated appreciably sooner than the kinetic energy (4 instead of 5). This can be explained by considering the enstrophies curves on Figure (VIII-3-a): at this Prandtl number of one, a temperature-variance dissipation rate equal to $\epsilon(t_*) = 2\nu D(t_*)$ will be obtained at the time t_*^T such that

$$D_T(t_*^T) = D(t_*) \quad ,$$

that is, approximately $t_*^T = 4$ for $t_* = 5$. Hence, this lag of about 20% in the respective variances decay is not caused by a difference between the enstrophies blow-up times, but indicates the fact that the temperature enstrophy diverges faster than the velocity enstrophy at the critical time.

Finally, Figure VIII-4 (from Lesieur et al., 1987) shows in this calculation the evolution with time of the temperature spectrum:

• for $t < t_*^T$, $E_T(k,t)$ is rapidly decreasing at large k
• at $t = t_*^T$ a $k^{-5/3}$ inertial-convective range establishes (in fact, the slope is closer to $k^{-3/2}$ than $k^{-5/3}$).

• for $t > t_*^T$, the spectrum will decay self-similarly, as will be explained in the next section.

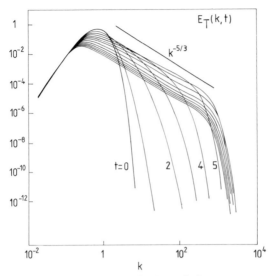

Figure VIII-4: time-evolution of the temperature spectra in the same calculation as in Figure VIII-3, showing the appearance of the inertial-convective range (from Lesieur et al., 1987).

In this calculation, in fact, the build up of the inertial-convective range is more rapid than the Kolmogorov cascade for the kinetic energy, since the latter establishes between 5 and 6 unit of time $1/v_0 k_i(0)$ (compare Figures VIII-4 and VII-3). This specifies how temperature cascades faster[4] than the velocity to high k. But, for a perfect fluid, this is not in contradiction with the possibility of forming simultaneously at t_c infinite ultra-violet $k^{-5/3}$ kinetic-energy and temperature spectra.

At this point, we have to mention that the already quoted large-eddy simulations of isotropic decaying turbulence performed by Lesieur

---

[4] If the conductivity is increased with respect to the viscosity, this phenomenon may be inhibited, as shown in the $E.D.Q.N.M.$ calculation of Lesieur et al. (1987) at a Prandtl number $P_r = 0.1$, where the velocity enstrophy now grows faster than the temperature enstrophy. In the same reference, a calculation at $P_r = 10$ shows unambiguously the formation of both the $k^{-5/3}$ inertial-convective range and the k^{-1} viscous-convective range.

and Rogallo (1989), Lesieur et al. (1989), and Métais and Lesieur (1990), show the establishment in the energy-containing eddies range of an *anomalous* range[5] following approximately the law:

$$E_T(k,t) = 0.1\frac{\epsilon_T}{\epsilon}\, v_0^2\, k^{-1} \quad . \qquad (VIII-3-24)$$

This calculation is shown on Figure VIII-5 (taken from Lesieur and Rogallo, 1989). An associated anomalous behaviour of the spectral eddy-conductivity has been noted in the same works. If valid[6], this range might be explained in the following way: one considers a temperature-flux equation analogous to (VIII-2-12), but with a shearing time determined by the energy-containing eddies and equal to $\tau(k_i) = 1/v_0 k_i$. This yields

$$\epsilon_T \sim k\, E_T(k)\, v_0\, k_i \quad , \qquad (VIII-3-25)$$

and a k^{-1} temperature spectrum. At this stage, the Reynolds number may be moderate[7]. Now, we assume a high Reynolds number, and that $\epsilon = v_0^3 k_i$ in (VIII-3-25). This permits to recover eq. (VIII-3-24), in which the constant 0.1 has been determined in the numerical experiment previously quoted. In this calculation, the enstrophy reaches a maximum at about $t = 4/v_0(0)k_i(0)$, corresponding to the appearance of a Kolmogorov cascade at high wave numbers. Meanwhile, the temperature enstrophy reaches its maximum at $2.4/v_0 k_i(0)$, corresponding to a much faster ultra-violet cascade of scalar. This is qualitatively in agreement with what was observed above for the closures, except for the exponent of the scalar spectrum (−1 instead of −1.5): the difference might be related to the large-scale intermittency of the temperature already mentioned in Chapter VI, which is certainly badly treated by the closures. Comparisons with laboratory experiments will be done below. Remark finally that the above large-scale k^{-1} temperature range does not seem to be related to Batchelor's small-scale viscous-convective range obtained at high Prandtl number. It has to be stressed that the anomalous k^{-1} regime in the large scales might be only transient, and that the temperature spectrum could eventually return to the self-similar decaying states studied in the next section.

---

[5] anomalous with respect to the $k^{-5/3}$ inertial-convective range scalar spectrum

[6] The verification of this prediction requires higher-resolution simulations.

[7] This scalar k^{-1} range has been found in direct-numerical simulations done by Métais and Lesieur (1988, 1990).

Turbulence in fluids

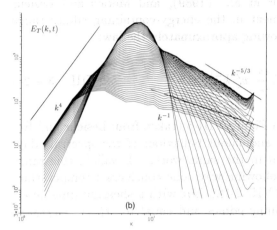

Figure VIII-5: decay with time of the temperature spectrum in the large-eddy simulation corresponding to the calculation of Figure VII-9. The temperature spectrum, initially rapidly decreasing, builds up an inertial-convective range in the small scales, and develops an approximate k^{-1} range in the energy-containing range (from Lesieur and Rogallo, 1989, courtesy The Physics of Fluids).

4 - The decay of temperature fluctuations

Once the temperature variance is dissipated at a finite rate, one can wonder about the existence of asymptotic (when ν and κ go to zero) temperature decay laws, and look for an exponent α_T such that

$$\frac{1}{2} <T(\vec{x},t)^2> \propto t^{-\alpha_T} \quad . \qquad (VIII-4-1)$$

4.1 Phenomenology

The problem depends in fact on the relative initial location of the temperature and velocity integral scales, as was shown experimentally by Warhaft and Lumley (1978)[8] An analytical study of this problem will be given here, based both on the phenomenology and on the *E.D.Q.N.M.* theory. It will try to explain some of the anomalous decay results found in the above quoted experiments: we will carry out the study by assuming first that both temperature and velocity scales are of the same order, and second, that the temperature is initially injected in scales much

---

[8]   in these experiments, another important parameter is the the grid heating intensity

smaller than the velocity. Such a study has important practical applications, for its results are generally used to fix some of the adjustable coefficients in the so-called "one-point closure" modelling methods employed for engineering purposes, even in compressible situations.

4.1.1 Non-local interactions theory

We first need to write down the non-local temperature fluxes and transfers, in the same way as for the kinetic energy in Chapter VII-6: the non-linear temperature transfer term in the r.h.s. of (VIII-3-14) can be symmetrized with respect to p and q, under the form $\iint_{\Delta_k} dp \, dq S_T(k,p,q)$, with $S_T(k,p,q) = S_T(k,q,p)$. Then we expand $(kq/p)S_T(k,p,q)$ [or $(kp/q)S_T(k,p,q)$] with respect to the small parameter q/k (or k/p). The technique of these expansions is described in (VII-6). To the first order with respect to the small parameter, we have, following VII-6 and without expanding θ :

a)**For** $q \ll k$:

$$p = k - qy \; ; \quad p^2 = k^2 - 2kqy \; ;$$

$$E(p) = E(k) - qy\frac{\partial E}{\partial k} \; ; \quad E_T(p) = E_T(k) - qy\frac{\partial E_T}{\partial k} \; ;$$

$$\frac{kq}{p}S_T(k,p,q) = \frac{1}{2}\theta^T_{kkq}(1-y^2)$$

$$[kqy\{2E_T(k) - k\frac{\partial E_T}{\partial k}\}E(q) + q^2 E_T(q)E(k) - \frac{q^4}{k^2}E_T(k)E(k)]$$
$$(VIII - 4 - 2)$$

and since

$$\int_0^1 d\phi \int_\phi^1 (1-y^2)dy = \frac{1}{4} \; ; \quad \int_0^1 d\phi \int_\phi^1 y(1-y^2)dy = \frac{2}{15}$$

the non local temperature flux π^{T+}_{NL} is written

$$\pi^{T+}_{NL}(k,t) = \frac{2}{15}\int_0^{ak} \theta^T_{kkq}q^2 E(q)dq \; [2kE_T(k) - k^2\frac{\partial E_T}{\partial k}]$$

$$+ \frac{1}{4}\int_0^{ak} \theta^T_{kkq}q^3 E_T(q)dq \; E(k)$$

$$- \frac{1}{4}\int_0^{ak} \theta^T_{kkq}q^5 dq \; E(k)\frac{E_T(k)}{k^2} \qquad (VIII - 4 - 3)$$

b)**For** $k \ll p$:

$$q = p - kz \; ; \quad q^2 = p^2 - 2kpz \; ;$$

$$E(q) = E(p) - kz\frac{\partial E}{\partial p} \; ; \quad E_T(q) = E_T(p) - kz\frac{\partial E_T}{\partial p} \; ;$$

$$\frac{kp}{q}S_T(k,p,q) = \theta^T_{kpp}\frac{k^2}{p^2}(1-z^2)[k^2E(p)E_T(p) - p^2E(p)E_T(k)]$$

$$(VIII-4-4)$$

and since $\int_0^1(1-z^2)dz = 2/3$, the non local temperature flux π^{T-}_{NL} is

$$\pi^{T-}_{NL}(k,t) = -\frac{4}{3}\int_0^k k'^2 E_T(k')dk' \int_{sup(k,k'/a)}^{\infty} \theta^T_{k'pp}E(p)dp$$

$$+\frac{4}{3}\int_0^k k'^4 dk' \int_{sup(k,k'/a)}^{\infty} \theta^T_{k'pp}\frac{E(p)}{p^2}E_T(p)dp \quad (VIII-4-5)$$

The first term in the r.h.s. of (VIII-4-3) corresponds to interactions responsible for the k^{-1} viscous-convective range, since the corresponding flux is constant for such a spectrum. More precisely one may write in this range from this equation (see also Kraichnan, 1968 b, Newman and Herring, 1979, and Herring et al., 1982):

$$E_T(k) = C_B \; \epsilon_T(\frac{\nu}{\epsilon})^{1/2}k^{-1} \quad , \qquad\qquad (VIII-2-23')$$

to be compared with (VIII-2-23). C_B is the Batchelor constant, here equal to (with $a_2 = 0$)

$$C_B = \frac{1}{3P}\left(\frac{\epsilon}{\nu}\right)^{1/2} \; , \quad P = \frac{2}{15}\int_0^{+\infty}\frac{q^2\;E(q)\;dq}{a_3\sqrt{\int_0^q p^2 E(p)dp + \nu q^2}} \quad ,$$

which can be checked to be a constant by assuming a kinetic energy spectrum of the form (VI-6-10).

Otherwise, the interesting non-local temperature transfers as far as the infrared dynamics is concerned come from eq. (VIII-4-5): when $k << k_i^T$, and taking $a = k/k_i^T$, it yields a non local temperature transfer equal to (to the lowest order)

$$T^T_{NL}(k,t) = -\frac{4}{3}k^2 E_T(k)\int_{k_i^T}^{\infty}\theta^T_{0pp}E(p)dp$$

$$+\frac{4}{3}k^4\int_{k_i^T}^{\infty}\theta^T_{0pp}\frac{E(p)}{p^2}E_T(p)dp \quad . \qquad (VIII-4-6)$$

This allows us to define an eddy-diffusivity in spectral space, analogous to the eddy-viscosity (VII-10-2), and equal to

$$\kappa_t = \frac{2}{3}\int_{k_i^T}^{\infty}\theta^T_{0pp}E(p)dp \quad . \qquad\qquad (VIII-4-7)$$

If one takes $k_i = k_i^T$ and a schematic energy spectrum equal to zero for $k < k_i$ and to $k^{-5/3}$ for $k > k_i$, with the asymptotic values (neglecting the time exponential contribution) for θ and θ^T, it is easy to check that the turbulent Prandtl number ν_t/κ_t is equal to $(a_2 + a_3)/6a_1$, as mentioned in section 3.

The second term of the r.h.s. of (VIII-4-6) is responsible for a positive k^4 temperature transfer, and the same remarks as for the k^4 kinetic energy transfer can be applied: if in particular the temperature spectrum is stationary, due to an external thermal forcing acting at the fixed wave number k_i^T, the temperature spectrum will be (for $k \ll k_i^T$) a k^2 equipartition spectrum resulting from a balance between the k^4 positive temperature transfer coming from resonant interactions of wave numbers of the order of k_i^T , and the turbulent diffusive action of these modes on k.

4.1.2 Self-similar decay

The study of the temperature decay will be made assuming that the kinetic energy spectrum decays self-similarly according to the laws (VII-10-8), the kinetic energy and the integral scale following power laws of time given by (VII-10-11) and (VII-10-12). In a similar way to eq. (VIII-4-1), we introduce α_E and α_l, such that

$$\frac{1}{2}v^2 \propto t^{-\alpha_E}, \quad l(t) \propto t^{\alpha_l} \qquad (VIII-4-8)$$

and remark that

$$\alpha_l = 1 - \frac{\alpha_E}{2} \quad , \qquad (VIII-4-9)$$

the precise value of α_E and α_l being given by eqs (VII-10-11) and (VII-10-12).

Firstly the temperature integral scale l_T is evaluated as shown below: we calculate the temperature dissipation rate ϵ_T as the ratio of the temperature variance $<T^2>$ divided by a characteristic dynamical time at scales of order l_T. This local time at l_T is l_T/v_T, where v_T is a velocity characteristic of the eddies of size l_T, that is $(\epsilon l_T)^{1/3}$. This yields[9]

$$\epsilon_T = <T^2> \epsilon^{1/3} l_T^{-2/3} \qquad (VIII-4-10)$$

In fact, (VIII-4-10) expresses only a proportionality. We have chosen to consider it as an equality, which defines precisely here the temperature integral scale l_T. This scale is of the order of $(k_i^T)^{-1}$, wave number

---

[9] This assumes however that l_T is in the kinetic energy spectrum $k^{-5/3}$ inertial range, which will not be valid if $l_T \gg l$. The latter case has been looked at in Herring et al. (1982).

where the temperature spectrum peaks. Eq. (VIII-4-10) will be valid
even in the case where $l_T \ll l$. As stressed above, it has no physical
significance when $l < l_T$: indeed, the local dynamical time at l_T will be
l_T/v , and (VIII-4-10) will have to be replaced by

$$\epsilon_T = <T^2> \epsilon^{1/3} l^{-2/3} \frac{l}{l_T} \ . \qquad (VIII-4-10')$$

We also assume, however small l_T/l, that $E_T(k,t)$ is given by

$$E_T(k,t) = C_{co}\epsilon_T \epsilon^{-1/3} k^{-5/3} \text{ for } k > k_i^T(t) \qquad (VIII-4-11)$$

$$E_T(k,t) = C_{s'}(t)k^{s'} \text{ for } k < k_i^T(t) \ . \qquad (VIII-4-12)$$

The justification of (VIII-4-11) comes from an inertial-convective range
assumption for the temperature spectrum[10] (the constant C_{co} is the
Corrsin-Oboukhov constant, that experiments show to be of the order of
0.7). Matching eqs (VIII-4-11) and (VIII-4-12) for $k = k_i^T \approx l_T^{-1}$ yields,
using (VIII-4-10)

$$C_{s'}(t) \approx <T^2> l_T^{s'+1} \ . \qquad (VIII-4-13)$$

The same reasoning as in Chapter VII-10 leads to, using (VIII-4-6)

$$\frac{dC_{s'}(t)}{dt} = 0 \text{ for } s' < 4 \qquad (VIII-4-14)$$

$$\frac{dC_{s'}(t)}{dt} \sim \int_{k_i^T}^{\infty} [p^3 E(p)]^{-\frac{1}{2}} \frac{E(p)}{p^2} E_T(p)dp \text{ for } s' = 4 \qquad (VIII-4-15)$$

and hence[11], from (VIII-4-11)

$$\frac{dC_{s'}(t)}{dt} \sim \epsilon_T l_T^5 \text{ for } s' = 4 \ . \qquad (VIII-4-16)$$

All these relations are general for any $l_T/l \leq 1$. A last interesting relation
can be easily derived (for $s' \leq 4$), i.e. (Lesieur et al., 1987)

$$\frac{1}{2}\frac{d}{dt} l_T^2 \sim \epsilon^{1/3} l_T^{4/3} \qquad (VIII-4-17)$$

[10] A similar analysis has been carried out by Métais and Lesieur (1986)
for the error spectrum in the predictability problem; it will be presented
in Chapter XI.
[11] The constant of proportionality in (VIII-4-15) has not been expressed, but is not needed for the present analysis.

which shows that l_T satisfies a Richardson-type law analogous to VII-10-13, but with possibly a different numerical constant.

Looking then for time power law solutions $t^{-\alpha_T}$ and $t^{\alpha_{l_T}}$ respectively for the temperature variance and integral scale, (VIII-4-17) yields

$$\alpha_{l_T} = 1 - \frac{\alpha_E}{2} \qquad (VIII-4-18)$$

which shows (from (VIII-4-9)) that α_l and α_{l_T} are equal. This, together with eqs (VIII-4-13), (VIII-4-14) and (VIII-4-16), gives

$$\alpha_T = (s'+1)\alpha_l - \gamma' = \frac{s'+1}{s+3}(2+\gamma) - \gamma' \qquad (VIII-4-19)$$

with $C_{s'}(t) \propto t^{\gamma'}(\gamma' = 0$ for $s' < 4)$, as predicted in Larchevêque et al. (1980) and Herring et al. (1982). When $s' = 4, \gamma'$ has been found numerically with the aid of the *E.D.Q.N.M.* to be equal to 0.06 (Chollet, 1983). Thus α_T is equal to 1.48 for $s = s' = 4$, in good agreement with the value of 1.5 found in Larchevêque et al. (1980).

In fact, these time power laws are not compatible with any value of the ratio l_T/l: indeed, let us write the equivalent expression of (VIII-4-10) for the velocity

$$\epsilon = v^2 \epsilon^{1/3} l^{-2/3} \qquad (VIII-4-20)$$

which can be used as a definition of l as well. Eq. (VIII-4-20) shows that

$$\alpha_E = 2t\frac{\epsilon}{v^2} = 2t \ \epsilon^{1/3} l^{-2/3} .$$

In the same way (VIII-4-10) allows us to define an "instantaneous" temperature variance decay exponent α'_T, such that the temperature variance should locally be tangent to a $t^{-\alpha'_T}$ law, with

$$\alpha'_T = 2t\frac{\epsilon_T}{<T^2>} = 2t \ \epsilon^{1/3} l_T^{-2/3} \quad . \qquad (VIII-4-21)$$

Eqs (VIII-4-20) and (VIII-4-21) show that

$$r = \frac{\alpha'_T}{\alpha_E} = (\frac{l}{l_T})^{2/3} \quad . \qquad (VIII-4-22)$$

In (VIII-4-22), α'_T/α_E can also be interpreted as the velocity and scalar time scales ratio. This expression is from Corrsin (1964). With the particular definitions taken here for l_T and l, (VIII-4-22) is always valid (for $l \geq l_T$) with a numerical constant equal to one, and even with a moderate Reynolds number. One may wonder how this relationship is modified with the classical definitions of the integral scales. It is claimed in Herring et al. (1982), from a high Reynolds number $E.D.Q.N.M.$ calculation ($l \geq l_T$), that a numerical constant of 1.63 has then to be introduced in front of the r.h.s. of (VIII-4-22). The same work stresses that, at a moderate Reynolds number, r is equal to l/l_T (even in the case $l \leq l_T$), a law which can immediately be derived from eq. (VIII-4-10') in this latter case. This agrees well with the experiment carried out by Sreenivasan et al. (1980).

When the temperature follows an actual time power law, α'_T is equal to α_T, and given by (VIII-4-19). This implies from (VIII-4-22) that one can have a time power law dependence only when l/l_T has the particular value predicted by (VIII-4-19) and (VII-10-11). This value depends on s and s', but in any case is very close to one. This demonstrates the point that the temperature variance decays as a power of time only if the temperature and velocity integral scales are of the same order.

4.1.3 Anomalous temperature decay

If l_T is initially much smaller than l, eq. (VIII-4-22) shows that the apparent temperature time decay exponent α'_T is much greater than α_E, which might explain the anomalous temperature decay exponents found experimentally in this case by Warhaft and Lumley (1978). It has been shown in Lesieur et al. (1987), using the above phenomenology, that the detailed time-evolution of the temperature variance and integral scales can be obtained analytically with these particular initial conditions. This analysis, which is an extension to an arbitrary $k^{s'}$ temperature spectrum (when $k \rightarrow 0$) of an analysis done by Nelkin and Kerr (1981) in the case $s' = 2$, is based on the integration of the Richardson equation (VIII-4-17). It is found

$$l_T(t) = (\frac{\alpha_E}{\alpha_T})^{3/2} \, l(t) \, [1 + B \, (\frac{t}{t_0})^{-2\alpha_l/3}]^{3/2} \quad , \qquad (VIII-4-23)$$

where B is a negative constant, such that

$$B = -1 + \frac{\alpha_T}{\alpha_E} \, [\frac{l_T(t_0)}{l(t_0)}]^{3/2} \quad ,$$

and initially ≈ -1. Therefore l_T/l increases with time. Hence $l_T(t)$ is going to grow and asymptotically (for t going to infinity) catch up with the time power law solution described above. Following eq. (VIII-4-22) the instantaneous temperature decay exponent will decrease with time and eventually reach the asymptotic value α_T given by (VIII-4-19). We again emphasize that these asymptotic values are fixed by the spectral exponents s and s' for $k \rightarrow 0$, and impose a particular ratio for l/l_T. If l/l_T is initially larger than this asymptotic ratio, it will decrease to it,

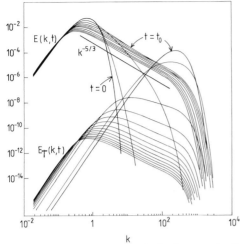

Figure VIII-6: same calculation as in Figure VIII-4 for the kinetic energy spectrum, but with an initial temperature integral scale 70 times smaller than the velocity integral scale (from Lesieur et al., 1987).

and so will the temperature decay exponent. The temperature variance decays as (Lesieur et al., 1987)

$$<T^2> \propto [B + (\frac{t}{t_0})^{2\alpha_l)/3}]^{-3\alpha_T/2\alpha_l} \quad . \qquad (VIII-4-24)$$

This is equivalent to

$$<T^2> \propto l_T^{-\alpha_T/\alpha_l} \quad . \qquad (VIII-4-25)$$

In the case $s' = 2$ one recovers the above quoted results of Nelkin and Kerr (1981). Notice finally that expansions of eqs (VIII-4-24) (VIII-4-25) for short times $(t - t_0)$ yield

$$l_T(t) \propto (t - t_0)^{3/2} \quad .$$

Hence, the temperature integral scale will see its instantaneous growth exponent increase from $3/2$ at short times to α_l at high times[12]. If one accepts to associate the mean thermal wake δ_T of a line source with l_T (see next section), this may be compared with the experimental findings of Warhaft (1984) and Stapountzis et al. (1986), who find a $\delta_T \propto (t - t_0)$ "turbulent-convective" range, and at higher times $\delta_T \propto (t - t_0)^{0.34}$ "turbulent-diffusive" range.

---

[12] The same behaviour will be found for the "error-length" characterizing the decorrelation front in the predictability problem (see Chapter XI).

To illustrate these results, Figure VIII-6 (taken from Lesieur et al., 1987) shows the evolution of the kinetic energy and temperature spectra in a high Reynolds number $E.D.Q.N.M.$ calculation where the initial ratio $l(0)/l_T(0) = 70$. Here $s = s' = 4$. One sees the rapid increase of l_T which tends to catch up with l. In this calculation the tangential temperature decay exponent goes from an initial value of 20 to a value of 3.5, after which the calculation is no longer significant since there remains a negligible amount of temperature variance. It is nevertheless to be expected that a calculation performed with $k^{-5/3}$ ranges extending to infinity would eventually yield a value of α'_T equal to α_T. In this calculation, as well as in Herring et al.'s results, the temperature integral scale does satisfy the Richardson law when $l_T \leq l$. One may also notice in the calculation of Figure VIII-6 that the temperature spectrum develops initially a range close to k^{-1}, as in Figure VIII-5, and then eventually reorganizes towards a $k^{-5/3}$ inertial-convective range.

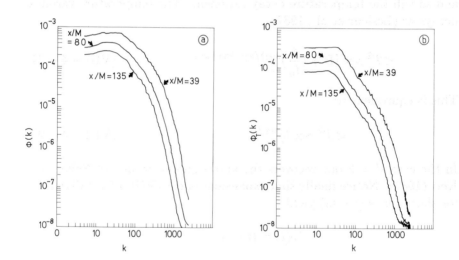

Figure VIII-7: kinetic energy (Figure VIII-7-a) and temperature spectra (Figure VIII-7-b) at three locations dowstream of the grid in the experiment by Warhaft and Lumley (1978), when $l_T \approx l$ initially. M is the grid mesh (Courtesy J. Fluid Mech.)

4.2 Experimental temperature decay data

The next results concern moderate Reynolds numbers: Figure VIII-7, taken from Warhaft and Lumley (1978), shows the time evolution of the kinetic energy and temperature spectra in a heated-grid turbulence experiment at a Reynolds number $v_0 l/\nu$ of the order of 150 . Here, and since the grid is heated, the temperature fluctuations are produced initially in the same scales as the velocity fluctuations. It is not surprising therefore, from the point of view of the above phenomenology, that the temperature decay exponent is found to be of same order ($\alpha_T = 1.41$) as the kinetic energy exponent[13] ($\alpha_E = 1.34$). At this moderate Reynolds number, a scalar range of slope $-3/2$ appears, although no inertial range is apparent for the corresponding kinetic energy spectrum. This point had already been noted by Yeh and Van Atta (1973) from heated-grid turbulence experiments [14], and could be due to the direct shearing of the scalar by the large-scale velocity gradients already mentioned in the theory leading to eq. (VIII-3-25).

When the temperature is injected at scales such that initially $l_T/l = 1/2$, the experimental temperature decay exponent (averaged during the experiment) is found to be 3.2 , in good agreement with another experiment of Sreenivasan et al. (1980). Figure VIII-8, from Lesieur et al. (1987), shows an $E.D.Q.N.M.$ calculation under conditions close to these experiments: the kinetic energy decreases as $t^{-1.32}$, and the temperature variance as $t^{-3.45}$. One can also see the velocity enstrophy peak at the time when the kinetic energy starts being dissipated. It is quite remarkable to see how well the $E.D.Q.N.M.$ theory fits the experiment in this "two-scale" (velocity and temperature) problem. It also seems that this moderate Reynolds number situation is widely influenced by the high Reynolds number theoretical predictions. As checked in the same reference, the spectra start building short $k^{-5/3}$ inertial ranges.

4.3 Discussion of the $L.E.S.$ results

In Lesieur and Rogallo's (1989) large-eddy simulation starting with identical rapidly decreasing velocity and temperature spectra, the ratio k_i^T/k_i of the respective peaks of the temperature and kinetic energy spectra has shifted from 1 initially to 0.64 at the end of the run, with a temperature decay exponent $\alpha_T = 3$ of the order of twice the velocity decay exponent α_E. In the large-eddy simulations of Lesieur et al. (1989), with different initial conditions, one gets at the end of the run $k_i^T/k_i = 0.56$, $\alpha_E = 1.37$ and $\alpha_T = 1.85$. In these two cases however, where the temperature spectra are close to k^{-1} up to the cutoff wave number k_c, an appreciable part of the temperature variance may have shifted to the subgrid scales: hence, the calculated temperature variance decay rate (corresponding to

---

[13] In a more recent experiment, Warhaft (1984) shows some anisotropic effects in the kinetic-energy decay: the longitudinal velocity energy decays as $t^{-1.40}$, and the transverse velocity energy as $t^{-1.32}$.

[14] with $\alpha_T = 1.33$

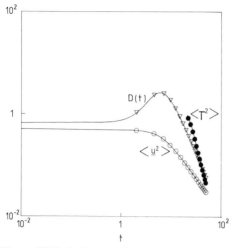

Figure VIII-8: time evolution of the kinetic energy, velocity enstrophy and temperature variance in an $E.D.Q.N.M.$ moderate Reynolds number calculation $(v_0/\nu k_i(0) = 140)$, where $l_T(0)/l(0) = 0.4$ (from Lesieur et al., 1987).

the explicit scales) may just be the signature of the fast scalar transfer towards subgrid scales, the total[15] temperature variance decaying at rates more comparable with the $E.D.Q.N.M.$ and the experimental predictions.

How then can we reconcile the $E.D.Q.N.M.$, $L.E.S$ and experimental results? The faster temperature transfer towards small scales, due to the lack of pressure term in the passive scalar equation, seems to be well observed in both $E.D.Q.N.M.$ and $L.E.S.$ calculations, if the scalar is injected in the energy-containing range or in the inertial range. This may produce transient scalar spectra determined by the larger-scale eddies deformation. As already stressed, it is possible that these are transient effects, and that, at high Reynolds and Peclet numbers, the temperature will eventually decay following the Corrsin-Oboukhov inertial-convective range.

4.4 Diffusion in stationary turbulence

Finally, it is of interest to look at the case where the kinetic energy spectrum is maintained stationary by external forces, while the temperature is decaying. We use the above phenomenological theory, where the Richardson law (VIII-4-17) is still valid, but things are simpler since ϵ is now a constant. For large times, l_T is proportional to $(\epsilon t^3)^{1/2}$. Eq. (VIII-4-13) is also valid, with the same distinction between the cases $s' < 4$ (where $dC_{s'}(t)/dt = 0$) and $s' = 4$. So for $s' < 4$ the temperature variance decays, for large times, proportionally to $(\epsilon\ t^3)^{-3(s'+1)/2}$. If

---

[15] calculated on the whole temperature spectrum fro $k = 0$ to $k = \infty$

for instance $s' = 2$, the temperature decay exponent is $-9/2$, as will be retrieved in the next section using another method.

5 - Lagrangian particle pair dispersion

Up to now we have mainly examined the statistics of a homogeneous passive scalar. But situations closer to reality are often the case when the scalar is locally injected in the homogeneous turbulent field, and then spreads out under the action of turbulent diffusion. It is, for instance, of interest to predict the evolution in time of the average size of a cloud of chemical or radioactive contaminant accidentally released in atmospheric turbulence[16].

The formalism of the Lagrangian tracers pair relative dispersion problem allows us to give a first answer to these questions: let $R(t_0)$ be a properly defined average diameter of the cloud at a given time t_0, and consider an ensemble of pairs of particles a distance $R(t_0)$ apart, randomly distributed and oriented in space. It is reasonable to accept the idea that the subsequent mean dispersion in time of the Lagrangian tracers pairs will give information on the spreading of the cloud. More precisely, one arbitrarily associates the pair separation variance $R(t)$ to the diameter of the cloud. Such an analogy does not take into account the influence of molecular diffusion, and will be valid only for isotropic turbulence.

The homogeneous formulation of the pair dispersion problem has been given by Batchelor (1952), and the problem has been studied with the aid of various statistical theories by Roberts (1961) using *D.I.A.*, Kraichnan (1966) using *L.H.D.I.A.*, Larchevêque and Lesieur (1981) using *E.D.Q.N.M.*, and Lundgren (1981). The p.d.f. $P(\vec{r}, t)$ that the two Lagrangian tracers of a pair are separated by the vector $\vec{r}$ can be easily shown to be equal to

$$P(\vec{r}, t) = <T(\vec{x} + \vec{r}, t)T(\vec{x}, t)> -\delta(\vec{r}) \qquad (VIII - 5 - 1)$$

where $T(\vec{x}, t)$ is a homogeneous passive scalar field. It is easy to show that $P(\vec{r}, t)$ satisfies the same equation as $\Phi(\vec{r}, t)$, the spatial correlation of T. The normalization conditions satisfied by $P(\vec{r}, t)$ impose

$$\int P(\vec{r}, t)d\vec{r} = 1 \qquad (VIII - 5 - 2)$$

---

[16] When the pollutant has reached scales of several kilometers, atmospheric turbulence is no longer isotropic, and such a study is beyond the present chapter. Diffusion in two-dimensional turbulence will be looked at in Chapter IX.

and the pair separation variance is

$$R^2 = \int r^2 P(\vec{r}, t) d\vec{r} \quad . \qquad (VIII - 5 - 3)$$

The problem has thus been reduced to the isotropic study of section 4, with the further condition (VIII-5-2), which corresponds to a passive scalar spectrum proportional to k^2 for $k \to 0$ (Larchevêque and Lesieur, 1981). One could then simply apply the results of section 4 with $s' = 2$, taking into account the difficulty that the relationship between $R(t)$ and $l_T(t)$ is not known[17]. When $R(t)^{-1}$ corresponds to Kolmogorov inertial-range eddies, it is possible to solve exactly eqs (VIII-3-17) and (VIII-3-18) for $P(r, t)$ (Larchevêque and Lesieur, 1981), looking for self-similar solutions of the same genre as proposed by Kraichnan (1966): one seeks a solution of the form

$$P(r, t) = F[R(t)]\, f\left[\frac{r}{R(t)}\right]) \quad , \qquad (VIII - 5 - 4)$$

and eqs (VIII-5-2) and (VIII-5-3) imply

$$F(R) = R^{-3} \quad ,$$

with two normalization conditions for the non dimensional function $f(x)$. The result is that $R(t)$ satisfies a Richardson law (with ϵ function of time in a decaying turbulence), and from Larchevêque and Lesieur (1981),

$$P(r, t) \sim R^{-3} \exp - \left[5.44(\frac{r}{R})^{2/3}\right] \quad . \qquad (VIII - 5 - 5)$$

One retrieves in particular the result that, for stationary forced turbulence, $P(0, t) \sim R^{-3}$ decays like $t^{-9/2}$ as shown in section 4. More generally, an expansion of eq. (VIII-5-5) for $r << R$ yields

$$P(0, t) - P(r, t) \sim \epsilon_T \epsilon^{-1/3} r^{2/3} \qquad (VIII - 5 - 6)$$

with

$$\epsilon_T = -\frac{1}{2}\frac{d}{dt}P(0, t) \sim R^{-4}\frac{dR}{dt} \sim \epsilon^{1/3}R^{-11/3} \quad . \qquad (VIII - 5 - 7)$$

This is, in the particular case studied here, the derivation of the Corrsin-Oboukhov law (VIII-2-14) for the second order scalar structure function, since, for any passive scalar

$$<[T(\vec{x} + \vec{r}, t) - T(\vec{x}, t)]^2> = 2[\Phi(\vec{r}, t) - \Phi(\vec{0}, t)] \quad .$$

[17] They turn out to be proportional.

To conclude these two sections, it seems that, in three-dimensional isotropic turbulence, the various lengths characteristic of diffusion (scalar integral scale, pair separation, scalar cloud diameter) follow a Richardson law provided they are smaller or of the order of the velocity integral scale l (and neglecting of course molecular viscous and diffusive effects). This is valid for stationary-forced or decaying-isotropic turbulence. It justifies a very simple mixing-length argument expressing the eddy-dispersion coefficient as

$$\frac{1}{2}\frac{d}{dt}R^2 = CR\, v_R \qquad\qquad (VIII-5-8)$$

where $R(t)$ is one of the diffusing lengths, $v_R = (\epsilon R)^{1/3}$ an associated characteristic velocity, and C a numerical constant to be determined. This is, however, valid only when $R(t)$ is smaller or equal to the velocity integral scale $l(t)$. When $R(t)$ is much greater than $l(t)$, turbulence can be approximated as a Brownian motion for what concerns its diffusion at scales of the order of R, with an eddy-dispersion coefficient proportional to lv , where v is the r.m.s. velocity. Matching both laws at $R = l$ gives an eddy-dispersion coefficient (proposed by Larchevêque, 1983) which can be written as

$$\frac{1}{2}\frac{d}{dt}R^2 = C\ \inf(R, l)\ v[\inf(R, l)] \qquad\qquad (VIII-5-9)$$

with

$$v[R] = (\epsilon R)^{1/3} \quad . \qquad\qquad (VIII-5-10)$$

The value of the constant C depends on the particular diffusing length one is interested in. If $R(t)$ corresponds to the particle pair dispersion problem, it has been shown in Larchevêque and Lesieur (1981) that C is approximately equal to $(3/2)C_{co}^{-1}$, where we recall that C_{co} is the Corrsin-Oboukhov constant in the inertial-convective range. Whit $C_{co} = 0.7$, $C = 2.14$. With the same $E.D.Q.N.M.$ calculation in spectral space, it has been found in Herring et al. (1982, see also Lesieur et al., 1987), when R is the scalar integral scale, a value of the order of $C = 0.3$, with a scalar integral scale defined either classically or by (VIII-4-10). This value varies slightly with the exponent s' and with the Prandtl number. It may be possible that the laws (VIII-5-7) (VIII-5-8) should be valid in situations other than the strictly homogeneous isotropic forced or decaying turbulence, at least in certain space directions.

6 - Single-particle diffusion

6.1 Taylor's diffusion law

Let us first recall the classical Taylor's law (1921) concerning the diffusion of a single particle in homogeneous stationary turbulence: let $\vec{X}(\vec{a}, t)$

be the Lagrangian position of a tracer located in $\vec{a}$ at $t = 0$. Let $\vec{v}(\vec{a}, t)$ be the Lagrangian velocity, such that

$$\frac{D}{Dt}\vec{X}(\vec{a}, t) = \vec{v}(\vec{a}, t) \quad , \qquad (VIII - 6 - 1)$$

which yields

$$\vec{X}(\vec{a}, t) = \vec{X}(\vec{a}, 0) + \int_0^t \vec{v}(\vec{a}, \tau) \, d\tau \quad . \qquad (VIII - 6 - 2)$$

Taking the average on an ensemble of velocity realizations ($\vec{a}$ is not random) leads to, if the turbulence is homogeneous with zero mean velocity, $<\vec{X}(\vec{a}, t)> = \vec{a}$. Now, let us consider the Lagrangian diffusion coefficient

$$K(t) = \frac{1}{2}\frac{D}{Dt}[\vec{X}(\vec{a}, t) - \vec{a}]^2 = <\vec{X}(\vec{a}, t).\vec{v}(\vec{a}, t)> \quad ,$$

which writes

$$K(t) = <[\vec{a} + \int_0^t \vec{v}(\vec{a}, \tau) \, d\tau].\vec{v}(\vec{a}, t)> = \int_0^t <\vec{v}(\vec{a}, \tau).\vec{v}(\vec{a}, t)> \, d\tau \quad .$$

Since turbulence is stationary,

$$<\vec{v}(\vec{a}, \tau).\vec{v}(\vec{a}, t)> = R_{ii}(t - \tau) \quad ,$$

where $R_{ij}(t)$ is the Lagrangian two-time velocity correlation tensor. Hence it is found:

$$K(t) = \int_0^t R_{ii}(t - \tau) \, d\tau = \int_0^t R_{ii}(\tau) \, d\tau \quad , \qquad (VIII - 6 - 3)$$

equation which is due to Taylor (1921).

Now, we recall a result due to Batchelor (1949): assume that turbulence is isotropic, and that $X_i(\vec{a}, t)$ is, at a given t, a gaussian random variable (the X_i are assumed to be independent). The p.d.f. $P(x_i, t)$ that $X_i(\vec{a}, t)$ is equal to x_i at time t is

$$P(x_i, t) = \frac{1}{\sqrt{2\pi\sigma_i^2}} \, e^{-(x_i - a_i)^2 / 2\sigma_i^2} \quad ,$$

with

$$\sigma_i^2 = <[X_i(\vec{a}, t) - a_i]^2> = \frac{1}{3} <[\vec{X}(\vec{a}, t) - \vec{a}]^2> = \frac{\sigma(t)^2}{3} \quad .$$

This allows to show that

$$\frac{\partial}{\partial t} P(x_i, t) = \frac{1}{2} \frac{D\sigma_i^2}{Dt} \frac{\partial^2 P(x_i, t)}{\partial x_i^2} \quad , \qquad (VIII - 6 - 4)$$

without summation upon the i. Now, let

$$Q(\vec{x}, t) = P(x_1, t) \, P(x_2, t) \, P(x_3, t) \qquad (VIII - 6 - 5)$$

be the p.d.f. that the tracer is, at time t, located at $\vec{x}$. It satisfies the diffusion equation

$$\frac{\partial}{\partial t} Q(\vec{x}, t) = \frac{K(t)}{3} \nabla^2 Q(\vec{x}, t) \quad . \qquad (VIII - 6 - 6)$$

We have already seen that the gaussianity assumption is questionable, particularly for the scalar. However, a subsequent $E.D.Q.N.M.$ analysis will be shown to give a similar equation. Furthermore, eq. (VIII-6-4) turns out to be valid at short and large times (see Batchelor, 1949).

At short times, when the diffusion distance is short compared with the integral scale, it is obtained from (VIII-6-3):

$$K(t) \approx v_0^2 \, t \quad , \quad \sigma(t) \approx v_0 \, t \quad , \qquad (VIII - 6 - 7)$$

with

$$v_0^2 = < \vec{v}(\vec{a}, 0)^2 > \quad .$$

The diffusion is *coherent*, in the sense that the r.m.s. displacement is proportional to the time. At high times, it is found, still from eq. (VIII-6-3):

$$K(t) = \frac{1}{2} \frac{D\sigma^2}{Dt} \int_0^{+\infty} R_{ii}(\tau) \, d\tau = v_0^2 \, T_L \quad ,$$

relation which defines the Lagrangian correlation time T_L. Therefore:

$$\sigma(t)^2 = 2 \, v_0^2 \, T_L \, t \quad . \qquad (VIII - 6 - 8)$$

The r.m.s. displacement is now proportional to $t^{1/2}$, which characterizes an *incoherent* diffusion, as in the Brownian motion or the random walk.

6.2 $E.D.Q.N.M.$ approach to single-particle diffusion

Let

$$\psi(\vec{x}, t) = \delta[\vec{x} - \vec{X}(\vec{a}, t)] \quad . \qquad (VIII - 6 - 9)$$

This scalar quantity is obviously conserved following the motion. Let us calculate now the mean value of ψ. By definition of the p.d.f. $Q(\vec{x}, t)$, it is obtained:

$$<\psi(\vec{x}, t)> = \int_{-\infty}^{+\infty} \delta(\vec{x} - \vec{y}) \, Q(\vec{y}) \, d\vec{y} = Q(\vec{x}, t) \quad . \qquad (VIII - 6 - 10)$$

This allows to relate the preceding Lagrangian quantities to the Eulerian formalism of passive-scalar diffusion. We are now interested in the diffusion of an inhomogeneous scalar $\psi(\vec{x}, t)$ in homogeneous isotropic turbulence[18]. We are going to write a diffusion equation satisfied by $Q = <\psi>$, using *E.D.Q.N.M.*-type techniques. By averaging eq. (IV-2-8) with $\kappa = 0$, it is obtained

$$\frac{\partial}{\partial t} <\hat{\psi}(\vec{k}, t)> + i\, k_j \int <\hat{u}_j(\vec{p}, t)\hat{\psi}(\vec{q}, t)> \delta(\vec{k} - \vec{p} - \vec{q})d\vec{p}d\vec{q} = 0 \ .$$

$$(VIII - 6 - 11)$$

This equation is not closed, since we need to determine the scalar-velocity correlation. For this, we use again eq. (IV-2-8) written for a wave vector $\vec{k}$ and eq. (IV-2-7) written for a wave vector $\vec{k}'$: multiplying the former by $\hat{u}_i(\vec{k}', t)$ and the latter by $\hat{\psi}(\vec{k}, t)$ and adding yields, since $<\hat{f}\hat{\psi}>= 0$:

$$(\frac{\partial}{\partial t} + \nu k'^2) <\hat{\psi}(\vec{k})\hat{u}_i(\vec{k}')> + ik_j \int <\hat{u}_j(\vec{p})\hat{u}_i(\vec{k}')\hat{\psi}(\vec{q})> \delta(\vec{k} - \vec{p} - \vec{q})d\vec{p}d\vec{q}$$

$$+ ik'_m P_{in}(\vec{k}') \int < \hat{u}_n(\vec{p}')\hat{u}_m(\vec{q}')\hat{\psi}(\vec{k}) > \delta(\vec{k}' - \vec{p}' - \vec{q}')d\vec{p}'d\vec{q}' = 0 \ .$$

$$(VIII - 6 - 12)$$

Eqs (VIII-6-11) and (VIII-6-12) are the two equations of the moments hierarchy with which we are going to work.

Let $\hat{\psi}'$ be the fluctuation of $\hat{\psi}$ with respect to its mean value. One can write

$$<\hat{\psi}(\vec{k})\hat{u}_i(\vec{p})\hat{u}_j(\vec{q})>=<\hat{\psi}'(\vec{k})\hat{u}_i(\vec{p})\hat{u}_j(\vec{q})> + <\hat{\psi}(\vec{k})><\hat{u}_i(\vec{p})\hat{u}_j(\vec{q})> \ .$$

$\hat{\psi}'$ and $\hat{u}$ are random fonctions of zero mean. If they were gaussian, the triple moment $<\hat{\psi}'(\vec{k})\hat{u}_i(\vec{p})\hat{u}_j(\vec{q})>= 0$ would be zero, and hence

$$<\hat{\psi}(\vec{k})\hat{u}_i(\vec{p})\hat{u}_j(\vec{q})>=<\hat{\psi}(\vec{k})><\hat{u}_i(\vec{p})\hat{u}_j(\vec{q})> \ . \qquad (VIII - 6 - 13)$$

The corresponding "Quasi-Normal approximation" permits in (VIII-6-12) to write:

$$<\hat{u}_j(\vec{p})\hat{u}_i(\vec{k}')\hat{\psi}(\vec{q})>= \hat{U}_{ji}(\vec{k}')\ \delta(\vec{p} + \vec{k}')\ <\hat{\psi}(\vec{q})>$$

$$<\hat{u}_n(\vec{p}')\hat{u}_m(\vec{q}')\hat{\psi}(\vec{k})>= \hat{U}_{nm}(\vec{q}')\ \delta(\vec{p}' + \vec{q}')\ <\hat{\psi}(\vec{k})> \ ,$$

and the second non-linear term in (VIII-6-12) is zero (since interactions imply $\vec{k}' = \vec{0}$). Therefore it is obtained, within this quasi-normal assumption:

$$(\frac{\partial}{\partial t} + \nu k'^2) <\hat{\psi}(\vec{k})\hat{u}_i(\vec{k}')> + ik_j\ \hat{U}_{ji}(\vec{k}') <\hat{\psi}(\vec{k} + \vec{k}')>= 0 \ ,$$

---

[18] but we do not need to assume the stationarity here

or, after a time integration:

$$< \hat{\psi}(\vec{k},t)\hat{u}_i(\vec{k}',t) >= -ik_j \int_0^t e^{-\nu k'^2(t-\tau)}\hat{U}_{ji}(\vec{k}',\tau) <\hat{\psi}(\vec{k}+\vec{k}'),\tau)> d\tau \; ,$$

which is equivalent to

$$< \hat{\psi}(\vec{q},t)\hat{u}_i(\vec{p},t) >= -iq_l \int_0^t e^{-\nu p^2(t-\tau)}\hat{U}_{lj}(\vec{p},\tau) <\hat{\psi}(\vec{p}+\vec{q}),\tau)> d\tau \; .$$

Finally, the quasi-normal diffusion equation writes:

$$\frac{\partial}{\partial t} < \hat{\psi}(\vec{k},t)>$$
$$+ k_j q_l \int_{\vec{p}+\vec{q}=\vec{k}} \int_0^t e^{-\nu p^2(t-\tau)}\hat{U}_{lj}(\vec{p},\tau) <\hat{\psi}(\vec{p}+\vec{q}),\tau)> d\tau d\vec{p} = 0 \; .$$

Noticing also, due to incompressibility, that

$$q_l \hat{U}_{lj}(\vec{p},\tau) = k_l \hat{U}_{lj}(\vec{p},\tau) \quad ,$$

the quasi-normal diffusion equation writes finally

$$\frac{\partial}{\partial t} < \hat{\psi}(\vec{k},t)>= -k_i k_j \int_{\vec{p}+\vec{q}=\vec{k}} \int_0^t e^{-\nu p^2(t-\tau)}\hat{U}_{ij}(\vec{p},\tau) <\hat{\psi}(\vec{k},\tau)> d\vec{p}d\tau \; .$$
$$(VIII-6-14)$$

Now, an "Eddy-Damped" procedure will consist in replacing νp^2 in (VIII-6-14) by $\nu p^2 + \mu_1(p) + \mu_2(q)$, so that the *E.D.Q.N.* diffusion equation is

$$\frac{\partial}{\partial t} < \hat{\psi}(\vec{k},t)>= - k_i k_j \int_{\vec{p}+\vec{q}=\vec{k}} \int_0^t e^{-[\nu p^2+\mu_1(p)+\mu_2(q)](t-\tau)}$$
$$\hat{U}_{ij}(\vec{p},\tau) <\hat{\psi}(\vec{k},\tau)> d\vec{p}d\tau \; .$$
$$(VIII-6-15) \; .$$

Finally, a "markovianization" will consist in replacing (VIII-6-15) by:

$$\frac{\partial}{\partial t} < \hat{\psi}(\vec{k},t)>= -k_i k_j \int_{\vec{p}+\vec{q}=\vec{k}} \Theta_{pq}^{(\psi u)}\hat{U}_{ij}(\vec{p},t)d\vec{p} <\hat{\psi}(\vec{k},t)>$$
$$(VIII-6-16)$$

with

$$\Theta_{pq}^{(\psi u)} = \int_0^t e^{-[\nu p^2+\mu_1(p)+\mu_2(q)](t-\tau)} \; d\tau = \frac{1 - e^{-[\nu p^2+\mu_1(p)+\mu_2(q)]t}}{[\nu p^2 + \mu_1(p) + \mu_2(q)]} \; .$$
$$(VIII-6-17)$$

The choice of μ_1 and μ_2 will be made as for the *E.D.Q.N.M.* applied to the velocity field, that is

$$\mu_1(p) = \lambda_1 [\int_0^p y^2 \, E(y,t) \, dy]^{1/2}$$

$$\mu_2(q) = \lambda_2 [\int_0^q y^2 \, E(y,t) \, dy]^{1/2} \quad , \qquad\qquad (VIII - 6 - 18)$$

where λ_1 and λ_2 are constants which have to be adjusted.

For diffusion times small in front of the turbulence large-eddy turn-over time, it is straightforward that

$$\Theta_{pq}^{(\psi u)} \approx t \quad .$$

Noticing then that

$$\int \hat{U}_{ij}(\vec{p},t)d\vec{p} = <u_i(\vec{x},t)u_j(\vec{x},t)> = \frac{<\vec{u}^2>}{3} \, \delta_{ij} \quad ,$$

it is obtained:

$$\frac{\partial}{\partial t} <\hat{\psi}(\vec{k},t)> = -k^2 \, t \, \frac{<\vec{u}^2>}{3} <\hat{\psi}(\vec{k},t)> \quad .$$

Coming back to the physical space, one recovers the diffusion equation (VIII-6-6) with the diffusion coefficient (VIII-6-7):

$$\frac{\partial}{\partial t} Q(\vec{x},t) > = \, t \, \frac{<\vec{u}^2>}{3} \, \nabla^2 Q(\vec{x},t) > \quad . \qquad\qquad (VIII - 6 - 19)$$

For larger times, the knowledge of the two constants is required. By analogy with section 3-1, we will consider a simplified case where $\lambda_2 = 0$, which allows to write eq. (VIII-6-16) as

$$\frac{\partial}{\partial t} <\hat{\psi}(\vec{k},t)> = -k_i k_j [\int \Theta_p^{(1)}(t)\hat{U}_{ij}(\vec{p},t)d\vec{p}] <\hat{\psi}(\vec{k},t)> \quad ,$$

where $\Theta_p^{(1)}$ is given from eq. (VIII-6-17) with $\lambda_2 = 0$. Using the isotropic form of the velocity spectral tensor, it is found:

$$k_i k_j P_{ij}(\vec{p}) = k^2 \sin^2 \gamma$$

where γ is the angle formed by $\vec{k}$ and $\vec{p}$. After an integration in spherical coordinates, one gets:

$$k_i k_j [\int \Theta_p^{(1)} \hat{U}_{ij}(\vec{p},t) \, d\vec{p}] = \frac{2}{3} \, k^2 \int_0^{+\infty} \Theta_p^{(1)}(t) \, E(p,t) \, dp \quad ,$$

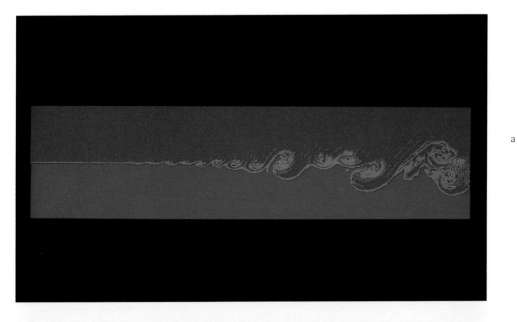

a

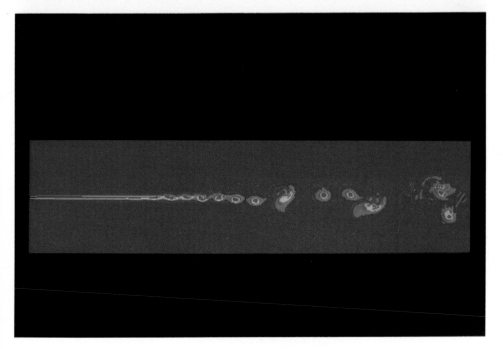

b

Plate 1: two-dimensional numerical simulation of Brown and Roshko's experiment shown in Figure I-4: a) passive dye contours; b) vorticity contours (courtesy X. Normand, Institut de Mécanique de Grenoble)

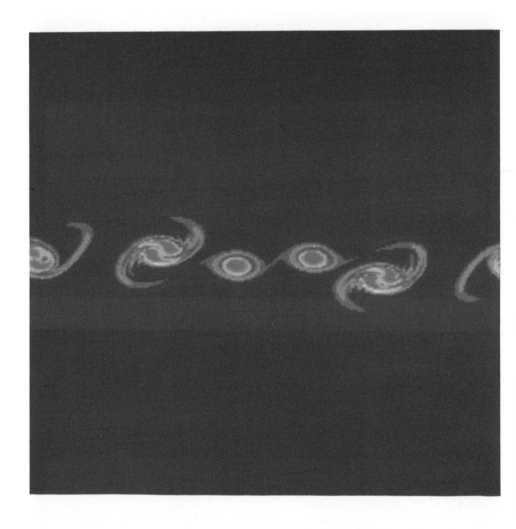

Plate 2: vorticity contours in the two-dimensional numerical simulations of the mixing layer reported in Lesieur et al. (1988) and Comte (1989)

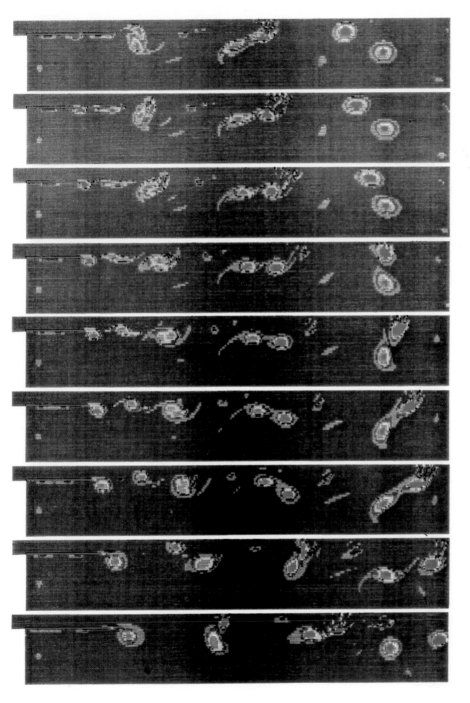

Plate 3: evolution with time of the vorticity field in a two-dimensional direct-numerical simulation of the flow above a backwards-facing step (courtesy A. Silveira, C.E.N.G. and I.M.G.)

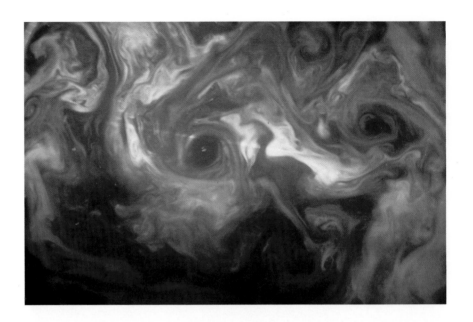

Plate 4: visualization of a horizontal section of turbulence in a tank rotating fastly about a vertical axis: the eddies shown are quasi-two-dimensional, due to the effect of rotation (courtesy E.J. Hopfinger, Institut de Mécanique de Grenoble)

Plate 5: satellite picture of the phytoplancton field on the surface of the Alboran sea (courtesy A. Morel, Laboratoire de Physique et Chimie Marine, Paris)

Plate 6: turbulence on Jupiter, in the vicinity of the great red spot (courtesy Jet Propulsion Laboratory, Pasadena)

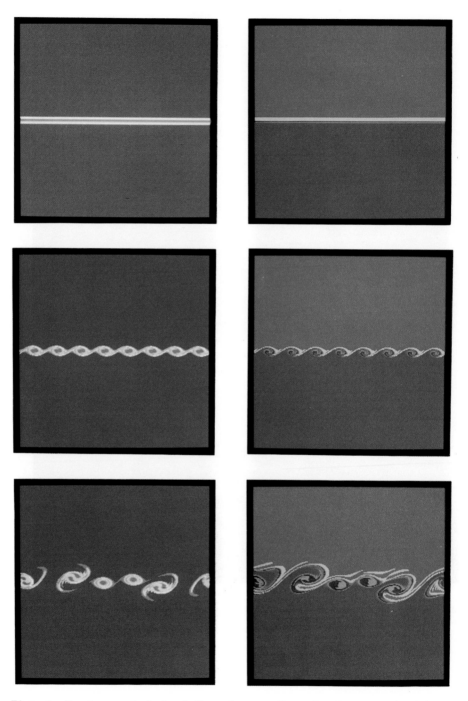

Plate 7: direct-numerical simulation of a two-dimensional temporal mixing layer: left, vorticity field; right, passive scalar field; one can see the formation of the primary vortices, and the subsequent pairings; (from Comte, 1989)

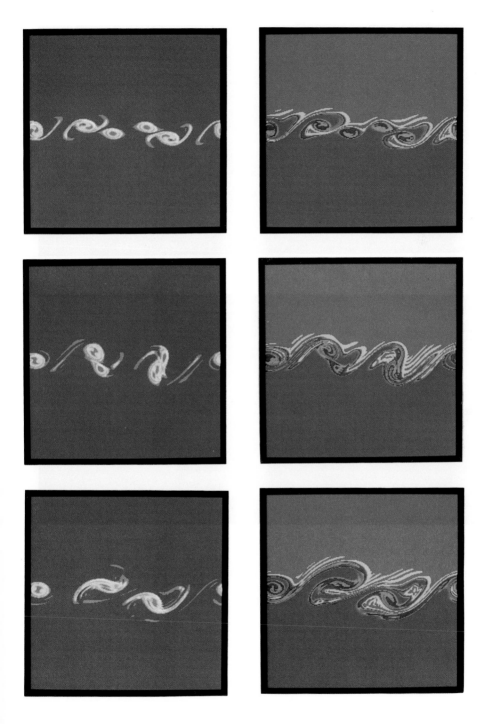

Plate 8: same calculation as in Plate 7: end of the evolution

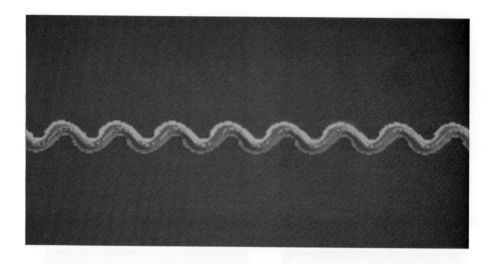

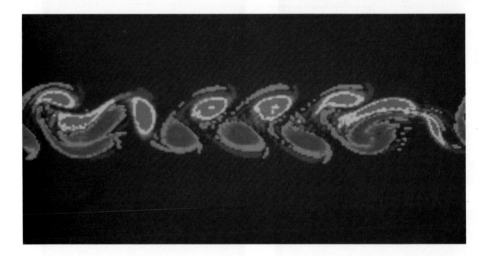

Plate 9: vorticity field in a direct-numerical simulation of a two-dimensional temporal Bickley jet: one can see the growth of the sinuous instability, and the formation of a Karman street, with alternate eddies of positive (red) and negative (blue) vorticity; (from Comte et al., 1987).

Plate 10: visualization by numerical-dye techniques of a temporal Karman
street resulting from a three-dimensional direct-numerical simulation of a periodic wake; (courtesy E. David and M.A. Gonze, I.M.G.)

Plate 11: experimental wake behind a splitter plate in a hydrodynamic tunnel;
(courtesy H. Werlé, ONERA)

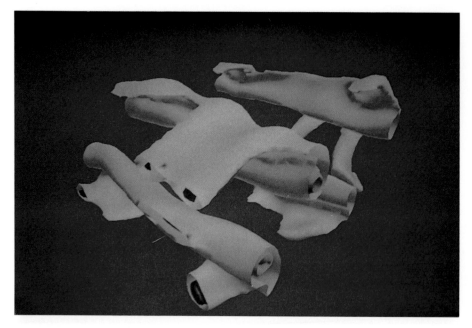

Plate 12: same calculation as in Plate 10, at a later time; the scalar field shows the formation of longitudinal hairpin streaks reconnecting the Karman billows.

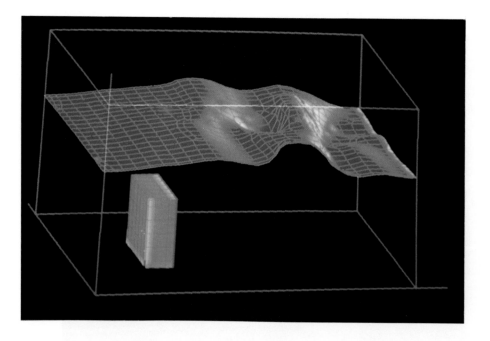

Plate 16: isopycnal surface in a finite-volume direct-numerical simulation of a strongly-stratified flow above an obstacle (courtesy of H. Laroche, I.M.G).

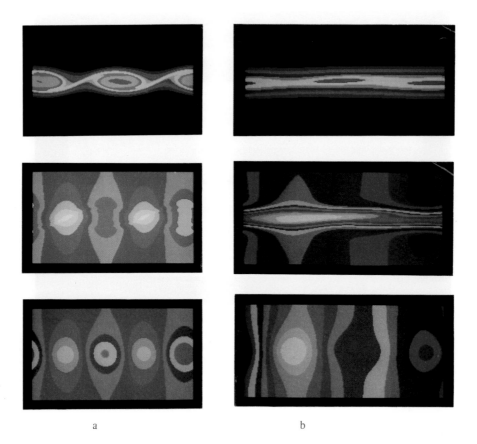

a   b

Plate 13: direct-numerical simulation of a two-dimensional periodic compressible mixing layer. The vorticity, temperature and pressure fields are shown for two convective Mach numbers: a) $M_c = 0.5$; b) $M_c = 0.7$ (courtesy X. Normand, I.M.G.)

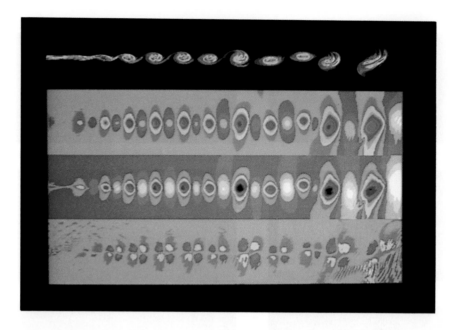

Plate 14: direct-numerical simulation of a two-dimensional spatially-growing compressible mixing layer. The Mach numbers of the two streams are 2 and 1.2. The vorticity, pressure, density and divergence fields are shown (courtesy Y. Fouillet et X. Normand, I.M.G.)

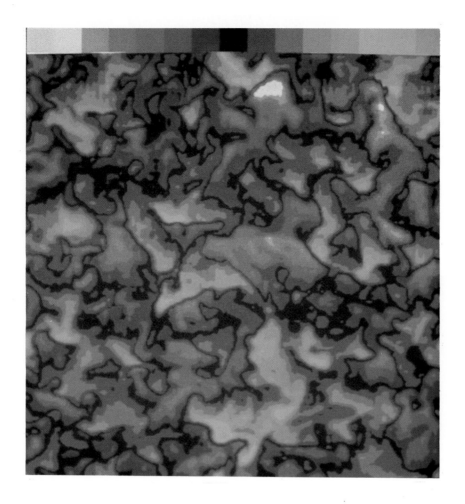

Plate 15: temperature distribution in the direct-numerical simulation of isotropic turbulence done by Métais and Lesieur (1990); the resolution is 128^3 Fourier modes.

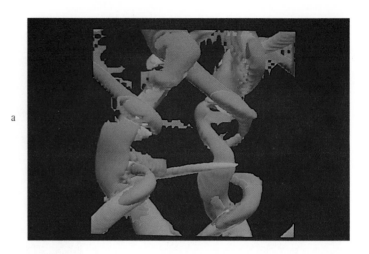

a

b

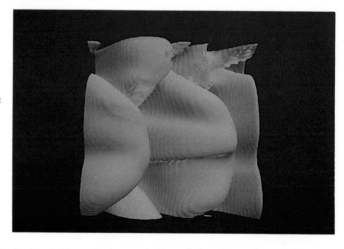

c

Plate 17: direct-numerical simulation of the periodic mixing layer forced by a small random three-dimensional perturbation done by Comte and Lesieur (1990); a) vortex structures; b) vortex lines; c) passive scalar at the interface. The resolution is 128^3 Fourier wave vectors, and the Reynolds number $U\delta_i/\nu = 100$.

Plate 18: same calculation as in Plate 17: vertical (in the x, y plane) cross section of the interface (courtesy P. Comte, IMG).

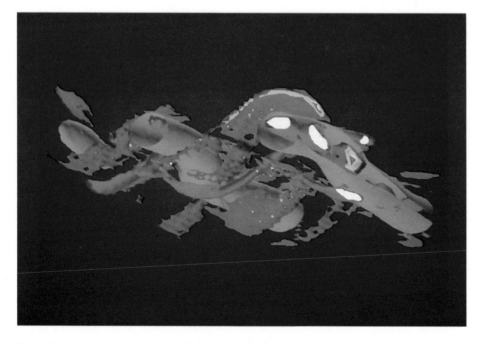

Plate 19: vortex structures in the same calculation as in Plate 17, but with a quasi-two-dimensional perturbation (courtesy P. Comte, IMG).

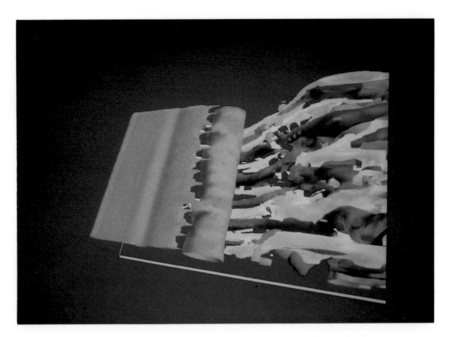

Plate 20: longitudinal (yellow and green) and spanwise (violet) vorticity components in the flow behind a backwards-facing step. The numerical simulation is due to A. Silveira (IMG and CENG).

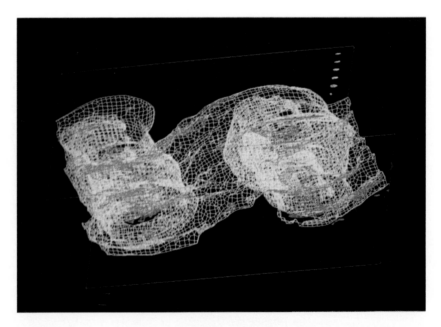

Plate 21: large-eddy simulation of the temporal mixing layer: interface at the end of the rollup, visualized with a numerical dye; in red is shown the positive longitudinal vorticity. From Comte (1989).

which, by inverse Fourier transform, may be written as:

$$\frac{\partial}{\partial t} Q(\vec{x}, t) = \frac{2}{3} [\int_0^\infty \Theta_p^{(1)}(t) \ E(p, t) \ dp] \ \nabla^2 Q(\vec{x}, t) \ . \qquad (VIII - 6 - 20)$$

It is again a diffusion equation, to be compared with (VIII-6-6), and we will associate to it the diffusion coefficient

$$K(t) = 2 \int_0^\infty \Theta_p^{(1)}(t) \ E(p, t) \ dp \ , \qquad (VIII - 6 - 21)$$

which has the same structure as the eddy-diffusivity (VIII-4-7).

Chapter IX

TWO-DIMENSIONAL
AND
QUASI-GEOSTROPHIC TURBULENCE

1 - Introduction

Let us begin by considering a fluid of uniform density ρ_0 in a frame which may be rotating with a constant rotation $\vec{\Omega}$. It obeys eq (II-5-7), where we recall that the "modified pressure" P also contains the gravity and centrifugal effects. This equation governs the motion of a rotating (or not) non-stratified flow in a laboratory experiment. Let us assume that the $\vec{z}$ axis of coordinates is directed along $\vec{\Omega}$, and look for two-dimensional solutions $\vec{u}(x, y, t)$ and $P(x, y, t)$. Let $u(x, y, t), v(x, y, t)$ and $w(x, y, t)$ be respectively the "horizontal" (that is perpendicular to $\vec{\Omega}$) and vertical components of the velocity. The continuity equation implies that the velocity field is horizontally non divergent

$$\frac{\partial u}{\partial x} + \frac{\partial v}{\partial y} = 0 \qquad (IX - 1 - 1)$$

and hence there exists a stream function $\psi(x, y, t)$ such that

$$u = \frac{\partial \psi}{\partial y} \quad ; \quad v = -\frac{\partial \psi}{\partial x} \quad . \qquad (IX - 1 - 2)$$

The Coriolis force in the equation of motion is then equal to $-2\Omega \vec{\nabla}\psi$ and can also be included in the pressure term, modified as $P'(x, y, t) = P(x, y, t) + 2\rho_0\Omega\psi$. The result is that the "horizontal" velocity field $\vec{u}_H(x, y, t)$ of components $(u, v, 0)$ satisfies a two-dimensional Navier-

244

Turbulence in fluids

Stokes equation

$$\frac{\partial u}{\partial t} + u\frac{\partial u}{\partial x} + v\frac{\partial u}{\partial y} = -\frac{1}{\rho_0}\frac{\partial P'}{\partial x} + \nu\nabla_H^2 u$$

$$\frac{\partial v}{\partial t} + u\frac{\partial v}{\partial x} + v\frac{\partial v}{\partial y} = -\frac{1}{\rho_0}\frac{\partial P'}{\partial y} + \nu\nabla_H^2 v$$

$$(IX-1-3)$$

with the incompressibility condition (IX-1-1). In the following, the suffix H refers to the horizontal coordinates: for instance $\nabla_H^2 = \partial^2/\partial x^2 + \partial^2/\partial y^2$ is the horizontal Laplacian operator. The vertical coordinate $w(x,y,t)$ obeys a two-dimensional passive scalar equation[1]

$$\frac{D_H w}{Dt} = \frac{\partial w}{\partial t} + u\frac{\partial w}{\partial x} + v\frac{\partial w}{\partial y} = \nu\nabla_H^2 w \quad . \qquad (IX-1-4)$$

This shows that the assumption of two-dimensionality does not imply a purely horizontal motion: a fluid particle will conserve during the motion (modulo the viscous dissipation) its initial vertical velocity. The components of the vorticity $\vec{\nabla}\times\vec{u}$ are: $\partial w/\partial y, -\partial w/\partial x, -\nabla_H^2\psi$. It can be easily checked, either directly from eq. (IX-1-3) or from the general vorticity equation (II-6-3) which is written

$$\frac{D_H\vec{\omega}}{Dt} = (\vec{\omega} + 2\vec{\Omega}).\vec{\nabla}_H\vec{u} + \nu\nabla_H^2\vec{\omega}$$

$$= \frac{\partial w}{\partial y}\frac{\partial\vec{u}}{\partial x} - \frac{\partial w}{\partial x}\frac{\partial\vec{u}}{dy} + \nu\nabla_H^2\vec{\omega} \quad ,$$

that the vertical vorticity $-\nabla_H^2\psi$ (which is also the vorticity of the horizontal velocity $\vec{u}_H$) obeys the same equation as (IX-1-4) . However, the vertical vorticity is no longer a passive scalar, since a perturbation brought to it would affect ψ and hence $\vec{u}_H$. The equation of motion for the stream function is then

$$[\frac{\partial}{\partial t} + J(.,\psi)]\nabla_H^2\psi = \nu\nabla_H^2(\nabla_H^2\psi) \qquad (IX-1-5)$$

where the Jacobian operator $J(A,B)$ is defined by

$$J(A,B) = \frac{\partial A}{\partial x}\frac{\partial B}{\partial y} - \frac{\partial A}{\partial y}\frac{\partial B}{\partial x} \quad . \qquad (IX-1-6)$$

Consider for example a two-dimensional purely horizontal flow: this demonstrates that a constant rotation has no effect on the dynamics

---

[1] "Passive", in the sense that any external perturbation modifying w will have no effect on the convective field $\vec{u}_H$.

of such a flow[2]. But we have already seen in Chapter III, and new examples will be given in the present chapter, that a strong rotation prevents small vertical velocity fluctuations from developing, and hence plays a stabilizing role with respect to the two-dimensional solutions.

Is it then possible to speak of *two-dimensional turbulence*? Clearly, the conservation of the vertical vorticity (modulo viscous diffusion) following the motion of the fluid particle is a severe constraint which seems to prevent all the vortex-stretching effects associated to the finite inviscid kinetic energy dissipation features of three-dimensional turbulence. However, a lot of weakly-viscous two-dimensional flows share the *mixing* and *unpredictability* properties, proposed in Chapter I as characteristic of turbulence. It will be seen below that two-dimensional turbulence is characterized (for low but non-zero viscosity) by a conservation of kinetic energy, a finite dissipation of enstrophy, and an exponential increase of the palinstrophy[3], already introduced in Chapter VI-7, and here equal to

$$P(t) = \frac{1}{2} < [\vec{\nabla} \times (\vec{\nabla} \times \vec{u}_H)]^2 > \quad . \qquad (IX - 1 - 7)$$

From a mathematical point of view, there is therefore no difficulty in studying turbulent solutions of the two-dimensional Navier-Stokes equations considered above. The problem lies in the physical possibility of realizing and maintaining such flows: indeed, and if one accepts the well-known concept of "return to three-dimensionality" (see e.g. Herring, 1974), a purely two-dimensional flow in an infinite domain will become three-dimensional if there is no external action tending to maintain the two-dimensionality. The first possibility is thus to consider in a laboratory a flow constrained between two planes of distance D : at scales much larger than D , one may expect the flow to be horizontal and two-dimensional. Nevertheless the boundary layers along the planes will develop, interact, and be responsible for active three-dimensional turbulence at scales smaller than D , which could rapidly dissipate the energy of the large two-dimensional scales. It is therefore necessary to limit the development of these boundary layers. This may be done with the aid of a rapid[4] rotation $\vec{\Omega}$ perpendicular to the boundaries (Colin de Verdière, 1980, Hopfinger et al., 1982), or (in a *M.H.D.* flow) by imposing a magnetic field $\vec{B}$ also perpendicular to the boundaries (Somméria, 1986, see

---

[2] If the boundary conditions concerning ψ are unchanged with respect to the non-rotating case.

[3] This word was introduced first in Pouquet et al. (1975). It is constructed with the aid of Greek derivatives: *strophy* stands for *rotation*, and *palin* for *again*, so that *palinstrophy* characterizes the *curl* of the *curl*.

[4] "Rapid", that is with a low Rossby number.

also Moreau, 1990). In both cases, the interior horizontal velocity field
obeys eq. (IX-1-5) with a further large-scale damping due to the action
of the boundary Ekman layers (in the case of rotation) and the Hartman
layers (in the $M.H.D.$ case). In geophysical situations, the shallowness
of the atmosphere or of the oceans (with respect to horizontal planetary
scales) and the rotation of the earth corresponds to the same situation.
In fact, the concept of two-dimensional turbulence and unpredictability
was developed by meteorologists who could not predict the evolution
of these planetary motions for more than a few days (Thompson, 1957,
Leith, 1968, Lorenz, 1969).

Let us mention also the experiments on two-dimensional turbulence
and two-dimensional shear flows done in liquid soap films by Couder
(1984) and Gharib and Derango (1989). The latter authors show in
particular impressive visualizations of Kelvin-Helmholtz vortices in the
flow over a backwards-facing step, which resemble very much the two-
dimensional direct-numerical simulations of this flow done by Silveira et
al. (1989) and presented on Plate 3.

Finally, Chapter XIII will discuss how the influence of a stable
stratification in an infinite fluid might lead in some cases to quasi-two-
dimensional flows organized in horizontal layers possessing some of the
features of two-dimensional turbulence, but with a strong vertical variability.

In section 2 , we will study the quasi-geostrophic theory, allowing one to write quasi-two-dimensional evolution equations for the large
scales of a stably-stratified shallow flow on a rapidly-rotating sphere.
The following sections will review the dynamics of two-dimensional turbulence, and look at how the results are modified in the quasi-geostrophic
case.

2 - The quasi-geostrophic theory

The mathematical details of this theory, proposed by Charney (1947),
are given in Pedlosky (1979). Here we will give the main physical ingredients of the approximation. Consider a rotating sphere (of rotation $\vec{\Omega}$)
of radius a. A given point of the flow M is defined by its vertical projection O on the sphere, and by its "altitude" z. To O is associated the
local frame $(\vec{x}, \vec{y}, \vec{z})$, $\vec{x}$ being directed along a parallel of latitude φ and
$\vec{y}$ along a meridian of longitude λ (see Figure IX-1). The components of
the velocity field $\vec{u}$ in the local frame are (u, v, w). The point O will be
assumed to be close to a reference point O_0 of longitude and latitude λ_0
and φ_0, in order to neglect the sphericity corrections. The "horizontal"

coordinates of M are defined by

$$x = a(\lambda - \lambda_0)\cos\varphi_0$$
$$y = a(\varphi - \varphi_0) \quad .$$

$$(IX - 2 - 1)$$

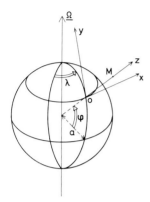

Figure IX-1: local frame on the rotating sphere.

Let D be the average depth of the fluid, L and U characteristic horizontal lengths and velocities, W a characteristic vertical velocity. We utilize the Boussinesq approximation on a rotating sphere, within the same approximation as done in eq. (II-9-14), that is:

$$\frac{Du}{Dt} = -\frac{\partial\tilde{p}}{\partial x} + fv + \nu_e\nabla^2 u$$

$$\frac{Dv}{Dt} = -\frac{\partial\tilde{p}}{\partial y} - fu + \nu_e\nabla^2 v$$

$$\frac{Dw}{Dt} = -\frac{\partial\tilde{p}}{\partial z} - \tilde{\rho}g + \nu_e\nabla^2 w \quad .$$

$$(IX - 2 - 2)$$

In fact, the velocity field in (IX-2-2) represents a filtered velocity field, averaged over a horizontal box of size δL, which is the smallest characteristic scale for which the following geostrophic approximation will be valid. Therefore, ν_e represents an eddy-viscosity corresponding to momentum exchanges with the "sub-geostrophic scales" smaller than δL. This is a very rough way of modelling the effect of these small scales, but ν_e will not play any role in the following geostrophic approximation. However, this coefficient will be essential for the understanding of the Ekman layer (see below). The continuity equation is approximated as

$$\frac{\partial u}{\partial x} + \frac{\partial v}{\partial y} + \frac{\partial w}{\partial z} = 0 \quad .$$

$$(IX - 2 - 3)$$

The scope of the present study is to write an approximate evolution equation for the quasi-horizontal large-scale velocity field, using expansions with respect to the small parameters D/L, the local Rossby number R_o, and the Froude number. In fact, the theory will consist of two major ingredients: the first one is called the geostrophic approximation, and the second is the application of Ertel's potential vorticity conservation in the special case of the geostrophic approximation.

2.1 The geostrophic approximation

The smallness of D/L justifies the hydrostatic assumption along the vertical, as was done in Chapter II-10. This leads to

$$\tilde{\rho} = -\frac{1}{g}\frac{\partial \tilde{p}}{\partial z} \quad . \tag{IX $-2-4$}$$

In dimensional variables, this corresponds to

$$\rho' = -\frac{1}{g}\frac{\partial p'}{\partial z} \quad , \tag{IX $-2-5$}$$

for the density and pressure fluctuations about the basic hydrostatic profile $\bar{\rho}(z), \bar{p}(z)$. This implies that the instantaneous density $\rho = \bar{\rho}+\rho'$ and pressure $p = \bar{p} + p'$ are also hydrostatically related:

$$\rho = -\frac{1}{g}\frac{\partial p}{\partial z} \quad . \tag{IX $-2-6$}$$

Now in the horizontal velocity equations, the Coriolis force is of the order of $f_0 U$, the non-linear terms $D\vec{u}_H/Dt$ of the order of U^2/L, and the dissipative term of the order of $\nu_e U/L^2$, where f_0 the value of the Coriolis parameter (defined by eq. (II-7-6)) at φ_0. From the Ekman layer study done below, $\nu_e \sim f_0 D_E^2$, where $D_E << D$ is the vertical thickness of the Ekman layers between which the fluid evolves. Hence the dissipative term is $\sim f_0 U D_E^2/L^2$. The ratio of the non-linear to the Coriolis term is of the order of the Rossby number $U/f_0 L$, and that of the dissipative to the Coriolis term of the order of $(D_E/L)^2 << 1$. Then, as soon as the Rossby number $U/f_0 L$ is small compared with one, and if O is close to O_0, the momentum equation in the rotating shallow layer on the sphere reduces to

$$-\vec{\nabla}_H \tilde{p} = f_0 \, \vec{z} \times \vec{u}_H \quad . \tag{IX $-2-7$}$$

Since $\bar{p}$ is a function of z only, eq. (IX-2-7) is equivalent[5] to

$$-\frac{1}{\rho_0}\vec{\nabla}_H p = f_0 \vec{z} \times \vec{u}_H \quad , \tag{IX $-2-8$}$$

---

[5] These relations might also have been obtained using the Navier-Stokes equations directly.

where ρ_0 is the average value of the density across the fluid layer. Eq. (IX-2-8) corresponds to the geostrophic balance between the pressure gradient and the Coriolis force. It shows that the horizontal flow follows the isobaric lines in a horizontal plane, in the cyclonic direction (that is the positive rotation imposed by Ω) in the vicinity of a pressure trough, and in the anticyclonic direction around a pressure peak. Eq (IX-2-8) implies also that

$$\vec{u}_H = -\vec{z} \times \vec{\nabla}_H \; \psi(x,y,z,t)$$
$$\psi(x,y,z,t) = -\frac{1}{\rho_0 f_0} \; p(x,y,z,t) \qquad (IX-2-9)$$

where ψ acts like a stream function for the horizontal motion, but depends on the vertical coordinate z. Let us remark also, from eq. (IX-2-6), that

$$\rho = \frac{\rho_0 f_0}{g} \frac{\partial \psi}{\partial z} \qquad (IX-2-10)$$

We will define the *geostrophic velocity* $\vec{u}_G$ by eq. (IX-2-9). Introducing a fluctuating stream function

$$\psi'(x,y,z,t) = -\frac{1}{\rho_0 f_0} \; p'(x,y,z,t) \quad , \qquad (IX-2-11)$$

with

$$\psi = -\frac{1}{\rho_0 f_0} \; \bar{p}(z) + \psi' \quad ,$$

the above relations write

$$\vec{u}_H = -\vec{z} \times \vec{\nabla}_H \; \psi'(x,y,z,t)$$
$$\tilde{\rho} = \frac{f_0}{g} \frac{\partial \psi'}{\partial z} \quad . \qquad (IX-2-12)$$

Hence, ψ' may also play the role of a stream function for the geostrophic field. We remark also that the geostrophic velocity is horizontally non divergent. This has important consequences for the vertical velocity of the actual flow: indeed, let us write the real velocity field as

$$\vec{u}(\vec{x},t) = \vec{u}_G(\vec{x},t) + \frac{U}{f_0 L} \; \vec{u}_1(\vec{x},t) + O[\frac{U}{f_0 L}]^2 \qquad (IX-2-13)$$

where $\vec{u}_1$ has a modulus of the order of U. Since $\vec{u}_G$ is horizontal, the vertical characteristic velocity W is of the order of $R_o W_1$, where W_1 is a typical vertical component of $\vec{u}_1$. Since the three-dimensional incompressibility condition (IX-2-3) is valid at any order of the expansion, it implies

$$\frac{\partial w_1}{\partial z} = -(\frac{\partial u_1}{\partial x} + \frac{\partial v_1}{\partial y}) \quad ,$$

which yields

$$\frac{W_1}{D} = O[\frac{U}{L}] \quad ; \quad \frac{W}{U} = O[(\frac{D}{L})(\frac{U}{f_0 L})] \quad . \qquad (IX-2-14)$$

These results show that a rapid rotation diminishes the ratio W/U from the value D/L (imposed by the shallowness of the layer and the continuity equation) to $(U/f_0 L)(D/L)$. This is a generalization of the Proudman-Taylor theorem to a shallow flow on a rotating sphere, and goes in favour of the limitation of the vertical fluctuations of the flow under the action of a rapid rotation, as seen in Chapter III for a flow of uniform density in a layer of arbitrary depth.

The local Rossby number is infinite at the equator, and so the geostrophic balance holds only in the medium and high latitudes. More specifically, the Rossby number is of the order of 0.3 in the Earth's atmosphere in medium latitudes ($f = 10^{-4}$ rd/s) for tropospheric jets of velocity 30 m/s and length scales of 1000 km . Since the geostrophic balance theory is valid at the lowest order with respect to the Rossby number, the latter fixes the precision of the determination of the actual velocity field using the geostrophic velocity inferred from the pressure field. In the ocean, a velocity of 5 cm/s and a length scale of 100 km yields a Rossby number of 5.10^{-2} at the same latitude.

By differentiating eq. (IX-2-9) with respect to z , and using eq. (IX-2-6), one obtains the so-called thermal wind equation

$$\frac{\partial \vec{u}_H}{\partial z} = -\frac{g}{\rho_0 f_0} \vec{z} \times \vec{\nabla}_H \rho \quad . \qquad (IX-2-15)$$

This shows a tendency for the horizontal density gradients to induce, under the action of a rapid rotation, zonal currents: in the Earth's atmosphere, for instance, the meridional density gradients are directed towards the poles, and the thermal-wind equation is therefore in good agreement with the zonal westerly jet streams in medium latitudes. Notice also that eq. (IX-2-15) is equivalent to

$$\frac{\partial \vec{u}_H}{\partial z} = -\frac{g}{f_0} \vec{z} \times \vec{\nabla}_H \tilde{\rho} = -\frac{g}{f_0} \vec{z} \times \vec{\nabla}_H \rho_* \quad . \qquad (IX-2-16)$$

2.2 The quasi-geostrophic potential vorticity equation

The geostrophic approximation is a "diagnostic" equation, allowing one to approximate the velocity by a horizontal field, the geostrophic velocity, and to calculate it from the pressure field, using eq. (IX-2-9). With the aid of the hydrostatic approximation, the velocity and the temperature can thus be completely determined from the pressure, with a precision of the order of the Rossby number. Therefore we need a "prognostic" equation allowing one to predict the evolution in time of the pressure field, or the vertical geostrophic vorticity $\omega_G = -\nabla_H^2 \psi$. Such an equation may be derived with the aid of systematic expansions with respect to the small parameters (Rossby number R_o, Froude number and D/L), as was done by Charney (1947, see also Pedlosky, 1979). Here, we will derive this equation using the *potential-vorticity conservation principle*, expanding the potential-vorticity equation.

One still works within the Boussinesq approximation, neglecting molecular diffusion. As shown in Chapter II-8, the potential vorticity

$$\zeta = (\vec{\omega} + 2\vec{\Omega}).\vec{\nabla}\rho_* \qquad (IX-2-17)$$

is conserved following the three-dimensional motion. We will expand this equation to the first order with respect to the above mentioned small parameters. Therefore, the vorticity $\vec{\omega}$ in (IX-2-17) may be replaced by the geostrophic vorticity $\vec{\omega}_G$, since it is of order R_o with respect to $2\vec{\Omega}$. Writing

$$\vec{\nabla}\rho_* = \frac{\partial\rho_*}{\partial z}\,\vec{z} + \vec{\nabla}_H\rho_* \quad ,$$

the potential vorticity writes:

$$\zeta = (\omega_G + f)\frac{\partial\rho_*}{\partial z} + (\vec{\omega}_G + 2\vec{\Omega}).\vec{\nabla}_H\rho_* \quad . \qquad (IX-2-18)$$

Let us consider the orders of magnitude of the second term in the r.h.s. of (IX-2-18), which is, due to the thermal-wind equation (IX-2-16), of the order of or smaller than $(f_0^2/g)\,\|\partial\vec{u}_H/\partial z\| \sim (f_0^2 U/gD)$. The first term in (IX-2-18) is minored (in modulus) by $\sim (U/L)(d\bar{\rho}_*/dz)$. Hence, the ratio of the second to the first term is of the order of $(f_0/N)^2\,(L/D)$, where N is the Brunt-Waisala frequency. The magnitude of this parameter can be specified by noticing that

$$\frac{f_0}{N} = \frac{F}{R_0}\frac{D}{L} \quad , \qquad (IX-2-19)$$

where F is the Froude number. Hence

$$(\frac{f_0}{N})^2\frac{L}{D} = (\frac{F}{R_o})^2\frac{D}{L}$$

will be small if the Froude number is of the order of or smaller than the Rossby number. In this case, the potential vorticity (IX-2-18) writes

$$\zeta = (\omega_G + f)\frac{\partial\rho_*}{\partial z} \quad . \qquad (IX-2-20)$$

We write

$$\rho_* = \bar{\rho}_*(z) + \tilde{\rho} \quad , \quad f = f_0 + f'$$

and assume that

$$\left| \frac{\tilde{\rho}}{\bar{\rho}_*} \right| , \left| \frac{\partial \tilde{\rho}}{\partial z} \middle/ \frac{d\bar{\rho}_*}{dz} \right| , \frac{f'}{f_0}$$

are small parameters of the order or smaller than R_o. To the first order, it is obtained

$$\zeta = f_0 \frac{d\bar{\rho}_*}{dz} + (\omega_G + f') \frac{d\bar{\rho}_*}{dz} + f_0 \frac{\partial \tilde{\rho}}{\partial z} \quad . \tag{IX – 2 – 21}$$

Dividing by $d\bar{\rho}_*/dz$, and using (II-9-9) and (IX-2-12), eq. (IX-2-21) writes:

$$\zeta = \frac{d\bar{\rho}_*}{dz} \left[-\nabla_H^2 \psi' + f - \left(\frac{f_0}{N} \right)^2 \frac{\partial^2 \psi'}{\partial z^2} \right] \quad . \tag{IX – 2 – 22}$$

The conservation of potential vorticity is written

$$\frac{D}{Dt} \zeta = \frac{D_H}{Dt} \zeta + w \frac{\partial}{\partial z} \zeta = 0 \quad , \tag{IX – 2 – 23}$$

where w is the vertical velocity. In (IX-2-23), the magnitude of $w \partial \zeta / \partial z$ relative to $D_H \zeta / Dt$ is, from (IX-2-14), equal to the Rossby number. Then it is easy to check that, to the lowest order, (IX-2-23) reduces to

$$\frac{D_H}{Dt} \zeta + f_0 w \frac{d^2 \bar{\rho}_*}{dz^2} = 0 \quad . \tag{IX – 2 – 24}$$

We assume for simplicity that N^2 is a constant. Hence the potential vorticity equation writes:

$$\frac{D_H}{Dt} \omega'_P = 0 \quad , \tag{IX – 2 – 25}$$

with

$$\omega'_P = -\nabla_H^2 \psi'(x, y, z, t) + f - \left(\frac{f_0}{N} \right)^2 \frac{\partial^2 \psi'}{\partial z^2} \quad . \tag{IX – 2 – 26}$$

At this point, and since $\psi - \psi'$ is a function of z only, the quasi-geostrophic potential vorticity equation may be replaced by:

$$\frac{D_H}{Dt} \omega_P = 0$$

$$\omega_P = -\nabla_H^2 \psi(x, y, z, t) + f - \left(\frac{f_0}{N} \right)^2 \frac{\partial^2 \psi}{\partial z^2} \quad . \tag{IX – 2 – 27}$$

More details may be found in Pedlosky (1979)[6]. As a matter of fact this equation is completely "Geostrophic", since it has been established within the geostrophic balance approximation.

The three terms of the r.h.s. of (IX-2-27 b) correspond to three different physical processes: $\omega_{BT} = -\nabla_H^2 \psi$, the barotropic potential vorticity, will give rise to the two-dimensional turbulence introduced in section 1; the variations of f with the latitude will generate Rossby waves; finally the baroclinic potential vorticity $\omega_{BC} = -(f_0/N)^2 \partial^2 \psi/\partial z^2$, which, from (IX-2-10), is proportional to $-\partial \rho/\partial z$, will be responsible for the so-called "baroclinic instability", due to the simultaneous action of a rotation and a density gradient.

There are several interpretations of this instability. We first give the explanation proposed by Rhines (1979): a horizontal density gradient[7] will result into a vertical density gradient (of opposite sign compared to the horizontal gradient) and generate baroclinic vorticity, of the order of ψ/r_I^2 , where r_I is the internal radius of deformation defined in (II-9-12). This baroclinic vorticity will be larger than the barotropic vorticity $\omega_{BT} \sim \psi/L^2$ for horizontal motions of wave length $L > r_I$. If one accepts the concept of geostrophic turbulence proposed by Charney (see section 5), non-linear interactions will tend to an equipartition between the baroclinic and barotropic components of vorticity. Therefore, the baroclinic vorticity will be a source of barotropic vorticity of same sign as the horizontal density gradient, that we may expect to condense into coherent vortices. This is why intense cyclonic disturbances form in medium latitudes in the atmosphere.

A more classical and linear point of view of the baroclinic instability is given by the study of the Eady model, which may be recovered from a linearization of eqs (IX-2-25) (IX-2-26), where it is assumed that the velocity fluctuates about a zonal westerly wind with constant vertical gradient $u = Az$, where A is a constant. Neglecting differential rotation, it is obtained

$$(\frac{\partial}{\partial t} + Az\frac{\partial}{\partial x})[\nabla_H^2 \psi'(x,y,z,t) + (\frac{f_0}{N})^2\frac{\partial^2 \psi'}{\partial z^2}] = 0 \quad , \qquad (IX-2-27')$$

equivalent to eq. (45-28) of the Eady model given in Drazin and Howard (1981), where a resolution by normal modes in the case of a rotating stratified channel of width L and depth D (with appropriate boundary conditions) is given: it is found that unstable modes appear when the Burger number $(DN/Lf_0)^2 = (r_I/L)^2$ is smaller than 0.58, equivalent

---

[6] The reader is warned that our stream function has an opposite sign compared to Pedlosky's.

[7] due for instance to the differential heating between the poles and the equator

to $L > 1.31 \, r_I$. This confirms the fact that the baroclinic instability develops at horizontal wave lengths larger than r_I.

There ought to be a sub-geostrophic diffusion term in the r.h.s. of (IX-2-27 a), due both to the small-scale turbulent diffusion of momentum and temperature. The simplest form for such an operator could be an eddy-viscous dissipation $\nu_e \nabla_H^2 \omega_G$ (which however ignores the temperature diffusion). Actually, oceanographers prefer to utilize a biharmonic diffusion operator proportional to $-(\nabla_H^2)^2 \omega_G$ (Holland, 1978). Higher-order Laplacian operators have been proposed by Basdevant and Sadourny (1983). In the numerical large eddy simulations of quasi-geostrophic turbulence, these higher order turbulent dissipation operators cause the dissipative effects to shift towards the smallest resolved scales, leaving the large scales unaffected by viscosity (see Chapter XII).

The quasi-geostrophic equation derived here can be generalized to a variable $d\bar{\rho}/dz$, with N function of z (see Pedlosky, 1979). However, it is valid within the interior of the geostrophic fluid, and requires knowledge of boundary conditions at the bottom and the top of the fluid layer. In the Earth's atmosphere, the bottom boundary conditions are due to the orography, and to the existence of a turbulent Ekman layer[8]. In the Ocean, there is also an upper Ekman layer due to wind stress. In order to understand the role of these boundary conditions, it may be useful to consider the particular case of a flow composed of n quasi-horizontal layers of fluids of uniform densities $\rho_1, ..., \rho_n$.

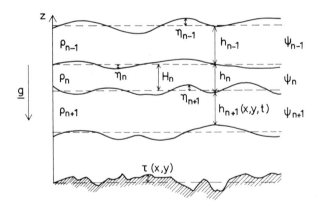

Figure IX-2: schematic vertical section of a n-layer geostrophic flow: η_n and η_{n+1} are the deviations of the interface of the layer n with respectively the layers $(n-1)$ and $(n+1)$.

---

[8] where an appreciable part of the geostrophic kinetic energy is dissipated

2.3 The n-layer quasi-geostrophic model

More details on the following derivation can be found in Pedlosky (1979). The fluid is assumed to be a superposition of homogeneous layers of density ρ_n (increasing with n), instantaneous thickness $h_n(x, y, t)$, and average thickness H_n (see Figure IX-2). The hydrostatic approximation is done on the vertical, permitting to calculate the instantaneous pressure at an altitude z as:

$$p(x, y, z, t) = p_A + g(\rho_1 h_1 + \rho_2 h_2 + ... + \rho_{n-1} h_{n-1})$$
$$+ \rho_n g(\eta_n + H_n + H_{n+1} + ... - z) \quad , \qquad (IX - 2 - 28)$$

where p_A is the pressure at a free surface, if any, and η_n the denivellation of the upper surface of the layer n. Using the following relations

$$\sum_i h_i(x, y, t) + \tau(x, y) - \eta_1 = \sum_i H_i = H$$

$$h_n - \eta_n + \eta_{n+1} = H_n$$

$$\bar{p}(z) = p_A + g(\rho_1 H_1 + \rho_2 H_2 + ... + \rho_{n-1} H_{n-1})$$
$$+ \rho_n g(H_n + H_{n+1} + ... - z) \, ,$$

where $\tau(x, y)$ is the bottom topography, and $\bar{p}(z)$ the hydrostatic pressure, it is obtained for the fluctuating[9] pressure $p'_n(x, y, t)$ in the layer n:

$$p'_n(x, y, t) = g[\rho_1 \eta_1 + (\rho_2 - \rho_1)\eta_2 + ... + (\rho_n - \rho_{n-1})\eta_n \quad . \quad (IX - 2 - 29)$$

Afterwards, the geostrophic balance is assumed in each of the homogeneous layers: this enables to associate to the n-th layer the stream function of the horizontal velocity field $\vec{u}_n(x, y, t)$ in the layer. It is equal to:

$$\psi_n(x, y, t) = -\frac{p'_n(x, y, t)}{\rho_n f_0} \quad .$$

Therefore, the conservation of the potential vorticity reduces here to the conservation of $(-\nabla_H^2 \psi_n + f)/h_n$ following the horizontal motion $\vec{u}_n$, as can be seen from a derivation paralleling that of the Barré de Saint Venant equation in II-10, or by referring to the original derivation of Ertel's theorem. Let the constant H_n be the mean thickness of the layer n. The potential vorticity can be approximated by

$$\frac{1}{H_n}[-\nabla_H^2 \psi_n + f - (\eta_n - \eta_{n+1})\frac{f_0}{H_n}] \, .$$

---

[9] with respect to $\bar{p}(z)$

Note that the quasi-geostrophic equation for the n th layer can also be written as

$$\frac{D_H^{(n)}}{Dt}[-\nabla_H^2 \psi_n(x,y,t) + f] = \frac{f_0}{H_n}(w_n - w_{n+1})$$

where $D_H^{(n)}/Dt$ stands for the derivative following the horizontal motion $\vec{u}_n$, and w_n and w_{n+1} are respectively the vertical velocities at the top and bottom of the layer.

In order to calculate η_n, one writes ψ_n and ψ_{n-1} using eq. (IX-2-29), which gives

$$\eta_n(x,y,t) = \frac{f_0}{g(\rho_n - \rho_{n-1})}(\rho_{n-1}\psi_{n-1} - \rho_n\psi_n) \quad . \qquad (IX-2-30)$$

In the upper layer, $\eta_1 = -f_0\psi/g$.

Let us consider first a one-layer fluid over a topography $\tau(x,y)$. The quasi-geostrophic potential vorticity equation is thus

$$\frac{D_H}{Dt}[-\nabla_H^2 \psi(x,y,t) + f + \frac{f_0}{H}(\tau - \eta)] = 0 \quad ,$$

that is

$$\frac{D_H}{Dt}[-\nabla_H^2 \psi(x,y,t) + f + \frac{f_0}{H}\tau + \frac{f_0^2}{gH}\psi] = 0 \quad . \qquad (IX-2-31)$$

This very simple case allows one to understand the role of the bottom topography, which can be included in the potential vorticity of the bottom. Another remark concerns the topographic Rossby waves, which will be generated by the variations of τ with y in the same way as Rossby waves are generated by the variations of f (see Pedlosky (1979). Notice also that the contribution to the potential vorticity due to the free surface is proportional to ψ/r_E^2 , where r_E is the external Rossby radius of deformation: if the horizontal wavelength L is much smaller than r_E, this term is negligible in front of $\nabla_H^2 \psi$, which constitutes the rigid-lid approximation.

For a two-layer fluid above a topography τ, with a free surface of denivellation η, it is obtained

$$\eta_1 = -\frac{f_0}{g}\psi_1 \; ; \; \eta_2 = \frac{f_0}{g\delta\rho}(\rho_1\psi_1 - \rho_2\psi_2) \; ; \; \eta_2 - \eta_1 = \frac{f_0}{g\delta\rho}\rho_2(\psi_1 - \psi_2) \; ,$$

where $\delta\rho = \rho_2 - \rho_1$ is the density difference (positive, since we are in a stable situation and the heaviest fluid is at the bottom). Because of (IX-2-10), $\psi_1 - \psi_2$ is characteristic of the temperature difference between the two layers. The two potential vorticity equations are (within the rigid lid approximation)

$$\frac{D_H^{(1)}}{Dt}[-\nabla_H^2\psi_1(x,y,t)+f+\frac{\psi_1-\psi_2}{R_1^2}]=0$$

$$\frac{D_H^{(2)}}{Dt}[-\nabla_H^2\psi_2(x,y,t)+f+\frac{\psi_2-\psi_1}{R_2^2}+f_0\frac{\tau}{H_2}]=0$$

$$(IX-2-32)$$

where $R_1 = N_1 H_1/f_0$ and $R_2 = N_2 H_2/f_0$ are two internal radii of deformation analogous to (II-9-12), the local Brunt-Vaisala frequencies being defined by

$$N_1^2 = \frac{g}{\rho_2}\frac{\delta\rho}{H_1} \; ; \; N_2^2 = \frac{g}{\rho_2}\frac{\delta\rho}{H_2} . \qquad (IX-2-33)$$

This two-layer model was first introduced by N.A. Phillips (1954). It can be interpreted in a different way displaying the interaction between a mean horizontal field (the "barotropic" mode)

$$\psi_{BT} = \frac{\psi_1+\psi_2}{2} \qquad (IX-2-34)$$

and the "baroclinic" mode

$$\psi_{BC} = \frac{\psi_1-\psi_2}{2} \qquad (IX-2-35)$$

which has been seen to characterize the temperature. To simplify matters, let us assume that there is no topography, and that both internal radii of deformation are equal

$$\frac{1}{R_1^2} = \frac{1}{R_2^2} = \frac{\mu^2}{2} . \qquad (IX-2-36)$$

A good exercise is to check that these modes satisfy the two following equations (where ∇^2 stands for the operator ∇_H^2, and $J(.,.)$ for the Jacobian)

$$\frac{\partial}{\partial t}\nabla^2\psi_{BT}+J(\nabla^2\psi_{BT},\psi_{BT})+J(\nabla^2\psi_{BC},\psi_{BC})+\beta\frac{\partial\psi_{BT}}{\partial x}=0$$

$$\frac{\partial}{\partial t}[(\nabla^2-\mu^2)\psi_{BC}]+J[(\nabla^2-\mu^2)\psi_{BC},\psi_{BT}]+J(\nabla^2\psi_{BT},\psi_{BC})$$

$$+\beta\frac{\partial\psi_{BC}}{\partial x}=0$$

$$(IX-2-37)$$

where the "β-plane approximation"

$$f = f_0 + \beta y \qquad (\text{IX} - 2 - 38)$$

has been made (β is assumed to be constant). An alternative way of obtaining these n-layer equations is to perform a spectral vertical expansion of the original quasi-geostrophic equation (IX-2-27) (with a constant N^2), which involves a three-dimensional fluid with a continuous density profile: for a vertical two-mode truncation of a layer of thickness H extending from $z = -H/2$ to $z = H/2$ for instance, one can look for solutions of the form (c.f. Hoyer and Sadourny, 1982)

$$\psi(x,y,z,t) = \psi_{BT}(x,y,t) + \sqrt{2}\psi_{BC}(x,y,t)\sin\frac{\pi z}{H} \qquad (\text{IX} - 2 - 39)$$

where the barotropic mode corresponds to a horizontal two-dimensional basic flow, and the baroclinic mode is the amplitude of a vertical sine perturbation describing the departure from two-dimensionality. The potential vorticity (IX-2-27) is thus equal to

$$\omega_p = -\nabla_H^2\psi_{BT} + f + \sqrt{2}(-\nabla_H^2 + \mu^2)\psi_{BC}\sin\frac{\pi z}{H} \qquad (\text{IX} - 2 - 40)$$

with

$$\mu = \pi\frac{f_0}{NH} \quad .$$

The spectral expansion can then be performed in the following way: one integrates the potential vorticity conservation equation successively from $z = -H/2$ to $+H/2$, then from $z = 0$ to $H/2$: one obtains exactly for ψ_{BT} and ψ_{BC} the same equations as (IX-2-37). It is remarkable to see here how a two-mode vertical spectral expansion of a fluid with a continuous density can be identified with a flow involving two homogeneous fluids of different densities. Let us consider for instance a two-dimensional "eddy" of more or less cylindric shape and of axis parallel to $\vec{z}$: if this eddy is perturbed by the sine perturbation considered above, the horizontal amplitude of this three-dimensional perturbation will be proportional to the baroclinic mode, and the baroclinic instability will consist in a transfer of energy from this three-dimensional perturbation to the two-dimensional motion.

The physical interpretation of this two-layer (or two-mode) model in terms of quasi-geostrophic turbulence will be given in section 5. To conclude this section we will look at the interaction of a quasi-geostrophic flow with an Ekman layer, and also study some barotropic and baroclinic quasi-geostrophic waves.

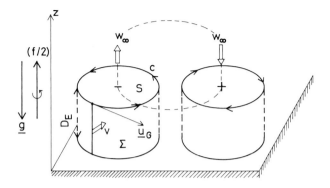

Figure IX-3: schematic view of the bottom Ekman layer.

2.4 Interaction with an Ekman layer

When a quasi-geostrophic flow $\vec{u}_G$ is close to a boundary, the latter will tend to diminish the velocity (which must be zero on the boundary), in such a way that the geostrophic balance will be lost. The consequence will be a rotation of the horizontal velocity in the boundary layer (the so-called Ekman spiral), and a vertical fluid flux at the top of the boundary layer, due to the loss of horizontal non divergence. To study this important problem which controls the dissipation of a geostrophic flow in the presence of a boundary, the simplest model to consider is that of a fluid of constant density ρ_0 rotating with a constant rotation $f_0/2$ about a vertical axis (see Figure IX-3)

$$\frac{Du}{Dt} = -\frac{1}{\rho_0}\frac{\partial P}{\partial x} + f_0 v + \nu_e \nabla^2 u \qquad (IX-2-41)$$

$$\frac{Dv}{Dt} = -\frac{1}{\rho_o}\frac{\partial P}{\partial y} - f_0 u + \nu_e \nabla^2 v \qquad (IX-2-42)$$

$$\frac{Dw}{Dt} = -\frac{1}{\rho_0}\frac{\partial P}{\partial z} + \nu_e \nabla^2 w \qquad (IX-2-43)$$

where the gravity and centrifugal force have been included in the pressure. The velocity field is a filtered velocity defined as in (IX-2-2). This permits one to assume that the fluid is geostrophic and equal to $\vec{u}_G$ of components u_G and v_G at the top or the bottom of the layer. Furthermore, the hydrostatic approximation is made in (IX-2-43), showing that the pressure P is a function of x and y only, and is then equal to its geostrophic value such that

$$f_0 \, v_G = \frac{1}{\rho_0}\frac{\partial P}{\partial x} \; ; \; -f_0 \, u_G = \frac{1}{\rho_0}\frac{\partial P}{\partial y} \; . \qquad (IX-2-44)$$

It justifies the above argument that the geostrophic balance will be lost in the boundary layer if the velocity modulus decreases. Thus the equations of motion become

$$\frac{Du}{Dt} = f_0(v - v_G) + \nu_e \nabla^2 u$$
$$\frac{Dv}{Dt} = -f_0(u - u_G) + \nu_e \nabla^2 v$$

$$(IX - 2 - 45)$$

In the outer geostrophic region, the inertial and viscous terms were neglected, so that the equation reduced to $v = v_G$ and $u = u_G$. In the Ekman layer on the contrary, dissipation is significant and cannot be neglected, while the inertial terms D/Dt are still discarded, since it is still assumed that the Rossby number is small. Let $\vec{\tilde{u}} = \vec{u}_H - \vec{u}_G$ of components $\tilde{u}$ and $\tilde{v}$, and the complex velocity fields $Z = \tilde{u} + i\tilde{v}$, $Z_G = u_G + iv_G$. The equation of motion reduces, when taking into account only the z-dependence, to

$$\frac{d^2 Z}{dz^2} = i \frac{f_0}{\nu_e} Z \ .$$

$$(IX - 2 - 46)$$

2.4.1 Geostrophic flow above an Ekman layer

It is the case envisaged in Figure IX-3: the boundary conditions are $Z(0) = -Z_G$ (no-slip condition on the lower boundary) and $Z(\infty) = 0$ (geostrophic flow on the top), and the solution of eq. (IX-2-46) is

$$Z = -Z_G \left(\exp -\sqrt{\frac{f_0}{2\nu_e}} z\right) \left(\exp -i\sqrt{\frac{f_0}{2\nu_e}} z\right) \ .$$

$$(IX - 2 - 47)$$

This shows both the exponential decrease of the horizontal velocity modulus, as well as the rotation of the velocity direction in the Ekman layer, and leads to the famous Ekman spiral for the horizontal velocity profile. More details can be found e.g. in Greenspan (1968) and Pedlosky (1979). The Ekman layer thickness is defined by

$$D_E = \pi \delta_E, \text{ with } \delta_E = \sqrt{\frac{2\nu_e}{f_0}}$$

$$(IX - 2 - 48)$$

and corresponds to a velocity equal to 96% of the geostrophic velocity. In practice, the Ekman layer thickness can be determined experimentally, and allows one to calculate the eddy-viscosity ν_e. D_E is about 1 km in the Earth's atmosphere, and a few tenths of a meter in the ocean.

One can then define the *Ekman number*

$$E_k = \frac{\nu_e}{f_0 D^2} = \frac{\delta_e^2}{2D^2} \ ,$$

$$(IX - 2 - 49)$$

which characterizes the relative importance of the viscous to the Coriolis forces in eqs (IX-2-41), (IX-2-42). If $E_k \ll 1$, one is back to the case of the above geostrophic analysis.

This analysis did not take into account the horizontal variation of the geostrophic velocity: let us consider at the top of the layer a closed isobaric contour C, to which the geostrophic velocity is tangential, and let Σ be a cylinder of section C and of generating lines parallel to $\vec{z}$ (see Figure IX-3). The fluid flux through Σ per unit length of C is equal, from (IX-2-47), to

$$\Phi = \int_0^{+\infty} v(z)dz = u_G \, \frac{\delta_E}{2}$$

since

$$v(z) = u_G \left(\exp -\sqrt{\frac{f}{2\nu_e}} \, z \right) \sin \left(\sqrt{\frac{f}{2\nu_e}} \, z \right) \quad .$$

Thus the total fluid flux through Σ is

$$\Phi_t = \frac{\delta_E}{2} \oint_C \vec{u}_G . \vec{\delta l} = \frac{\delta_E}{2} \iint_S \vec{\omega}_G . \vec{z} \, dS$$

where $\vec{\omega}_G$ is the geostrophic vorticity, and S the surface enclosed by C. If the variations of $\vec{\omega}_G$ on S are ignored, the horizontal flux across Σ is equal to $(\delta_E/2)S$ times the vorticity of the geostrophic motion at the top of the Ekman layer. Since, for the sake of continuity, this horizontal flux has to be balanced by a vertical flux of velocity w_∞ at the top of the layer, we obtain the important relation

$$w_\infty = \frac{\delta_E}{2} \omega_G \qquad\qquad (IX-2-50)$$

which fixes for the geostrophic flow the vertical velocity coming from the interaction with the Ekman layer. One can also notice that a cyclonic geostrophic motion will imply a positive w_∞ (Ekman pumping), while an anticyclonic motion will bring outer fluid into the Ekman layer. Hence, there is a secondary circulation within the geostrophic flow, which is induced by the Ekman layer.

One can for instance write the quasi-geostrophic potential-vorticity equation for one layer above an Eckman layer extending over a topography τ. Potential vorticity conservation is then written

$$\frac{D_H}{Dt}[-\nabla_H^2 \psi(x,y,t) + f] = \frac{f_0}{H}(w_s - w_i) \qquad (IX-2-51)$$

where w_s and w_i are the vertical velocities respectively at the top and bottom of the layer. It has been shown by Pedlosky (1979) that the effects of topography and free surface could be decoupled from the Ekman layer effects, in such a way that, from eq. (IX-2-50):

$$w_i = \frac{D_H \tau}{Dt} - \frac{\delta_E}{2} \nabla_H^2 \psi(x, y, t) \qquad (IX - 2 - 52)$$

$$w_s = \frac{D_H \eta}{Dt} \quad ,$$

where $\eta(x, y, t)$ is the elevation of the free surface, equal to $-(f_0/g)\psi$. It is found

$$\frac{D_H}{Dt}[-\nabla_H^2 \psi + f + \frac{f_0}{H}\tau + \frac{f_0^2}{gH}\psi] = \frac{f_0 \delta_E}{2H} \nabla_H^2 \psi . \qquad (IX - 2 - 53)$$

The new term introduced by comparison to (IX-2-31) is the Ekman layer damping term, which dissipates the vorticity linearly, and hence in the large eddies[10].

2.4.2 The upper Ekman layer

When a geostrophic ocean is driven by a wind exerting a strain $\vec{\sigma}$ on the surface, this strain is transmitted to the geostrophic circulation through an upper Ekman layer, to which the above calculation applies with appropriate boundary conditions. The Ekman layer is still described by eq. (IX-2-46), with boundary conditions which are now $Z = U - Z_G$ on the top $z = 0$ ($\vec{U} = U\vec{x}$ is the ocean surface velocity driven by the wind), and $Z = 0$ on the bottom $z \to -\infty$. It is found

$$Z = (U - Z_G) \exp \sqrt{\frac{f_0}{2\nu_e}} z \quad .$$

If one assumes that $U_G << U$, it is easily found that

$$\vec{\sigma} = \rho_0 \nu_e \left(\frac{\partial \vec{u}}{\partial z}\right)_{z=0} = \rho_0 \nu_e \frac{U}{\delta_E}(\vec{x} + \vec{y}) \quad , \qquad (IX - 2 - 54)$$

which shows that, due to the Coriolis force, the ocean surface is deviated 45^0 right[11] of the wind stress. On the other hand, the vertical integration

---

[10] There ought to be also in the r.h.s. of (IX-2-53) a turbulent damping due to sub-geostrophic scales, as discussed above: it will dissipate in the small scales the vorticity carried there along the enstrophy cascade (see below).

[11] in the northern hemisphere

of u and v show (see Pedlosky (1979) that the horizontal flux across the Ekman layer is:

$$\vec{m}_E = \frac{U\delta_E}{2}(\vec{x} - \vec{y}) = \frac{\vec{\sigma} \times \vec{z}}{\rho_0 f_0} \ .$$

Hence, a vertical integration of the 3D continuity equation yields, for the vertical velocity $w_{-\infty}$ at the bottom of the Ekman layer:

$$w_{-\infty} = \vec{\nabla}_H.\vec{m}_E = \frac{f_0}{\rho_0}\vec{z}.\vec{\nabla} \times \vec{\sigma} \ . \qquad (IX-2-55)$$

When applied to a quasi-geostrophic layer of an ocean forced by the wind, the complete Potential vorticity equation is:

$$\frac{D_H}{Dt}[-\nabla_H^2\psi + f + \frac{f_0\tau}{H} + \frac{f_0^2}{gH}\psi] =$$

$$\frac{f_0^2}{\rho_0 H}\vec{z}.\vec{\nabla} \times \vec{\sigma} + \frac{f_0\delta_E}{2H}\nabla_H^2\psi \ . \qquad (IX-2-56)$$

When a multi-layer model is considered (within the rigid-lid approximation), the forcing terms are introduced in the upper layer, and the orographic and Ekman dissipation terms in the bottom layer.

Let us mention finally that eq.(IX-2-54) permits an understanding of the phenomenon called *upwelling*: when a wind blows along a coast (with the water to the right in the northern hemisphere[12]), the warm ocean surface layer will be entrained offshore, and, by continuity, deep cold water will upwell to the surface. The result is a cold coastal current going in the same direction as the wind. Famous examples are the Californian current in the northern hemisphere, which goes southwards, and the Humboldt current in the southern hemisphere, which goes northwards: they are initiated by the anticyclonic winds west of the American Pacific coast. Due to the horizontal thermal gradients thus created, they may be subject to a baroclinic instability, which might explain the eddies observed in the Californian current[13]. A very famous (but not yet really understood) anomaly of the Humboldt current corresponds to the so-called *El Nino* phenomenon, where this current chaotically reverses its direction every 2 or 3 years, transforming into a southwards current of warm water. This catastrophic[14] event is accompanied with climatic anomalies such as the reversal of trade winds, and might involve extremely complicated non-linear couplings between the atmosphere and the oceans.

---

[12] and left in the southern hemisphere

[13] However, eddies found in the ocean may have other causes, such as horizontal shear instabilities or topographic trapping for instance

[14] for the local fish industries

2.5 Barotropic and baroclinic waves

Up to now, we have mentioned the propagation of inertial waves due to a constant rotation in a three-dimensional flow. Here we focus our interest on the Rossby waves, which are obtained within the geostrophic approximation applied to a shallow fluid on a rotating sphere, and are due to the variation in the latitude of the "effective" Coriolis force $-f\vec{z}\times$ $\vec{u}_H$. These waves are extremely significant in the Earth's atmosphere and oceans. The simplest model to study them is eq (IX-2-31) with no topography, using the rigid-lid approximation, and with the β-plane approximation (IX-2-38), that is

$$\frac{\partial}{\partial t}\nabla^2\psi + J(\nabla^2\psi, \psi) + \beta\frac{\partial\psi}{\partial x} = 0 \quad . \qquad (IX-2-57)$$

A linearization about a basic state at rest yields the famous Rossby equation (1939)

$$\frac{\partial}{\partial t}\nabla^2\psi + \beta\frac{\partial\psi}{\partial x} = 0 \qquad (IX-2-58)$$

which admits wave solutions of the form

$$\psi(x,y,t) = \psi_0 \exp i(\vec{k}.\vec{x} - \varpi t) \qquad (IX-2-59)$$

with $\vec{k} = (k_1, k_2, 0)$ and $\vec{x} = (x, y, 0)$, provided the dispersion relation

$$\varpi = -\frac{\beta k_1}{k_1^2 + k_2^2} \qquad (IX-2-60)$$

is satisfied. These waves travel to the west, due to the minus sign in (IX-2-60). The reader will find in Pedlosky (1979) all the details of their dynamics. A particularly important result is the conservation of the group velocity times the mean kinetic energy of the wave during the reflexion of a wave-packet on a boundary parallel to $\vec{y}$: since it may be shown that the reflexion decreases the group velocity if the frontier is to the west (and increases it in the opposite case), a western frontier in an oceanic basin will intensify the variability of the currents. This applies in particular to the Gulf Stream. Another remark is that the Rossby waves differ from the inertial waves, since the latter are generated by a constant rotation, while Rossby waves need a differential rotation. Such a differential rotation, due to the Earth's sphericity, does not seem to be related to the differential rotation effects encountered for instance on Jupiter or in rotating stars, and which correspond to differences of zonal velocities: the shears thus created may be responsible for Kelvin-Helmholtz type instabilities and formation of coherent eddies such as Jupiter's great red spot. Notice finally that a bottom topographic term in eq. (IX-2-31) corresponding to a slope directed in a certain direction (chosen as the y direction) will generate waves of the same type as the Rossby waves and travelling in the x direction. They are called "topographic Rossby waves", and can be easily obtained in rapidly rotating fluid experiments in a laboratory (Colin de Verdière, 1979).

It is interesting to perform the same analysis on a two-layer geostrophic fluid, with the linearized equations (IX-2-37):

$$\frac{\partial}{\partial t}\nabla^2\psi_{BT} + \beta\frac{\partial\psi_{BT}}{\partial x} = 0 \qquad (IX-2-61)$$

$$\frac{\partial}{\partial t}[(\nabla^2 - \mu^2)\psi_{BC}] + \beta\frac{\partial\psi_{BC}}{\partial x} = 0 \qquad (IX-2-62)$$

The linearized equation for the barotropic mode is the classical Rossby equation. The dispersion relation for the baroclinic equation is

$$\varpi_{BC} = -\frac{\beta k_1}{k_1^2 + k_2^2 + \mu^2} \quad . \qquad (IX-2-63)$$

For wave lengths much larger than the internal radius of deformation $\approx \mu^{-1}$, these "baroclinic Rossby waves" will have a frequency much smaller than the barotropic frequency given by (IX-2-60). Consequently the characteristic times will be much larger: this is particularly true in the oceans, where the typical velocity spin-up times are of 100 to 200 days, while the thermal spin-up time (corresponding to the baroclinic mode) is of several years.

Let us briefly summarize an analysis, taken from Pedlosky (1979), showing how the baroclinic instability can be understood in the frame of the two-layer model: we take a basic flow corresponding to a constant westerly zonal velocity U_0 , that is

$$\psi_{BT} = U_0 y \qquad (IX-2-64)$$

and a temperature ψ_{BC} decreasing (with a constant gradient) towards the north, that is

$$\psi_{BC} = -U_1 y \quad . \qquad (IX-2-65)$$

Then, after linearization of eqs (IX-2-37), one obtains, for the fluctuations $\tilde{\psi}_{BT}$ and $\tilde{\psi}_{BC}$ about the basic state, solutions of the form

$$\tilde{\psi}_{BT} = A \, \exp i\,(\vec{k}.\vec{x} - \varpi t)$$
$$\tilde{\psi}_{BC} = B \, \exp i\,(\vec{k}.\vec{x} - \varpi t)$$

with the dispersion relation

$$\varpi = k_1 \left(U_0 \pm U_1 \left[\frac{k^2 - \mu^2}{k^2 + \mu^2} \right]^{1/2} \right) \qquad (IX - 2 - 66)$$

where k is the modulus of the horizontal wave vector $\vec{k}$ of components k_1 and k_2. This shows that for wave lengths smaller than the internal radius of deformation, ϖ is real and waves can propagate. For scales larger than the internal radius, ϖ is a pure imaginary and the perturbations can amplify exponentially, which corresponds to the baroclinic instability. Thus, the ratio of the barotropic to the baroclinic mean energy of the perturbation is

$$\frac{E_{BT}}{E_{BC}} \sim \frac{\mu^2 + k^2}{\mu^2 - k^2}$$

which shows that it is in the vicinity of the internal radius of deformation that the instability is the more active. This is why it is generally taken for granted that the baroclinic instability arises at horizontal scales of the order of the internal radius of deformation.

3 - Two-dimensional isotropic turbulence

We now return to purely two-dimensional isotropic turbulence obeying eq. (IX-1-3)[15]. The essential characteristics of the dynamics of such a turbulence[16] are described in Kraichnan and Montgomery (1980) and Lesieur (1983). As was the case in three dimensions, one can introduce the two-dimensional spatial Fourier transform of the velocity and stream function

$$\hat{u}_i(\vec{k}, t) = (\frac{1}{2\pi})^2 \int [\exp -i\vec{k}.\vec{x}] \; \hat{u}_i(\vec{x}, t) dx dy \qquad (IX - 3 - 1)$$

$$\hat{\psi}(\vec{k}, t) = (\frac{1}{2\pi})^2 \int [\exp -i\vec{k}.\vec{x}] \; \psi(\vec{x}, t) dx dy \qquad (IX - 3 - 2)$$

with $\vec{k} = (k_1, k_2)$ and $\vec{x} = (x, y), i = (1, 2)$. We recall from Chapter V that for isotropic two-dimensional turbulence

$$< \hat{u}_i(\vec{k'}, t) \hat{u}_j(\vec{k}, t) > = \frac{E(k, t)}{\pi k} P_{ij}(\vec{k}) \; \delta(\vec{k} + \vec{k'}) \qquad (IX - 3 - 3)$$

---

[15] or equivalently eq. (IX-1-5)

[16] The interaction of two-dimensional turbulence with Rossby waves and baroclinic instability will be discussed in the last section.

$$< \hat{\psi}(\vec{k}',t)\hat{\psi}(\vec{k},t) >= \hat{\Psi}(k,t)\, \delta(\vec{k}+\vec{k}') \ . \qquad (IX-3-4)$$

The kinetic energy and the enstrophy are still respectively

$$\int_0^{+\infty} E(k,t)dk \quad \text{and} \quad \int_0^{+\infty} k^2 E(k,t)dk \ ,$$

with

$$E(k,t) = \pi k^3 \hat{\Psi}(k,t) \ . \qquad (IX-3-5)$$

As in three-dimensional turbulence, the kinetic energy is conserved by the non-linear terms of the equations, and (IX-1-3) yields in the unforced case

$$\frac{d}{dt}\int_0^{+\infty} E(k,t)dk = -2\nu \int_0^{+\infty} k^2 E(k,t)dk \ . \qquad (IX-3-6)$$

But since there is no vortex stretching, the enstrophy also obeys a conservation equation

$$\frac{d}{dt}\int_0^{+\infty} k^2 E(k,t)dk = -2\nu \int_0^{+\infty} k^4 E(k,t)dk \ . \qquad (IX-3-7)$$

This result will form the basis for all the phenomenological theories of two-dimensional turbulence. It has to be stressed however that an infinite family of "inviscid"[17] invariants can be constructed with the aid of the enstrophy: indeed, eq. (IX-1-5) expresses the "inviscid" conservation of the vorticity ω following the fluid motion, and hence of any functional $f(\omega)$. Since the motion is incompressible, the surface $\delta\Sigma$ of the horizontal sections of the fluid tubes is also conserved, and the integral $\int f(\omega)\delta\Sigma$ on the domain over which the fluid extends is conserved with time[18]. This implies the invariance of $< f(\omega) >$ with time, if one accepts identifying the ensemble and spatial averages. It is therefore possible that the existence of these invariants could modify the following dynamical conclusions, based on theories which generally preserve only the kinetic energy and enstrophy invariance.

3.1 Fjortoft's theorem

This theorem[19] (Fjortoft, 1953) serves as a basis for the two-dimensional turbulence phenomenology: let us consider the two-dimensional Euler

---

[17] "Inviscid" means here without considering the viscous dissipation terms, which however exist and may contribute to the dissipation of the quantities considered, even at small viscosity.

[18] This is valid however only if the domain is compact, or if periodicity holds on the boundaries: the latter condition is fulfilled for homogeneous turbulence (fluid in a "square").

[19] which has no relation with Fjortoft's inviscid instability criterion demonstrated in Chapter III

equation in Fourier space (Fourier representation in a square), truncated in order to retain only three parallel wave vectors $\vec{k}_1$, $\vec{k}_2$ and $\vec{k}_3$. For simplification we assume $\vec{k}_2 = 2\vec{k}_1$ and $\vec{k}_3 = 3\vec{k}_1$. Let $E(k_i, t)$ be the kinetic energy at the mode $\vec{k}_i$. Conservation of kinetic energy and enstrophy imply that between two times t_1 and t_2, the variation $\delta E_i = E(k_i, t_2) - E(k_i, t_1)$ satisfies two constraints

$$\delta E_1 + \delta E_2 + \delta E_3 = 0 \qquad (IX-3-8)$$

$$k_1^2 \, \delta E_1 + k_2^2 \, \delta E_2 + k_3^2 \, \delta E_3 = 0 \qquad (IX-3-9)$$

which yields

$$\delta E_1 = -\frac{5}{8} \, \delta E_2 \; ; \; \delta E_3 = -\frac{3}{8} \, \delta E_2 \qquad (IX-3-10)$$

$$k_1^2 \, \delta E_1 = -\frac{5}{32} k_2^2 \, \delta E_2 \; ; \; k_3^2 \, \delta E_3 = -\frac{27}{32} k_2^2 \, \delta E_2 \; . \qquad (IX-3-11)$$

Therefore, if the intermediate wave number k_2 loses kinetic energy ($\delta E_2 < 0$), more energy will go to k_1 than to k_3, and more enstrophy will go to k_3 than to k_1. These conclusions must, of course, be reversed if $\delta E_2 > 0$. But, it is the first case which has more physical significance when freely decaying two-dimensional turbulence is considered: indeed let us envisage a continuous range of wave numbers, and an initial kinetic energy spectrum peaked about a wave number $k_i(0)$. Due to non-linear interactions, it is expected that the peak will spread out towards other modes, and consequently the amount of kinetic energy in the vicinity of $k_i(0)$ will decrease; Fjortoft's result will then suggest that more kinetic energy (resp. less enstrophy) will go towards modes $k < k_i(0)$ than towards modes $k > k_i(0)$. Actually, this is what the direct numerical simulations show in this case, though it has to be stressed that Fjortoft's result is not always true for triads of wave vectors ($\vec{k}_1, \vec{k}_2, \vec{k}_3$) with an arbitrary relative orientation, as was shown by Merilees and Warn (1975).

3.2 The enstrophy cascade

Since the enstrophy tends to cascade towards large wave numbers, it has been proposed by Kraichnan (1967) and Leith (1968) that the enstrophy flux

$$Z(k) = \int_k^\infty k^2 T(k) dk \qquad (IX-3-12)$$

(where $T(k)$ is the kinetic energy transfer in the evolution equation of the energy spectrum) should be independent of k at high wave numbers and equal to β. Then a dimensional argument, similar to the one used in three-dimensional turbulence for the Kolmogorov energy cascade, assumes that $E(k)$ is a function of β and k only. This leads to the enstrophy cascade concept, where the kinetic energy spectrum is given by[20]

$$E(k) \sim \beta^{2/3} k^{-3} \quad . \qquad (IX - 3 - 13)$$

In fact, Kraichnan and Leith proposed this enstrophy cascade in the context of a turbulence forced at a fixed wave number k_i by a stationary forcing injecting kinetic energy at a rate ϵ and enstrophy at a rate $\beta = k_i^2 \epsilon$. The same k^{-3} enstrophy cascade was proposed by Batchelor (1969) in the context of a freely-decaying two-dimensional turbulence, where a self similar evolving spectrum of the form

$$E(k,t) = v^3 t \ F(k \ vt) \qquad (IX - 3 - 14)$$

was assumed: with such a spectrum (where the kinetic energy $(1/2)v^2$ is independent of time, as will be shown below), the enstrophy is proportional to t^{-2} provided the integral $\int^\infty x^2 F(x) \ dx$ converges; the enstrophy dissipation rate

$$\beta = -\frac{dD(t)}{dt} \qquad (IX - 3 - 15)$$

is proportional to t^{-3}, and the assumption of a range where $F(x) \propto x^\alpha$ with a coefficient independent of v leads to $\alpha = -3$ and a spectrum $E(k) \sim t^{-2} k^{-3}$ corresponding to (IX-3-13).

Physically, the enstrophy cascade concept corresponds to the fact that a fluid blob imbedded into a larger scale velocity shear will be elongated in the velocity direction, and will see its transverse characteristic dimension decrease; since at the same time the vorticity of each fluid point is conserved, the result will be a steepening of the vorticity gradients in the direction transverse to the velocity, with a flux of vorticity into the small scales. When the local vorticity gradients are high enough, the molecular viscosity dissipates the vorticity. Then an enstrophy dissipation wave number can be introduced, function of β and ν only. It has a different form from (VI-4-2) and is now equal to

$$k_d = (\frac{\beta}{\nu^3})^{1/6} \quad . \qquad (IX - 3 - 16)$$

On the other hand, a k^{-3} energy spectrum extending from k_i to k_d leads to (neglecting the logarithmic corrections) $k_i \sim \beta^{1/3}/v$ and

$$\beta^{1/3} \sim D(t)^{1/2}.$$

---

[20] The enstrophy flux β has of course nothing to do with the parameter β of Rossby waves introduced in the preceding section.

Estimating roughly D as $k_i^2 v^2$, one finally obtains

$$Re = \frac{v}{\nu k_i} \sim (\frac{k_d}{k_i})^2 \qquad (IX - 3 - 17)$$

which shows that the total number of degrees of freedom of two-dimensional turbulence is of the order of Re, at variance from the $Re^{9/4}$ value which has been determined for three-dimensional turbulence.

The first direct numerical simulations of the enstrophy cascade were carried out by Lilly both in the forced and freely-evolving cases (see Lilly, 1969, 1971, 1973). He found a spectrum very close to k^{-3} on about one decade. He confirmed also, by comparing the stream function and vorticity contours, the fact that vorticity tends to cascade towards small scales while kinetic energy goes to larger and larger scales, where it remains trapped if the fluid extends on a domain of finite extent. Higher-resolution calculations (with up to 128^3 modes in a spectral-method calculation) in the decaying case, using spectral methods or finite-differences methods as well, were done by Herring et al. (1974). They found a k^{-4} spectrum instead of the law (IX-3-13), and stressed that the extent of the spectral range was insufficient to describe properly the enstrophy cascade, due to the low resolution. They stress also that, at these resolutions, the dynamics of the large scales is Reynolds number independent. These calculations display also in the vorticity contours evidence of the presence of what will be called later on coherent structures.

Other interpretations of the vorticity transfers towards small scales in two-dimensional isotropic turbulence have been given, in particular by Saffman (1971), who proposes that the vorticity conservation following the motion will produce vorticity shocks, and that the resulting enstrophy spectrum consists of a random superposition of shocks, and is then proportional to k^{-2}; hence the energy spectrum should follow a k^{-4} inertial range. As a matter of fact, an inviscid numerical simulation done by Kida and Yamada (1984), using time power series expansions, displays an energy spectrum proportional to $k^{-4.6}$. However, such a theory does not take into account the role of viscosity, which will be seen to be needed if one wants to dissipate the enstrophy at a finite rate.

As already pointed out, a very impressive characteristic of two-dimensional isotropic turbulence is the early formation of Kelvin-Helmholtz type coherent eddies, which seem to result from inflectional instabilities of the initial velocity field. These structures have been clearly identified in the numerical simulations of Fornberg (1977) and McWilliams (1984) in decaying situations (see also Babiano et al., 1987, Brachet et al., 1986, 1988, and Farge and Sadourny, 1989), and also in some stationary forced situations with a forcing spectrum at high wave numbers (Basdevant et al., 1981, Herring and McWilliams, 1985). These coherent structures are visible on Figure IX-4, taken from Brachet et al. (1986). As was shown first by McWilliams (1984), and confirmed by the the above quoted references, they survive much longer after the small-scale turbulence has been dissipated by viscosity. Their influence on the two-dimensional kinetic-energy spectrum may be multiple: firstly, they produce spatial intermittency [21], and it has been argued by Basdevant et al. (1981) and Basdevant and Sadourny (1983) that this intermittency could restore the transfers localness and yield a kinetic energy spectrum steeper than the k^{-3} enstrophy cascade model presented above. Secondly, they create strong velocity gradients in the stagnation regions inbetween, as shown by Brachet et al. (1988): in the latter spectral calculation, performed in the decaying case at a very high resolution (800^2 modes), there are two distinct phases: during a first period (when coherent eddies form), the energy spectrum is not submitted to viscosity (which affects the small scales), and seems to be close to Saffman's law, with formation of intense vortex sheets strained between the large eddies. Once the dissipative scales are excited[22] there is a transition towards a k^{-3} energy spectrum, corresponding to a "packing" of these vortex sheets. Such a piling up, which is apparent on Figure IX-4, could be the result of spiral vortex distributions due to the coherent vortices[23]

In reviewing the various physical mechanisms which may contribute to the enstrophy cascade, let us mention the work of Staquet et al. (1985), Staquet (1985) and Lesieur et al. (1988) and Comte et al. (1989), concerning the longitudinal spatial energy spectra in the mixing layer (see Chapter XIII), which could shed some light on the build up of the enstrophy cascade: indeed it is shown by a numerical simulation that when two "fundamental" [24] eddies pair, the contribution to the longitudinal spectrum coming from the harmonics of the fundamental mode will collapse on the other modes to form a continuous spectrum [25]

---

[21] Indeed, if one considers the *E.D.Q.N.M.* model of two-dimensional turbulence (see below), it is possible to give in that frame an analytical derivation of the k^{-3} law (with a logarithmic correction), which also shows that the enstrophy interactions in the enstrophy cascade are not local, but semi-local in the sense that $Z(k)$ is dominated by triads (k', p, q) such that $q << k' \approx k \approx p$.

[22] time which corresponds to a maximum in the enstrophy-dissipation rate

[23] A theory due to Moffatt (1986) and Herbert (1988) proposes that a spiralling vortex should have a $k^{-11/3}$ spectrum, intermediate between k^{-3} and k^{-4}.

[24] "Fundamental" means here that the eddies result from the primary instability of the inflexional shear.

[25] longitudinal spatial spectrum

with an inertial range intermediate between k^{-4} and k^{-3}: therefore the pairing of large energetic eddies of same-sign vorticity seems essential to create an enstrophy cascade. The same pairing interactions[26] have been observed in the above quoted numerical simulations of decaying two-dimensional isotropic turbulence done by Babiano et al. (1987), Brachet et al. (1988) and Farge and Sadourny (1989).

These arguments show that the concept of enstrophy cascade is not a simple one and might correspond to some statistical averaging of various physical situations. It will however prove to be extremely useful towards understanding the two-dimensional turbulence dynamics. We believe that the main conclusions derived below will remain qualitatively correct, even if the actual spectral exponent departed from the -3 Kraichnan-Leith-Batchelor's exponent.

Figure IX-4: isovorticity lines in the isotropic two-dimensional direct numerical simulation of Brachet et al. (1986), for a pseudo-spectral calculation with a resolution of 512^2 (Courtesy M.E. Brachet)

3.3 The inverse energy cascade

This concept is from Kraichnan (1968), and holds only when the turbulence is forced at a fixed wave number k_i . Fjortoft's theorem has shown that the kinetic energy could be transferred more easily towards large scales than towards small scales. In fact it will be shown below that, within the *E.D.Q.N.M.* model, the kinetic energy flux through the enstrophy cascade is zero. So the kinetic energy injected at the

---

[26] that is, merging of two eddies of same sign which rotate around one another

rate ϵ at k_i can only cascade backwards towards small wave numbers. Kraichnan's argument is then the same as for the three-dimensional Kolmogorov kinetic energy cascade, except for the sign of the kinetic energy flux $\Pi(k)$ which is now negative. In this range ($k < k_i$) the kinetic energy spectrum is proportional to $\epsilon^{2/3}k^{-5/3}$. Figure IX-5 shows schematically the kinetic energy spectrum obtained with such a forcing.

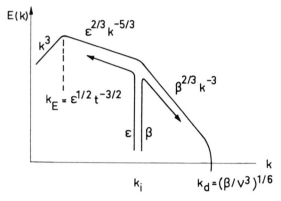

Figure IX-5: schematic double cascading spectrum of forced two-dimensional turbulence (from lesieur, 1983).

The inverse cascade range is not stationary at low wave numbers, since kinetic energy is continuously supplied at a rate ϵ, without any dissipation. Writing that the kinetic energy contained under the Kolmogorov spectrum is proportional to ϵ, it is then easy to show that the wave number k_E characteristic of the spectrum maximum will decrease like (Pouquet et al., 1975)

$$k_E(t) \sim \epsilon^{-1/2}t^{-3/2} \qquad (IX-3-18)$$

a Richardson law for the associated scale (which here is the integral scale of the turbulence, characteristic of the energetic eddies). For $k < k_E$, the spectrum is proportional to k^3, due to resonant non local interactions of two energetic wave numbers $\approx k_E$, as shown in Basdevant et al. (1978)[27]. Practically, there always exist in the numerical calculations or in the experiments a minimum non-zero wave number corresponding to the maximum extension of the domain: it has been shown in that case (forced turbulence) that the kinetic energy cannot increase indefinitely, as it would do for a $k^{-5/3}$ spectrum extending to $k = 0$, but will have an upper bound, due to the action of viscosity which will eventually play a role in the large scales when the latter will be extremely

---

[27] see also subsection **3.4**

energetic (Pouquet et al., 1975). In order to prevent this excessive accumulation of energy on the lower wave number, it is possible to add
to the r.h.s. of the equation of motion (IX-1-5) a damping term proportional to $-\nabla_H^2 \psi$, which will dissipate the inverse cascading kinetic
energy and limit the infrared extent of the inverse energy cascade: this
was done in particular in the numerical simulations of Lilly (1969, 1973).
Physically, such a damping is provided by the Ekman layer dissipation
in quasi-geostrophic turbulence (cf eq. (IX-2-53)), or by the Hartman
layers in the two-dimensional *M.H.D* turbulence between two planes
studied by Somméria (1986). In this case, where a stationary forcing is
produced by two-dimensional Taylor-Green vortices driven electrically,
a $k^{-5/3}$ inverse energy cascade has actually been measured in a certain
range of the parameters characterizing the dissipation, and this seems
to be the first experimental evidence of the inverse energy cascade of
two-dimensional turbulence (see Figure IX-6-A). Direct numerical calculations done by Lilly (1973), Fyfe et al. (1977), Frisch and Sulem
(1984) and Herring and McWilliams (1985) show some evidence of the
inverse cascade (Figure IX-6-B). Notice finally that there is no such cascade in the freely-decaying case, but only a stronger decrease $\propto t^{-1}$
(compared to the $t^{-0.3\approx0.5}$ law of three-dimensional turbulence) of the
wave number $k_i(t)$ where the spectrum peaks. This situation, presented
on the large-eddy simulation of Figure IX-7-A (from Staquet, 1984), will
be looked at in more detail below.

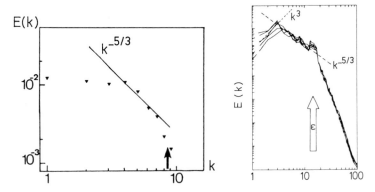

Figure IX-6: A- (left) experimental kinetic energy spectrum obtained by Sommeria (1986) in a forced two-dimensional turbulence in mercury. The arrow figures the energy injection wave number (courtesy J. Fluid Mech.). B- (right)
inverse energy cascade obtained in the numerical simulation of Frisch and
Sulem (1984). The calculation shows also the k^3 infrared spectrum due to
non local interactions (courtesy The Physics of Fluids).

Physically, the strong inverse transfers of decaying two-dimensional

isotropic turbulence result undeniably from the pairing of large energetic eddies of same vorticity sign and therefore correspond to subharmonic instabilities similar to those leading to the pairing of coherent
structures in the mixing layer[28]. However, things are less clear in the
forced case where, as pointed out by Herring and McWilliams (1985), a
strong random forcing may suppress the formation of the coherent structures. It may also, in the case where these structures form, inhibit the
various pairings, which are dependent upon the phases of the various
subharmonic instabilities.

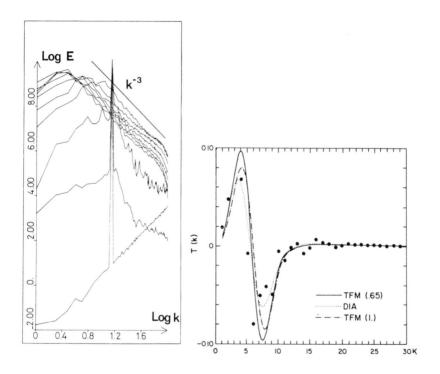

Figure IX-7: A-(left) freely-decaying kinetic energy spectrum obtained numerically in the large-eddy simulation of Staquet (1985). A self-similar k^{-3}
Batchelor spectrum establishes, with an enstrophy cascade extending above
$k_i(t)$, and no inverse $k^{-5/3}$ energy cascade. B-(right) comparison between
the two-dimensional turbulence kinetic energy transfers computed in a direct
numerical simulation (points), the $D.I.A.$ approximation , and the Test-Field-
Model with two different values of an adjustable parameter (from Herring et
al., 1974, courtesy J. Fluid Mech.).

---

[28] The latter were experimentally displayed by Winant and Browand
(1974).

3.4 The two-dimensional $E.D.Q.N.M.$ model

The application of $E.D.Q.N.M.$-type techniques to two-dimensional isotropic turbulence was made in particular in Kraichnan (1971), Leith (1971), Herring et al. (1974), Pouquet et al. (1975), Holloway (1976), Lesieur and Herring (1985), and Métais and Lesieur (1985). Using the spectral tensor expression (V-5-3) in the equation (VII-3-8), one obtains the following expression for the kinetic energy transfer $T(k,t)$

$$T(k,t) = \frac{2}{\pi} \iint_{\Delta_k} dp \ dq \ \theta_{kpq} \frac{k^2}{pq} b_2(k,p,q)$$
$$[kE(p,t)E(q,t) - pE(q,t)E(k,t)] \qquad (IX - 3 - 19)$$

with the same notations as in Chapter VII, the coefficient b_2 being given by

$$b_2(k,p,q) = 2\frac{p}{k}(xy - z + 2z^3)(1 - x^2)^{\frac{1}{2}}$$
$$= 2\frac{(k^2 - q^2)(p^2 - q^2)}{k^4}(1 - x^2)^{\frac{1}{2}} \ . \qquad (IX - 3 - 20)$$

The relaxation time for triple correlations θ_{kpq} has the same expression, in terms of the energy spectrum, as for three-dimensional turbulence. Only the numerical constant will slightly change from the value 0.36 proposed in Chapter VII to a value of 0.40, as stressed in Lesieur and Herring (1985). This corresponds to a "Kolmogorov" constant in the $k^{-5/3}$ inverse energy cascade of the order of $6 \approx 7$, as determined by Kraichnan (1971 a) and Pouquet et al. (1975). This value is of the order of magnitude of the experimental results of Sommeria (1986) and the direct-numerical simulations of Lilly (1969), Frisch and Sulem (1984) and Herring and McWilliams (1985), which yield values comprised between 4 and 9. This relatively high value proves that the two-dimensional Kolmogorov inverse energy cascade is far less efficient than its three-dimensional direct counterpart.

The transfer (IX-3-19) conserves kinetic energy and enstrophy, and it may be checked that

$$\int_0^{+\infty} T(k,t)dk = 0; \quad \int_0^{+\infty} k^2 T(k,t)dk = 0 \ . \qquad (IX - 3 - 21)$$

Such a transfer, as indicated by Fjortoft's theorem, is now positive in the large scales, as can be checked on Figure IX-7-B, which shows both a direct and a closure calculation of the two-dimensional turbulence isotropic transfer determined by Herring et al. (1974). This behaviour is completely different from the three-dimensional transfer of Figure (VI-2) which is negative in the large scales, indicating thus the tendency for the kinetic energy to cascade up to the small scales.

The same techniques as proposed in Chapters VII and VIII allow one to calculate the non-local flux of enstrophy (Kraichnan, 1971 a, Pouquet et al., 1975, Basdevant et al., 1978, Lesieur and Herring, 1985, Métais and Lesieur, 1986), which is equal to

$$
Z_{NL}(k,t) = -\frac{1}{4} \int_0^{ak} \theta_{kkq}\, q^2 E(q)dq\; k^3 \frac{\partial}{\partial k} k E(k)
$$

$$
+ \frac{1}{2} \int_0^k k'^4 E(k')dk' \int_{sup(k,k'/a)}^{\infty} \theta_{k'pp} \frac{\partial}{\partial p} p E(p)dp
$$

$$
- \int_0^k k'^5 dk' \int_{sup(k,k'/a)}^{\infty} \theta_{k'pp} \frac{E(p)^2}{p} dp \quad (IX-3-22)
$$

where a is the non-localness parameter. The non-local kinetic energy flux calculated in the same way is

$$
\Pi_{NL}(k,t) = -\frac{1}{4} \int_0^{ak} \theta_{kkq}\, q^2 E(q)dq\; k \frac{\partial}{\partial k} k E(k)
$$

$$
+ \frac{1}{2} \int_0^k k'^2 E(k')dk' \int_{sup(k,k'/a)}^{\infty} \theta_{k'pp} \frac{\partial}{\partial p} p E(p)dp
$$

$$
- \int_0^k k'^3 dk' \int_{sup(k,k'/a)}^{\infty} \theta_{k'pp} \frac{E(p)^2}{p} dp \;. \quad (IX-3-23)
$$

Let us first consider the infrared region $k \to 0$: it can be checked (see Basdevant et al., 1978) that $\Pi_{NL}(k)$ reduces to the third term on the r.h.s. of (IX-3-23), and that the corresponding energy transfer $-\partial \Pi_{NL}(k)/\partial k$ is itself of the order of

$$
k^3 \int_{k_i}^{\infty} \theta_{kpp} \frac{E(p)^2}{p} dp
$$

where k_i is the wave number where the energy spectrum peaks (i.e. $k_E(t)$ in a forced situation and $k_i(t)$ in a decaying situation). This term is therefore responsible for the infrared k^3 kinetic energy spectrum already mentioned on Figures IX-5 and IX-6-B.

In the enstrophy cascade, it turns out that the first term in the r.h.s. of (IX-3-22) gives the essential contribution to the enstrophy flux. As already mentioned, it involves wave number triads (k', p, q) such that $q << k' \approx k \approx p$, the same that are responsible for the k^{-1} viscous-convective range of the three-dimensional passive scalar at high Prandtl number, and which correspond to the creation of smaller scales through the straining of a larger-scale velocity shear. The enstrophy flux $Z(k)$ can then be approximated by (Kraichnan, 1971 a)

$$Z(k) = -\frac{1}{4}\int_0^k \theta_{kk0}\, q^2 E(q)dq\, k^3 \frac{\partial}{\partial k}kE(k) \qquad (IX-3-24)$$

which itself can be modified as (Pouquet et al., 1975), Basdevant et al., 1978)

$$Z(k) = -\frac{1}{4}k^3\frac{\partial}{\partial k}[kE(k)\theta_{kk0}\int_0^k q^2 E(q)dq] \qquad (IX-3-25)$$

which is not very different from (IX-3-24) in the enstrophy cascade, but offers the further advantage of being both conservative of energy and enstrophy, and therefore able to approximate the enstrophy flux whatever the value of k (and not only in the enstrophy cascade). Eq (IX-3-25) can be solved analytically, assuming a constant enstrophy flux β. It leads to

$$E(k) \sim \beta^{2/3}k^{-3}(A + 2\ln\frac{k}{k_i})^{-1/3} \qquad (IX-3-26)$$

where the constant A may be evaluated approximately by assuming a zero enstrophy flux (IX-3-25) for $k < k_i$. This yields $A = 1$ and a $k^{-5/3}$ inverse energy cascade for $k < k_i$. As noticed in Lesieur and Herring (1985), the slope of the energy spectrum in a neighbourhood k_i beyond k_i is thus

$$\frac{\partial \ln E(k)}{\partial \ln k} = -3 - 2\frac{A}{3} = -\frac{11}{3},$$

steeper than -3. Then the logarithmic correction (3-26), originally proposed by Kraichnan (1971 a) under a slightly different form, could also be a candidate in explaining the steepening with respect to k^{-3} of the enstrophy cascade.

Let us now turn our attention to the kinetic energy flux in the enstrophy cascade: from (IX-3-23) it is approximately equal to

$$\Pi(k) = \frac{Z(k)}{k^2} + 2\int_0^k \nu_t(k')k'^2 E(k')dk' \qquad (IX-3-27)$$

where

$$\nu_t = \frac{1}{4}\int_k^\infty \theta_{0pp}\frac{\partial}{\partial p}pE(p)dp\ . \qquad (IX-3-28)$$

Indeed, the third term in the r.h.s. of (IX-3-23) can be neglected. Contrary to the three-dimensional turbulence case, this eddy-viscosity, of the order of $-(1/4)kE(k)\theta_{kk0}$, is negative for any spectrum decreasing faster than k^{-1}. The total energy flux through the enstrophy cascade is composed of a positive term $Z(k)/k^2$ corresponding to the energy flux due to the semi-local interactions responsible for the enstrophy cascade, minus negative eddy-viscosity interactions which send back into the large scales the kinetic energy drained towards the small scales by the enstrophy cascade. The result turns out to be a zero kinetic energy flux for a k^{-3} kinetic energy spectrum. This implies for high Reynolds number two-dimensional turbulence a very peculiar dynamics, where the large scales (unaffected by molecular viscosity), conserve exactly their kinetic energy, while their vorticity is dissipated at a finite rate towards small scales through an enstrophy cascade. It is clear from that result that any system of inviscid dynamical equations which do not dissipate enstrophy cannot pretend to represent two-dimensional turbulence correctly.

Again one may wonder about the physical significance of this negative eddy-viscosity: it could for instance be due to the pairing of small scale eddies, or even to other types of subharmonic instabilities displayed by Sinai and co-workers in the case of the "Kolmogorov flow", a class of two-dimensional flows forced in the small scales by a periodic field (see Meshalkin and Sinai, 1961, Sivashinsky and Yakhot, 1985). More generally, one can widen the concept of negative eddy viscosity and include in it all the larger-scales formation effects. One must emphasize, however, that, in the context of two-dimensional numerical Large-eddy simulations for instance, the "negative eddy viscosity" effects described above are not represented at all by a negative eddy-viscosity, since the modelled scales need to conserve their kinetic energy, while dissipating their enstrophy at a finite rate in the subgrid scales (see Chapter XII).

3.5 Freely-decaying turbulence

A freely-evolving two-dimensional turbulence (no forcing) is now studied with a spectrum initially concentrated at a wave number $k_i(0)$. This problem has already been envisaged when writing eq. (IX-3-14) and its consequences. Here we will study the initial build-up period of the enstrophy cascade, and then show how one can derive the law $k_i(t) \sim (vt)^{-1}$ which underlies the Batchelor law (IX-3-14).

From eq. (IX-3-7) the enstrophy can only decay or remain constant, and is thus upper bounded by its initial value $D(0)$. Consequently, eq. (IX-3-6) implies that the rate of dissipation of kinetic energy will always tend to zero with the viscosity: hence the kinetic energy of two-dimensional turbulence is conserved for any time at vanishing viscosity, at variance with the result of finite dissipation at finite time occurring in the case of the closures applied to three-dimensional turbulence.

The question now arises of the time-evolution of the enstrophy, which, because of (IX-3-7), obliges one to look at the palinstrophy: from the expression (IX-3-19) of the kinetic energy transfer, and the calculations of Pouquet et al. (1974), the *E.D.Q.N.M.* evolution equation for the palinstrophy is found, with our notations, to be

$$\frac{dP(t)}{dt} = \frac{4}{\pi} \int_0^\infty \int_0^\infty p^4 q^2 E(p,t)E(q,t)\theta_{kpq}A(p,q)dpdq$$

$$- 2\nu \int_0^{+\infty} k^6 E(k,t)dk \qquad (IX-3-29)$$

with

$$A(p,q) = \begin{cases} (\pi/2)[1 - (q/p)^2], & \text{if } q \le p \\ (\pi 2)[1 - (q/p)^2](p/q)^2, & \text{otherwise.} \end{cases}$$

We may then symmetrize the non-linear term of (IX-3-29) with respect to p and q, and obtain

$$\frac{dP(t)}{dt} = 2 \int_0^{+\infty} dp \int_0^p q^2(p^4 - q^4)E(p,t)E(q,t)\theta_{kpq} \, dq$$

$$- 2\nu \int_0^{+\infty} k^6 E(k,t)dk \quad . \qquad (IX-3-30)$$

Since θ_{kpq} is majored by $(\mu_k + \mu_p + \mu_q)^{-1}$, itself majored by μ_q^{-1}, where μ_k is given by (VIII-3-9) with $a_1 = 0.4$, the non linear term of the r.h.s. of (IX-3-30) is upper bounded by

$$2 \int_0^{+\infty} dp \int_0^p p^4 q^2 E(p,t)E(q,t)\mu_q^{-1} dq =$$

$$\frac{4}{a_1} \int_0^{+\infty} dp \, p^4 E(p,t)[\int_0^p q^2 E(q,t)dq]^{\frac{1}{2}} \le \frac{4}{a_1}P(t)D(t)^{\frac{1}{2}} \ .$$

Finally (IX-3-30) leads to the following inequality

$$\frac{dP(t)}{dt} \le \frac{4}{a_1}P(t)D(t)^{\frac{1}{2}} - 2\nu \int_0^{+\infty} k^6 E(k,t)dk \qquad (IX-3-31)$$

which reduces to

$$\frac{dP(t)}{dt} \le \frac{4}{a_1}P(t)D(t)^{\frac{1}{2}} \qquad (IX-3-32)$$

in the limit of a zero viscosity. The palinstrophy majoration corresponding to (IX-3-32) is

$$P(t) \le P(0)\exp\frac{4}{a_1}D(0)^{\frac{1}{2}}t \qquad (IX-3-33)$$

and is better than the one obtained in Pouquet et al. (1975) which involved an exponential of t^2. The conclusion is, as already stressed in Pouquet et al. (1975), that for any fixed t the palinstrophy, even high, remains bounded when the viscosity goes to zero. This implies from (IX-3-7) that, in the same conditions, the enstrophy dissipation rate goes to zero, and the enstrophy is conserved for any time. There is therefore no finite inviscid enstrophy dissipation at a finite time, similar to the finite kinetic energy dissipation of three-dimensional turbulence.

It would however be an error to think that two-dimensional turbulence does not dissipate enstrophy when the viscosity is small but finite. This would contradict in particular the t^{-2} enstrophy dissipation law derived from (IX-3-14), and we stress again that the important property of two-dimensional turbulence is the finite dissipation of enstrophy. Actually, the *E.D.Q.N.M.* or direct calculations of freely-evolving two-dimensional turbulence all show that the palinstrophy starts increasing, following approximately an exponential of the type (IX-3-33), reaches a maximum, and then decreases under the action of viscosity. The maximum corresponds to a time t_c where the whole span of the energy spectrum, from $k_i(0)$ to $k_d(t_c)$ has been filled up. The enstrophy dissipation wave number $k_d(t_c)$ is defined from (IX-3-16), with $\beta = t_c^{-3}$:

$$k_d(t_c) = (\nu t_c)^{-1/2} \quad .\qquad (IX-3-34)$$

The palinstrophy $P(t_c)$ is thus approximately equal to

$$P(0)\exp\frac{4}{a_1}D(0)^{1/2}t_c$$

from the inviscid stage described by (IX-3-33), and to $t_c^{-2}k_d(t_c)^2$ from the enstrophy cascade assumption. Noticing finally that the total kinetic energy $(1/2)v^2$ is proportional to $t_c^{-2}k_i(t_c)^{-2}$, and the initial palinstrophy to $k_i(0)^4 v^2$, one obtains approximately

$$t_c \approx D(0)^{-1/2}\ln\frac{k_d(t_c)}{k_i(t_c)}\qquad (IX-3-35)$$

which indicates that the "critical" time t_c is proportional to the logarithm of the Reynolds number (IX-3-17). It is only after t_c that the enstrophy will be dissipated at a finite rate $\sim t^{-2}$. The existence of such a time was first conjectured by Tatsumi and Yanase (1981), and its correct determination $\propto \ln R$ was given by Basdevant (1981): before t_c, the palinstrophy grows "inviscidly" according to (IX-3-33), and the enstrophy should decrease like

$$D(t) = A - \nu B\exp[C\ D(0)^{1/2}t]\qquad (IX-3-36)$$

where A, B, and C are three constants. The enstrophy dissipation rate is thus proportional to $\nu \exp[CD(0)^{1/2}t]$. After t_c, the kinetic energy spectrum decays self-similarly according to (IX-3-14), and the enstrophy dissipation rate is proportional to t^{-3}. The critical time t_c goes to infinity with the Reynolds number, that is when the viscosity goes to zero: in this case, one recovers the above mentioned result of inviscid conservation of enstrophy for any time. Let us mention finally that the inviscid exponential evolution of $P(t)$ has been verified in the direct numerical Euler calculations of Kida and Yamada (1984).

Batchelor's analysis (IX-3-14) is based on a self-similar assumption of the type (VI-6-9), with an integral scale $l \approx k_i(t)^{-1}$ proportional to vt. It is possible to justify such an assumption (Lesieur and Herring, 1985) on the basis of two hypotheses which have been derived from the closures, i.e. the existence of an infrared k^3 energy spectrum, and of a k^{-3} enstrophy cascade. As shown in Lesieur and Herring (1985), the k^3 spectrum for $k \to 0$ will appear whatever the slope s of the initial spectrum, due to the strong k^3 energy transfer in low k. One assumes then that

$$E(k,t) = C(t)k^3, \ k < k_i(t);$$
$$E(k) = \beta^{2/3}k^{-3}, \ k > k_i(t) \ . \qquad \text{(IX}-3-37)$$

The enstrophy is negligibly contributed to by the k^3 range, which yields $\beta^{2/3} \approx D(t)$, and hence respectively a t^{-2} and t^{-3} behaviour for $D(t)$ and β. The total kinetic energy is proportional to $t^{-2}k_i(t)^{-2}$, which yields

$$k_i(t) \sim (vt)^{-1} \qquad \text{(IX}-3-38)$$

and demonstrates Batchelor's law. Finally, matching both spectral ranges at $k_i(t)$ gives

$$C(t) \sim v^6 t^4 \ . \qquad \text{(IX}-3-39)$$

The law (IX-3-38) has been verified by Rhines (1975) in a direct-numerical simulation, with a wave number $< k(t) >$ characteristic of the energy-containing eddies defined by

$$< k(t) > = \frac{\int_0^{+\infty} kE(k,t)dk}{\int_0^{+\infty} E(k,t)dk} \ . \qquad \text{(IX}-3-40)$$

This calculation led to

$$\frac{1}{v}\frac{d}{dt} < k >^{-1} = \frac{1}{30} \qquad \text{(IX}-3-41)$$

which shows that the "doubling time" of the "eddy" $< k >^{-1}$ (that is the time necessary to form an eddy of size $2/< k >$) is, in units of time

$(v <k>)^{-1}$, equal to 30. As stressed by Staquet (1985) on the basis of a similar two-dimensional isotropic calculation, this doubling time is 15 when evaluated with the aid of the wave number $k_i(t)$ where the kinetic energy spectrum peaks.

It has to be stressed that quantitative discrepancies in the transfers exist at moderate Reynolds numbers between the closure predictions and the direct-numerical simulations of two-dimensional isotropic turbulence, as shown by Herring and Kraichnan (1985). This is due to the intermittency caused by the coherent eddies, which are strongly non gaussian. It is however difficult to tell if these discrepancies will be important at very high Reynolds numbers[29]. In this respect, the work of Brachet et al. (1988) showing, at a high resolution, the transition from a k^{-4} to a k^{-3} spectrum is encouraging.

4 - Diffusion of a passive scalar

The two-dimensional passive scalar diffusion is a significant problem when one is interested in the large-scale diffusion of tracers or pollutants in the atmosphere or the ocean, or by temperature fluctuations in two-dimensional laboratory experiments. It also gives information about the way vorticity (or potential vorticity in geostrophic turbulence) is transported by the flow. Since the vorticity and the passive scalar both obey equation (IX-1-4), they have close analogies. However, the scalar, whose variance is an inviscid invariant, is not constrained to the double energy-enstrophy conservation like the velocity field. This will rule out the possibility of strong inverse scalar transfers and inverse scalar cascades.

The phenomenology of the two-dimensional passive scalar problem has been given in Lesieur et al. (1981) and Lesieur and Herring (1985). Let $T(\vec{x}, t)$ be the scalar, $E_T(k,t)$ its spectrum such that

$$\frac{1}{2} < T(\vec{x}, t)^2 >= \int_0^{+\infty} E_T(k, t)dk \quad . \qquad (IX - 4 - 1)$$

$\epsilon_T = \int_0^{+\infty} k^2 E_T(k,t)dk$ is the scalar dissipation rate and κ the scalar diffusivity. The scalar dissipation wave number is now (Lesieur and Herring, 1985)

$$k_c = (\frac{\epsilon_T}{\kappa^3})^{1/6} \quad . \qquad (IX - 4 - 2)$$

The corresponding inertial-convective, inertial-diffusive and viscous-convective ranges are shown in Figure IX-8, taken from Lesieur and Herring

---

[29] Indeed, it is for these situations that the closures have been developed.

(1985): if the scalar is injected in the enstrophy cascade, and for $k <$ inf(k_c, k_d), an Oboukhov type analysis, already employed in the three-dimensional case, shows that the scalar spectrum is proportional to the cascading enstrophy spectrum and is consequently of the form

$$E_T(k,t) \sim \frac{\epsilon_T}{\beta} k^2 E(k,t) \sim \epsilon_T \beta^{-1/3} k^{-1} \qquad (IX-4-3)$$

as also proposed in Mirabel and Monin (1982). In the inertial-diffusive range, it has already been mentioned in Chapter VIII that eq. (VIII-2-20) is still valid, that is

$$E_T(k,t) \sim \epsilon_T \kappa^{-3} k^{-4} E(k,t) \sim \epsilon_T \kappa^{-3} \beta^{2/3} k^{-7} \qquad (IX-4-4)$$

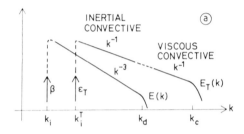

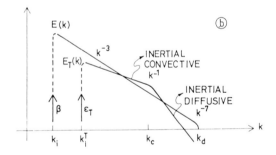

Figure IX-8: schematic inertial-ranges of the kinetic energy and scalar spectra in the enstrophy cascade. The enstrophy is injected at k_i with a rate β, and the scalar at k_i^T with a rate ϵ_T. a) Prandtl number > 1; b) Prandtl number < 1 (from Lesieur and Herring, 1985, courtesy J. Fluid Mech.).

As for the viscous-convective range, it is still proportional to k^{-1}. This phenomenology is supported by an *E.D.Q.N.M.* analysis, whose spectral equation is written (from VIII-3-12 which is still valid)

$$\left(\frac{\partial}{\partial t} + 2\kappa k^2\right) E_T(k,t) = \frac{4}{\pi} \iint_{\Delta_k} dp \, dq \, \theta_{kpq} \frac{p}{q} (1-x^2)^{\frac{1}{2}}$$

$$[kE_T(p,t)E(q,t) - pE(q,t)E_T(k,t)] . \qquad (IX-4-5)$$

Here we have taken the same triple-correlations relaxation time as for the velocity, which is certainly not justified but does not greatly influence the results. The use of a modified θ time of the form θ_{00q}, already considered in the three-dimensional case, has the advantage of allowing a Fourier return in the physical space, yielding then eq. (VIII-3-16) which is still valid (see Larchevêque and Lesieur, 1981, Lesieur and Herring, 1985).

Using the same non-local techniques already employed above, and in Chapters VII and VIII, one can calculate the non-local flux of scalar, equal to

$$\Pi_{NL}^T(k,t) = -\frac{1}{4}\int_0^{ak} \theta_{kkq}\, q^2 E(q)dq\, k^3 \frac{\partial}{\partial k}\frac{E_T(k)}{k}$$

$$+ \int_0^k k'^2 E_T(k')dk' \int_{sup(k,k'/a)}^\infty \theta_{k'pp} E(p)dp$$

$$- \int_0^k k'^3 dk' \int_{sup(k,k'/a)}^\infty \theta_{k'pp}\frac{E(p)E_T(p)}{p}dp \ .(IX-4-6)$$

This expansion shows again, as for the energy spectrum, the existence of a positive k^3 scalar transfer in the infrared region $k \to 0$. The second term in the r.h.s. of (IX-4-6) corresponds to an eddy-diffusivity

$$\kappa_t = \frac{1}{2}\int_{k/a}^\infty \theta_{0pp} E(p)dp \qquad (IX-4-7)$$

which, contrary to the eddy-viscosity, is always positive. The first term in the r.h.s. of (IX-4-6) is identical to the enstrophy flux in the enstrophy cascade if $E_T(k,t)$ is replaced by $k^2 E(k,t)$. This allows one to demonstrate the equality

$$E_T(k,t) = \frac{\epsilon_T}{\beta}E(k,t) \qquad (IX-4-8)$$

in the enstrophy cascade inertial-convective range (Lesieur and Herring, 1985), provided the local scalar transfer may, as for the enstrophy, be neglected[30]. Finally, when the scalar is forced into the $k^{-5/3}$ inverse kinetic energy cascade, it is going to cascade to higher wave numbers along a direct $k^{-5/3}$ inertial-convective range, as shown in Lesieur and Herring (1985).

Curiously, and when compared with the direct-numerical simulations which have been done for this passive scalar problem, the conclusions of this statistical analysis for the k^{-1} inertial-convective range seem to work quite well, though the enstrophy spectra determined in

---

[30] This assumption should however be verified, either with the aid of the *E.D.Q.N.M.* or using direct-numerical simulations.

these (low resolution) computations are steeper than k^{-1} (Babiano et al., 1987). It has also been noticed by Holloway and Kristmannsson (1984) that the scalar is more easily diffused by turbulence than is the vorticity. It would be interesting to verify the discrepancy between the enstrophy and scalar spectra on higher resolution computations. It might be that such an effect results simply from the absence of pressure term in the scalar equation[31], and that, as in the three-dimensional case, the k^{-1} scalar spectra observed in the numerical simulations are the result of a random shearing of the scalar by the large eddies, corresponding to eq. (VIII-3-25).

To end this section, let us consider the diffusion problem from a Lagrangian point of view, by looking for instance at the pair dispersion problem studied in Chapter VIII. As shown in Larchevêque and Lesieur (1981) and Lesieur and Herring (1985) with the aid of the *E.D.Q.N.M.* closure, the probability density $P(r, t)$ that the pair should be a distance $\vec{r}$ apart admits a similarity solution of the form

$$P(r, t) = R^{-2} f(\frac{r}{R}) \qquad (IX - 4 - 9)$$

where R is the r.m.s. pair separation. R satisfies a Richardson law $(dR^2/dt) \sim \epsilon^{1/3} R^{4/3}$ in the inverse energy cascade, and in the enstrophy cascade the dispersion law

$$\frac{1}{2} \frac{dR^2}{dt} \sim \beta^{1/3} R^2 \qquad (IX - 4 - 10)$$

which was proposed originally by Lin (1972) on a phenomenological basis. Such a law leads to an exponential dispersion for a stationary forced enstrophy cascade, but the derivation of Larchevêque and Lesieur (1981) leading to eqs (IX-4-9) and (IX-4-10) is not valid when the turbulence is decaying (which implies $\beta^{1/3} \sim t^{-1}$) if R^{-1} and $k_i(t)$ are not proportional. This situation occurs in a passive scalar *E.D.Q.N.M.* calculation done by Montmory (1986) for a decaying turbulence, where the scalar integral scale $1/k_i^T$ still increases exponentially when it corresponds to enstrophy-cascading eddies ($k_i^T > k_i$). The scalar variance decays then much more rapidly than the enstrophy. This is at variance with one of the conclusions of Lesieur and Herring (1985), stressing a t^{-2} scalar decay, and confirms the already-mentioned fact that the scalar is more efficiently diffused by two-dimensional turbulence than the enstrophy. Again, this might be due to the strong deformation of the scalar by the

---

[31] There is no pressure in the vorticity equation as well, but the deformation of vorticity by the velocity reacts on the velocity field and may inhibit the vorticity diffusion, compared with the scalar diffusion.

large-scale velocity gradient, and the absence of pressure in the scalar equation.

Eq (IX-4-10) also implies that the second-order velocity structure function is proportional to $\beta^{2/3} r^2$. It is then tempting to relate such a dispersion law (resp. structure function) to the kinetic energy spectrum, as was done for three-dimensional turbulence when the Richardson law was equivalent to the Kolmogorov law for the kinetic energy spectrum. This poses however some problems, as stressed by Babiano et al. (1985), where it has been shown that a r^2 structure function could correspond to spectra decreasing faster than k^{-3}. Thus, the determination of diffusion or dispersion statistics in a two-dimensional turbulent flow, as was for instance experimentally done for the EOLE experiment in the atmosphere (see Morel and Larchevêque, 1974), could not be sufficient to characterize the flow dynamically. It seems finally that, from the two-dimensional numerical simulation point of view, the vorticity gives a very efficient characterization of the large structures of the flow, as attest for instance the isovorticity lines of the mixing-layer calculations presented in Chapters I and III. Low pressure regions give also an accurate representation of the coherent vortices[32]

5 - Geostrophic turbulence

This is a turbulence satisfying the quasi-geostrophic potential vorticity conservation equation[33]. The main effects which can cause the dynamics of such a turbulence to differ from two-dimensional turbulence are the density stratification (responsible for baroclinic effects), the β-effect (responsible for Rossby wave propagation), the bottom topography, the dissipation by an Ekman layer, and possibly the existence of a finite external Rossby radius of deformation[34]. It is, of course, difficult to study the influence of these physical effects all together, and preferable to consider separately the influence of each on two-dimensional turbulence:

The influence of topography in geostrophic flows has been studied extensively, and a review can be found in Verron and Le Provost (1985): in order to conserve its potential vorticity $(\omega + f)/h$, a flow of initially zero relative vorticity will react to the decrease of h due to a topography of positive height by creating negative relative vorticity. This explains

---

[32] In three dimensions, the representation of iso-surfaces of various vorticity components, associated with vortex lines and pressure contours, allow to determine the topology of large and small-scale vortices.

[33] It may also be referred to as *quasi-geostrophic turbulence*.

[34] The effect of the latter is weak, and will not be considered here.

the trapping of anticyclonic vortices by a topography. The same reasoning would lead to the formation of cyclonic vortices over a trough. When the topography is complex and chaotic-like, numerical simulations of decaying geostrophic turbulence show that its spatial features eventually lock on the topography (see Holloway, 1986). We recall also the analogy between topographic and differential rotation Rossby waves.

The dissipation by a bottom Ekman layer has been seen to damp linearly the vorticity. This large-scale dissipation can limit or even prevent the formation of the inverse energy cascade, as occurs for instance in the Earth's atmosphere.

The interaction of Rossby waves with two-dimensional turbulence has been studied by Rhines (1975) with the aid of direct-numerical simulations, and by Holloway and Hendershott (1977), Legras (1980) and Bartello and Holloway (1990) using stochastic closures. An important length, introduced by Rhines (1975), corresponds in the potential vorticity equation to a balance between the relative vorticity $\omega = \partial v/\partial x - \partial u/\partial y$ and the βy term due to the differential rotation. This length is

$$l_R = (\frac{2U}{\beta})^{1/2} \qquad\qquad (IX - 5 - 1)$$

where U is a characteristic scale of turbulence. At scales smaller than l_R, $\beta y << \omega$, and the turbulence is not affected by the Rossby waves: if one imagines for instance a kinetic energy forcing at scales much smaller than l_R, the inverse energy cascade is going to build isotropic eddies up to the scale $\approx l_R$, where the Rossby waves start propagating. As shown by Rhines (1975), the wave propagation will induce a strong anisotropy by blocking the growth of the structures in the meridional direction, and hence producing elongated structures in the zonal direction[35]. This has been proposed as a possible explanation for the zonal jets observed on Jupiter (Williams, 1978, 1979) and perhaps Saturn: indeed the width of these jets is about 20000 km on Jupiter, of the same order as the Rhines's length which can be approximately measured. Such an explanation supposes however that there exists an energy forcing at scales smaller than 20000 km: this forcing is certainly not of baroclinic origin, since measurements have shown there is no meridional temperature gradient on Jupiter, but could be due to agitation from below arising from the turbulent thermal convection caused by the energy source in the interior of the planet. The situation would thus be a spherical-geometry analogue of the rotating turbulence experiment developed by Hopfinger

---

[35] This may be associated with the inhibition of the pairing in a mixing layer submitted to a differential rotation, when β exceeds a critical value, mentioned in Chapter III: thus, eq. (IX-5-1) is a statistical analogue of eq. (III-2-18) if it is replaced by an equality

et al. (1982), where quasi-two-dimensional eddies are forced by three-dimensional turbulence from below: if this conjecture happened to be true, Jupiter would be another example of a large-scale organization due both to a fixed scale (the Rhines length) and a small-scale agitation.

Another theory of Jupiter's zonal structure, put forward by Busse (1983), proposes that it is a surface manifestation of internal thermal convection columns parallel to the axis of rotation (see Figure IX-9). This poses however some problems, since the existence of such columns would imply meridional transports at the surface which are not apparent on the pictures of the planet.

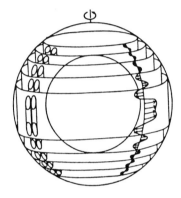

Figure IX-9: convective columns obtained in a rapidly rotating spherical atmosphere such as Jupiter, and the associated surface zonal flows, from Busse (1983) (courtesy Geophys. Astrophys. Fluid Dyn.)

In the Earth's atmosphere and oceans in medium latitudes, the Rhines's length is of 1000 km and 100 km respectively. This is of the order of the internal radius of deformation where the kinetic energy is provided by baroclinic instability. In the atmosphere, where this forcing can be considered as approximately stationary, this could be one of the reasons for the non existence of a $k^{-5/3}$ inverse energy cascade in the planetary scales[36].

A theory of geostrophic turbulence including baroclinic effects has been proposed by Charney (1971), with the concept of potential enstrophy cascade generalizing the two-dimensional enstrophy cascade: one considers the potential vorticity conservation equation (IX-2-27), and rescale x and y by the internal radius of deformation, z by the vertical height of the layer, and t using the horizontal scale and velocity.

---

[36] Another one being the Ekman dissipation in the atmospheric boundary layer.

Forgetting about the differential rotation term βy, one obtains

$$\frac{D_H}{Dt}(\frac{\partial^2\psi}{\partial x^2} + \frac{\partial^2\psi}{\partial y^2} + \frac{\partial^2\psi}{\partial z^2}) = \frac{1}{Re}\nabla_H^2(\nabla_H^2\psi) \ , \qquad (IX-5-2)$$

where Re is a horizontal turbulent Reynolds number representing the "sub-geostrophic diffusion" of potential vorticity already discussed above. The rescaled stream function $\psi(x, y, z, t)$ characterizes the horizontal velocity through its horizontal derivatives $\partial\psi/\partial y$ and $-\partial\psi/\partial x$, and the density through $\partial\psi/\partial z$. One can consider a random homogeneous three-dimensionally isotropic stream function, whose three-dimensional spatial Fourier transform is $\hat{\psi}(k_1, k_2, k_3, t)$. The potential enstrophy is defined by

$$D_p = \frac{1}{2} < (\frac{\partial^2\psi}{\partial x^2} + \frac{\partial^2\psi}{\partial y^2} + \frac{\partial^2\psi}{\partial z^2})^2 > \qquad (IX-5-3)$$

and is conserved with time [37] (modulo the horizontal turbulent diffusion). The other quadratic invariant is the total energy

$$E_t = \frac{1}{2} < (\frac{\partial\psi}{\partial x})^2 + (\frac{\partial\psi}{\partial y})^2 + (\frac{\partial\psi}{\partial z})^2 > \qquad (IX-5-4)$$

sum of the mean horizontal kinetic energy

$$\frac{1}{2} < (\frac{\partial\psi}{\partial x})^2 + (\frac{\partial\psi}{\partial y})^2 >$$

(barotropic energy) and of the mean "available potential energy"

$$\frac{1}{2} < (\frac{\partial\psi}{\partial z})^2 >$$

(baroclinic energy). Let $\hat{\Psi}(k, t)$ such as

$$< \hat{\psi}(\vec{k}', t)\hat{\psi}(\vec{k}, t) >= \hat{\Psi}(k, t)\delta(\vec{k} + \vec{k}') \qquad (IX-5-5)$$

$$E(k, t) = 2\pi k^4 \hat{\Psi}(k, t) \ . \qquad (IX-5-6)$$

Thus the total energy is $\int_0^{+\infty} E(k, t)dk$, and the potential enstrophy is $\int_0^{+\infty} k^2 E(k, t)dk$. As for two-dimensional turbulence, both quantities are conserved by the non-linear terms of the equations, the difference being here that $E(k, t)$ is a three-dimensional spectrum. Nevertheless, the conclusions are the same: if potential enstrophy is injected at a

---

[37] The conservation holds also for the squared potential vorticity spatially averaged on any horizontal plane.

given wave number k_i at a rate β, it will cascade towards large wave numbers along a $\beta^{2/3} k^{-3}$ potential enstrophy cascade. Since isotropy is assumed, this cascade will correspond to a simultaneous k^{-3} cascade of the enstrophy of the horizontal motion (barotropic enstrophy) and of the density $\partial \psi / \partial z$ (baroclinic enstrophy). The injection wave number corresponding to the alimentation of this potential enstrophy cascade will have non-dimensional components of the order of one, since r_I turns out to be the scale where the horizontal kinetic energy is, through the baroclinic instability, fed into the system from the available potential energy. Such a conversion curbs the development of the isotropy in scales larger than r_I: the behaviour in these scales has been studied by Salmon (1978) and Hoyer and Sadourny (1982) with a two-layer geostrophic model to which the *E.D.Q.N.M.* was applied. Such studies provide a stochastic version of the baroclinic instability: for instance, Hoyer and Sadourny (1982) show that if the energy is fed into the system under a baroclinic form at a scale[38] larger than r_I, it will cascade down a $k^{-5/3}$ spectrum[39] up to the internal radius of deformation, where a part of it will be transformed into barotropic energy: the potential enstrophy will cascade to higher wave numbers under a double barotropic and baroclinic form, according to Charney's theory; the barotropic energy produced at r_I^{-1} will cascade back to smaller wave numbers along a $k^{-5/3}$ inverse energy cascade, in the opposite direction of the baroclinic energy. The situation for the $k^{-5/3}$ inverse horizontal kinetic energy cascade is thus the same as for the two-dimensional turbulence, where we have shown that the temperature flux is positive in the inverse energy cascade. Therefore, the horizontal kinetic energy of geostrophic turbulence can be approximated by that of a two-dimensional turbulence which would be alimented in energy at the internal radius of deformation, at a rate corresponding to the conversion of baroclinic into barotropic energy.

Recent multi-layer numerical simulations of geostrophic turbulence with a high vertical resolution (up to 7 layers) (Hua and Haidvogel, 1986) have confirmed Charney's isotropy assumption, and shown also that a large part of the geostrophic dynamics is captured by two-layer models. Geostrophic turbulence, especially with a small number of vertical modes, is a very useful model towards understanding the physics of rapidly-rotating stably-stratified flows. Nevertheless, such an approximation discards a lot of important physical effects, such as the interaction of turbulence with the internal gravity waves considered in Chapter II on the basis of the Boussinesq approximation. It seems therefore that

---

[38] This scale, determined by the differential heating between the equator and the pole, is about 3000 km for the Earth's atmosphere, while one recalls that the internal radius of deformation is about 1000 km.

[39] k is here a horizontal wave number

a deeper understanding of the atmospheric and oceanic dynamics will require the recourse to three-dimensional equations such as Boussinesq approximation or full Navier-Stokes equations[40]. This will be rendered easier with the significant present development of scientific calculation facilities.

5.1 Rapidly-rotating stratified fluid of arbitrary depth

When deriving the Proudman-Taylor theorem in Chapter III-5, the same expansion of the velocity fields in terms of the Rossby number, assumed to be small, would have led for the Navier-Stokes equation (II-5-7) to the geostrophic balance

$$\frac{1}{\rho_0}\vec{\nabla}P = -2\vec{\Omega} \times \vec{u}_{2D}^{(0)} \quad , \qquad (IX-5-7)$$

where P is the pressure modified by gravity and centrifugal forces. Hence the leading-order two-dimensional solution obtained through the Proudman-Taylor analysis is geostrophic, but, because of eq. (III-5-4), this flow obeys a two-dimensional Navier-Stokes equation only if $\partial v_3/\partial z = 0$. One may look now at the combined effect of a strong rotation and stratification (stable or unstable) in a fluid of finite depth: one considers the Boussinesq approximation, with a hydrostatic basic state, with an arbitrary gravity distribution (the gravity includes centrifugal effects). The vorticity equation writes, from eqs (II-8-9) and (II-8-11), and including viscous dissipation

$$\frac{\partial\vec{\omega}}{\partial t} + \vec{u}.\vec{\nabla}\vec{\omega} = \vec{\omega}.\vec{\nabla}\vec{u} + 2\Omega\,\frac{\partial\vec{u}}{\partial z} + \vec{\nabla}\rho_* \times \vec{g} + \nu\nabla^2\vec{\omega} \quad . \qquad (IX-5-8)$$

As in Chapter III-5, we take U, D and D/U as velocity, length and time units. We normalize $\vec{\nabla}\rho_* \times \vec{u}$ by a typical value of $|\vec{g}.\vec{\nabla}\rho_*|$. Hence the vorticity equation in non-dimensional variables is

$$\frac{\partial\vec{\omega}}{\partial t} + \vec{u}.\vec{\nabla}\vec{\omega} = \vec{\omega}.\vec{\nabla}\vec{u} + R_o^{-1}\,\frac{\partial\vec{u}}{\partial z} - |R_i|\vec{\nabla}\rho \times \vec{\alpha} + R_e^{-1}\nabla^2\vec{\omega} \quad , \quad (IX-5-9)$$

where $R_o = U/2\Omega D$ is the Rossby number, $R_e = UD/\nu$ the Reynolds number, and

$$R_i = \vec{g}.\vec{\nabla}\rho_*\,\frac{D^2}{U^2} \quad , \qquad (IX-5-10)$$

the Richardson number. The unit vector $\vec{\alpha}$ is directed in the direction opposite to $\vec{g}$. Hence, for strong rotation and stratification ($R_o << 1$ and

---

[40] The use of the so-called "primitive equations", which are not geostrophic but where the hydrostatic approximation is done along the vertical, may not be sufficient, since they filter out internal-gravity waves.

$|R_i| \gg 1$), the R_o^{-1} and R_i terms dominate, and the vorticity equation reduces to (in dimensional variables)

$$\frac{\partial \vec{u}}{\partial z} = \frac{1}{2\Omega} \vec{g} \times \vec{\nabla} \rho_* \quad .$$

$$(IX - 5 - 11)$$

This is a generalization of the thermal-wind equation (IX-2-16). The latter is recovered if $\vec{g}$ is parallel to $\vec{\Omega}$. In this case, the consideration of the velocity equation shows that the horizontal flow is geostrophic, since $\vec{g}$ is vertical. In the more general case where $\vec{g}$ is non uniform (due for instance to a spherical repartition of mass, as in a star, and to centrifugal acceleration), the geostrophic balance is lost. For a basic density distribution with spherical symmetry for instance, no thermal wind is induced by eq. (IX-5-11)[41], as noted by Carrigan and Busse (1983).

---

[41] since $\vec{g}$ and $\vec{\nabla}\rho_*$ are aligned

Chapter X

ABSOLUTE EQUILIBRIUM ENSEMBLES

1 - Truncated Euler Equations

In this chapter, we will consider a turbulence within a box (that is to say which can be expanded into an infinite serie of discrete wave vectors $\vec{k}_n$ with velocity amplitudes $\hat{\underline{u}}(\vec{k}_n, t)$, as introduced earlier). Let us first consider the Navier Stokes equations in Fourier space, relating this infinite set of modes. The equation is truncated by retaining the modes lower than a cutoff wave number k_{max}, and properly normalized:

$$(\frac{\partial}{\partial t} + \nu k^2)\hat{u}_i(\vec{k}, t) = -\frac{i}{2}P_{ijm}(\vec{k}) \sum \hat{u}_j(\vec{p}, t)\hat{u}_m(\vec{q}, t) \qquad (X-1-1)$$

where the sum $\sum$ only keeps the modes such that $\vec{k} = \vec{p} + \vec{q}$ and whose modulus is smaller than k_{max}. If k_{max} is superior or equal to the Kolmogorov wave number (in three dimensions), it is expected that the truncated equations will correctly represent the turbulence, and this is precisely what is done in the so-called direct-numerical simulations of turbulence. If this condition is not fulfilled, and if we start initially with an energy spectrum sharply peaked at an initial energetic wave number k_0 (in a freely decaying turbulence), eq. (X-1-1) will properly describe the evolution of turbulence in the early stage, when the cascade has not yet reached k_{max}. But as soon as the Kolmogorov energy cascade forms (that is, from Chapter VII, at a time of about 5 initial large-eddy turnover times), energy will tend to go beyond k_{max} in the dissipative scales, which is not permitted by the truncated equation (X-1-1). Instead, energy will then accumulate at k_{max}, which may be a source of numerical instability, and in any case does not give an acceptable description of turbulence (cf Figure X-1). It is therefore a major difficulty for the numerical simulations in this case (called *Large-Eddy Simulations*, as already mentioned before) to model the kinetic energy transfer towards the *subgrid scales* $k > k_{max}$. This problem will be looked at in Chapter XII.

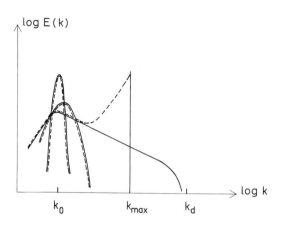

Figure X-1: schematic comparison of the evolution of a kinetic energy spectrum obeying Navier-Stokes equations (straight line) and truncated Navier-Stokes equations (dashed line).

In this chapter, we consider truncated Euler equations

$$\frac{\partial}{\partial t}\hat{u}_i(\vec{k},t) = -\frac{i}{2}P_{ijm}(\vec{k})\sum \hat{u}_j(\vec{p},t)\hat{u}_m(\vec{q},t) \qquad (X-1-2)$$

with the same conditions $k,p,q \leq k_{max}$. This system does not provide an accurate description of Euler equations, since the latter require an infinite set of wave numbers, except in the early stage of the evolution described in Figure 1. It will however be seen that the study of the equilibrium statistical mechanics of such a system will provide useful information for the study of turbulence, especially in two-dimensions. Further details on this analysis can be found for instance in Rose and Sulem (1978) and Orszag (1970 a-b, 1977).

2 - Liouville's theorem in the phase space

Let us first consider the three-dimensional case: to each mode $\hat{\underline{u}}(\vec{k}_n,t)$ one associate two real vectors, the real and imaginary part, which are both in the plane perpendicular to $\vec{k}_n$ and can therefore be represented each by their two real components in that plane; let

$$y_{n_1}(t), y_{n_2}(t), y_{n_3}(t), y_{n_4}(t)$$

be these real components. One can remark that

$$\hat{\underline{u}}(\vec{k}_n, t)^* . \hat{\underline{u}}(\vec{k}_n, t) = \sum_{i=1}^{4} y_{n_i}(t)^2 \quad . \tag{X-2-1}$$

Then if N wave vectors are retained in the truncation, the system can be represented by a point of $m = 4N$ coordinates $y_{n_i}(t)$ (i from 1 to 4) in a phase space determined by $y_a(t)$ (a from 1 to $4N$). In a similar way to what has been done in Chapter VI-2, it is easy to show that eq. (X-1-2) conserves the kinetic energy

$$\frac{1}{2} \sum_{\vec{k}_n} \hat{\underline{u}}(\vec{k}_n, t)^* . \hat{\underline{u}}(\vec{k}_n, t) = \frac{1}{2} \sum_{a=1}^{m} y_a(t)^2 \tag{X-2-2}$$

and then that the system evolves in the phase space on a sphere of radius determined by the initial kinetic energy. From (X-1-2) a system describing the evolution of the $y_a(t)$ can be written

$$\frac{d}{dt} y_a(t) = \sum_{b,c=1}^{m} A_{abc}\, y_b(t) y_c(t) \tag{2-3}$$

where the A_{abc} are complicated coupling coefficients which do not need to be written here: indeed, the kinetic energy conservation (X-2-3) implies, if only three "modes" (a, b, c) are considered, a detailed conservation property

$$A_{abc} + A_{bca} + A_{cab} = 0 \quad . \tag{X-2-4}$$

As mentioned in Rose and Sulem (1978), it can be shown also that A_{abc} is zero as soon as two of the indices (a, b, c) are equal. This allows to write an "incompressibility" condition for the generalized velocity field of components $dy_a(t)/dt$ in the phase space:

$$\frac{\partial}{\partial y_a} \frac{dy_a(t)}{dt} = 0 \quad . \tag{X-2-5}$$

The following analysis can be made either by considering the *micro-canonical ensemble* (systems all lying on a given energy sphere)[1] or the

---

[1] Such an analysis has been carried out by Basdevant and Sadourny (1975) in the case of two-dimensional turbulence.

canonical ensemble (systems of arbitrary energy). Here we adopt the latter point of view, considering in the phase space a collection of systems of density $\rho(y_1, ...y_m, t)$. Since the total number of systems is evidently preserved in the motion, and the volumes are preserved as well, a generalized continuity equation can be written

$$\frac{D\rho}{Dt} = \frac{\partial\rho}{\partial t} + \sum_{a=1}^{4N} \frac{dy_a}{dt} \frac{\partial\rho}{\partial y_a} = 0 \qquad (X-2-6)$$

which is Liouville's theorem for the problem considered.

We now look for *equilibrium solutions* of eq. (X-2-5), that is solutions $P(y_1, ...y_m)$ which do not depend explicitly on t. This is the case for any function of a conserved quantity. In particular the kinetic energy $(1/2) \sum_{a=1}^{m} y_a^2$ is conserved with the motion. By analogy with the statistical thermodynamics at the equilibrium, one will consider the particular Boltzmann-Gibbs equilibrium distribution

$$P(y_1, ...y_m) = \frac{1}{Z} \exp -\frac{1}{2}\sigma \sum_{a=1}^{m} y_a^2 \qquad (X-2-7)$$

where Z is the partition function of the system

$$Z = \int\!\!\int .. \int \exp -\frac{1}{2}\sigma \sum_{a=1}^{m} y_a^2 \; dy_1 dy_2 ... dy_{4N} \qquad (X-2-8)$$

and σ a positive constant[2]. Eq. (X-2-7) can be interpreted as a Gaussian probability distribution of the systems in the phase space. One then assumes that the ensemble average $< \rho(y_1, ...y_m, t) >$ of an ensemble of given systems $\rho(y_1, ...y_m, t)$ obeying eqs (X-2-3) and (X-2-6) will eventually relax towards the distribution (X-2-7). Such a behaviour has been numerically checked by Lee, J. (1982). This allows us to calculate the "mean energy" of the mode "a":

$$<y_a(t)^2> = \frac{1}{Z} \int\!\!\int .. \int y_a^2 [\exp -\frac{1}{2}\sigma \sum_{b=1}^{4N} y_b^2] dy_1 ... dy_m \qquad (X-2-9)$$

---

[2]   In statistical thermodynamics, σ is related to the temperature by the Boltzmann relation $\sigma = 1/KT, T$ being the temperature and K the Boltzmann constant. So in our problem, σ characterizes the inverse of the "temperature" of the inviscid truncated system. It has of course no relation with the physical temperature of turbulence.

which turns out to be independent of a. There is then equipartition of energy between the modes $a = 1, ..., 4N$, and also between the wave vectors $\vec{k}_n$. The kinetic energy spectrum, proportional to

$$2\pi k^2 < \vec{\tilde{u}}(\vec{k}, t)^* . \vec{\tilde{u}}(\vec{k}, t) >,$$

is then proportional to k^2, and we obtain a spectrum of equipartition of kinetic energy among the modes. This result is from Lee, T.D. (1952), and expresses in some way for the energy spectrum an accumulation of energy at the maximum wave number, characteristic of the ultraviolet energy transfers of three-dimensional turbulence (Figure X-2).

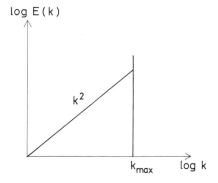

Figure X-2: equipartition kinetic energy spectrum of truncated three-dimensional Euler equations.

One might object to this analysis the fact that the kinetic energy is not the only invariant, and that the conservation of helicity might alter the equipartition k^2 spectrum: such a study has been made by Kraichnan (1973), but the qualitative conclusions (ultraviolet energy transfer) are not fundamentally modified. As stressed in Rose and Sulem (1978), this is no longer true for three-dimensional strong $M.H.D.$ helical turbulence, where a study of the equilibrium ensemble solutions (Frisch et al., 1975) led to the conjecture of an inverse cascade of magnetic helicity[3], which was later on verified in the frame of the $E.D.Q.N.M.$ closure (Pouquet et al., 1976).

---

[3] correlation of the vector potential with the magnetic field, and analogous to the kinetic helicity if the magnetic field is replaced by the vorticity

3 - The application to two-dimensional turbulence

The best way to build the phase space is to use the stream function $\hat{\psi}(\vec{k}, t)$, the y_a being now its real or imaginary parts. One has here two invariants to consider, the kinetic energy

$$\frac{1}{2} \sum_{\vec{k}_n} k_n^2 |\hat{\psi}(\vec{k}_n, t)|^2 = \frac{1}{2} \sum_{a=1}^m k_a^2 y_a(t)^2 \qquad (X - 3 - 1)$$

and the enstrophy

$$\frac{1}{2} \sum_{\vec{k}_n} k_n^4 |\hat{\psi}(\vec{k}_n, t)|^2 = \frac{1}{2} \sum_{a=1}^m k_a^4 y_a(t)^2 \qquad (X - 3 - 2)$$

which implies that the system moves on the intersection of the kinetic energy sphere and the enstrophy ellipsoid. The invariant to consider is a linear combination of the kinetic energy and enstrophy, and the equilibrium distribution (X-2-7) has now to be replaced by (Kraichnan, 1967, 1975 b)

$$P(y_1, ... y_m) = \frac{1}{Z} \exp -\frac{1}{2} \sum_{a=1}^m k_a^2 (\sigma + \mu k_a^2) y_a^2 \qquad (X - 3 - 3)$$

where σ and μ are two constants. Again the validity of such a distribution has been verified by Fox and Orszag (1973) and Basdevant and Sadourny (1975). The "mean" stream function variance in the mode "a" is now

$$<y_a(t)^2> = \frac{1}{Z} \iint \cdot\cdot \int y_a^2 \exp -\frac{1}{2} \sum_{b=1}^m k_b^2 (\sigma + \mu k_b^2) y_b^2 \; dy_1 ... dy_m$$

which reduces to

$$<y_a(t)^2> = \frac{1}{Y} \int_{-\infty}^{+\infty} y^2 \exp -\frac{1}{2} k_a^2 (\sigma + \mu k_a^2) y^2 \; dy \qquad (X - 3 - 4)$$

with

$$Y = \int_{-\infty}^{+\infty} \exp -\frac{1}{2} k_a^2 (\sigma + \mu k_a^2) y^2 \; dy \quad . \qquad (X - 3 - 5)$$

With the change of variable

$$y' = k_a(\sigma + \mu k_a^2)^{1/2} y \quad ,$$

(X-3-4) and (X-3-5) can easily be calculated, to yield

$$<y_a(t)^2> = [k_a^2(\sigma + \mu k_a^2)]^{-1} \qquad (X-3-6)$$

and hence

$$<|\hat\psi(k)|^2> = \frac{2}{k^2(\sigma + \mu k^2)} \qquad (X-3-7)$$

$$E(k) = \pi k^3 <|\hat\psi(k)|^2> \sim \frac{k}{\sigma + \mu k^2} \quad . \qquad (X-3-8)$$

The constants σ and μ can be determined from (X-3-8) in terms of the mean kinetic energy and enstrophy. If σ and μ are both positive, the wave number $(\sigma/\mu)^{1/2}$ can easily be shown to be of the order of the wave number

$$k_i = \left(\frac{\int_0^\infty k^2 E(k) dk}{\int_0^\infty E(k) dk} \right)^{1/2}$$

characteristic of an average wave number of the spectrum (Basdevant and Sadourny, 1975). Then if k_{max} is much larger than k_i, the energy spectrum will be a k^{-1} enstrophy equipartioning[4] spectrum for $k >> k_i$ and a k energy equipartition spectrum for $k \to 0$. The possibility of a negative σ (with a positive μ) is permitted if there exists a lower wave number bound k_{min} such that

$$k_{min} > (\frac{-\sigma}{\mu})^{1/2} \quad .$$

Therefore, and if k_{min} is very close to $(-\sigma/\mu)^{1/2}$, one can obtain an arbitrarily high kinetic energy spectrum in the vicinity of k_{min}. This is what Kraichnan called *negative temperature* states", since the "temperature" of the system is characterized by σ, which is here negative. Such a behaviour could of course be considered as an indication of the inverse energy cascade in the viscous problem.

Let us now consider the two-dimensional passive scalar: since it possesses only one quadratic invariant (the scalar variance), the same analysis will lead to equipartition of scalar variance among the wave vectors $\vec{k}$, and thus to a scalar spectrum proportional to k. If one is in the case where σ is positive, the scalar spectrum and the kinetic energy spectra will be both proportional to k in low wave numbers, while the enstrophy spectrum will be $\propto k^3$. Consequently, there will be more scalar than enstrophy in this infrared spectral range: such a behaviour was verified by Holloway and Kristmannsson (1984) with an

---

[4] indeed the equipartition of enstrophy yields $k^2 E(k) \propto k$

inviscid truncated direct numerical simulation, and shows the dynamical difference of the scalar and the enstrophy in this range.

4 - Two-dimensional turbulence over topography

One must first stress that the existence of a differential rotation (β-effect) does not modify the results of section 3, since the conservation of kinetic energy and enstrophy is not changed. It yields then isotropy, in contradiction with the anisotropic effects of differential rotation displayed in the viscous calculations of Rhines (1975) and Bartello and Holloway (1990)[5]: this is an example where the predictions of inviscid truncated systems are at variance with the real viscous flows dynamics.

An interesting problem, the solution of which is given in Holloway (1986), is that of a geostrophic turbulence over a random topography: from (IX-2-31), the potential vorticity is[6] $-\nabla^2\psi + h$, where $h(x,y)$ is the topography, properly normalized. Let $\hat{h}(\vec{k})$ be the spatial Fourier transform of $h(\vec{x})$, and let h_a be the real and imaginary parts of the various $\hat{h}(\vec{k}_n)$. With the same notations as in Section 3, the quadratic invariants are the kinetic energy (X-3-1) and the potential enstrophy

$$\frac{1}{2}\sum_{\vec{k}_n}|k_n^2\hat{\psi} + \hat{h}|^2 = \frac{1}{2}\sum_{a=1}^{m}(k_a^2 y_a + h_a)^2 \quad . \qquad (X-4-1)$$

The probability distribution corresponding to a linear combination of the kinetic energy and potential enstrophy is then

$$P(y_1,...y_m) = \frac{1}{Z}\exp-\frac{1}{2}\sum_{a=1}^{m}k_a^2(\sigma + \mu k_a^2)y_a^2 + 2\mu k_a^2 h_a y_a + \mu h_a^2 \quad .$$

This permits one to calculate $<y_a^2>$, equal to

$$<y_a(t)^2> = \frac{1}{Y'}\int_{-\infty}^{\infty} y^2 \exp-\frac{1}{2}[k_a^2(\sigma + \mu k_a^2)y^2 + 2\mu k_a^2 h_a y + \mu h_a^2] \; dy$$

$$(X-4-2)$$

with

$$Y' = \int_{-\infty}^{\infty} \exp-\frac{1}{2}[k_a^2(\sigma + \mu k_a^2)y^2 + 2\mu k_a^2 h_a y + \mu h_a^2] \; dy \quad . \quad (X-4-3)$$

---

[5] More precisely, it was shown by Shepherd (1987) that the ergodic property was lost if β is high.

[6] within the rigid-lid approximation

Expressing

$$k_a^2(\sigma + \mu k_a^2)y^2 + 2\mu k_a^2 h_a y + \mu h_a^2$$

under the form

$$k_a^2(\sigma + \mu k_a^2)[(y + \mu \frac{h_a}{\sigma + \mu k_a^2})^2 + \sigma\mu \frac{h_a^2}{k_a^2(\sigma + \mu k_a^2)^2}]$$

and with the change of variable

$$y' = y + \mu \frac{h_a}{\sigma + \mu k_a^2},$$

(X-4-2) can be written as

$$<y_a(t)^2> = \frac{1}{Y} \int_{-\infty}^{+\infty} y^2 \exp{-\frac{1}{2}k_a^2(\sigma + \mu k_a^2)y^2} \; dy$$

$$+ \mu^2 \frac{h_a^2}{(\sigma + \mu k_a^2)^2} \qquad\qquad (X-4-4)$$

where Y is still given by (X-3-5). Then, if one considers the topography as random, the modal vorticity variance at the wave vector $\vec{k}$ is

$$<\hat{\omega}(-\vec{k})\hat{\omega}(\vec{k})> = k^4 <|\hat{\psi}(k)|^2> = 2\frac{k^2}{\sigma + \mu k^2} + \mu^2 k^4 \frac{<\hat{h}(-\vec{k})\hat{h}(\vec{k})>}{(\sigma + \mu k^2)^2}$$

which shows that the resulting flow is the superposition of the classical two-dimensional turbulence equilibrium solution found in the preceding section, and of a contribution due to the topography. This topographic component can be clearly understood when looking at the vorticity topography correlation $<\hat{\omega}(-\vec{k})\hat{h}(\vec{k})>$ which turns out to be equal to $2 <h_a y_a>$ for an arbitrary a. The same changes of variable as above yield

$$<h_a y_a> = \frac{h_a}{Y} \int_{-\infty}^{+\infty} (y - \mu \frac{h_a}{\sigma + \mu k_a^2}) \exp{-\frac{1}{2}k_a^2(\sigma + \mu k_a^2)y^2} \; dy$$

$$= -\mu \frac{h_a^2}{\sigma + \mu k_a^2} \qquad\qquad (X-4-5)$$

and hence

$$<\hat{\omega}(-\vec{k})\hat{h}(\vec{k})> = - <\hat{h}(-\vec{k})\hat{h}(\vec{k})> \frac{k^2}{\sigma + \mu k^2} \; . \qquad\qquad (X-4-6)$$

The topographic component of the vorticity is then locked on to the topography, with a sign corresponding to the potential vorticity conservation phenomenology, that is in the anticyclonic direction above a positive topography (compression of relative vortex tubes), and in the cyclonic direction above a trough (stretching of relative vortex tubes).

This is a quite remarkable result, which has been observed also in the direct-numerical simulations in the viscous case (see Holloway, 1986). It seems therefore one might oppose a *maximum entropy principle* (leading to the Boltzmann-Gibbs equilibrium distribution), which yields in this case[7] the formation of organized eddies locked to the topography, to the *minimum enstrophy principle* of Bretherton and Haidvogel (1976, see also Leith, 1984) which assumes that the flow is going to evolve under the action of viscosity towards a state of minimal enstrophy. Though the example of inviscid truncated solutions for two-dimensional turbulence above topography is quite impressive, it has to be remembered that turbulence is an essentially *dissipative* phenomenon (of enstrophy in two dimensions), and that results closer to reality will certainly be obtained with theories allowing dissipation to act. But absolute equilibrium ensembles are a useful qualitative tool to explore the direction of the transfers among the various scales of motion.

---

[7] where one recalls that the system is *non dissipative*

Chapter XI

THE STATISTICAL PREDICTABILITY THEORY

1 - Introduction

The concept of predictability has been widely used throughout the preceding chapters, and is one of the major ingredients in our definition of turbulence. When turbulence is examined from the point of view of the "chaos" in dynamical systems, the predictability is studied by looking at the sign of the Liapounov exponent l, with $d(t) \propto e^{lt}$, where $d(t)$ is an appropriate measure of the distance between two figurative points initially very close in the phase space: a positive exponent means then an exponential separation, and implies loss of predictability.

When developed (and hence *dissipative*) turbulence is envisaged, the concept of predictability is a priori more vague, since no evident phase space then exists. It took about fifteen years for this concept to find a satisfactorily mathematical formulation, through the works of P.D. Thompson, E.A. Novikov, E.N. Lorenz, J. Charney, C.E. Leith and R.H. Kraichnan for instance. The reader is referred to the paper of Thompson (1984) for a historical overview of the subject: the problem was pointed out by meteorologists concerned with the amplification in the forecast models, as time was going on, of the errors contained in the initial conditions. The latter are due to the inaccuracy of the measurements and interpolation of the observing net data. Let us quote Thompson (1984): *"Suppose that the prediction model were perfect, and that the model equations could be integrated without error. Then there would still be a practical limitation on the accuracy of prediction, owing to the fact that the initial analysis is subject to (...) errors (...). The working hypothesis was (...) that the growth of error was a distinctively nonlinear phenomenon and that (...) the (...) prediction model (...) would amplify errors"*.

At the beginning of these studies, the problem was posed more as a stability problem, where one looks at the evolution of a "perturbation" (actually the initial departure from the real flow) superposed to the flow one desires to forecast. Then a statistical formalism was proposed by Novikov (1959), and used in the Fourier space by Lorenz (1969), Leith (1971) and Leith and Kraichnan (1972): instead of considering one given realization of the flow $\vec{u}^{(\alpha)}(\vec{x},t)$ where α stands for the particular realization of the random field studied, perturbed by $\delta\vec{u}^{(\alpha)}(\vec{x},t)$, one considers two ensembles of flows defined by the random functions

$$\vec{u}_1(\vec{x},t) = \{\vec{u}^{(\alpha)}(\vec{x},t)\} \qquad\qquad (XI-1-1)$$

$$\vec{u}_2(\vec{x},t) = \{\vec{u}^{(\alpha)}(\vec{x},t) + \delta\vec{u}^{(\alpha)}(\vec{x},t)\} \ . \qquad (XI-1-2)$$

The statistical formalism restricts the study to random functions $\vec{u}_1$ and $\vec{u}_2$ having the same statistical properties. This implies in particular that $<\delta\vec{u}>$ is zero, and that the two random flows have the same spectrum. The statistical predictability will study the statistical properties of $\delta\vec{u} = \vec{u}_1 - \vec{u}_2$. Of particular interest will be the relative energy of the error

$$r(t) = \frac{<(\vec{u}_1 - \vec{u}_2)^2>}{2 <\vec{u}_1^2>} \qquad\qquad (XI-1-3)$$

also called the error rate: when the two fields are initially very close, $|\delta\vec{u}^{(\alpha)}| << |\vec{u}^{(\alpha)}|$ in each of the realizations α considered, and $r(0)$ is consequently much smaller than 1. The two fields are then almost completely correlated. On the contrary, if the error between the two fields grows in such a way that they become decorrelated (and hence that $<\vec{u}_1.\vec{u}_2>= 0$), one will have $r = 1$.

In the Fourier space, and for isotropic turbulence, one considers the kinetic energy spectrum $E(k,t)$ of $\vec{u}_1$ and $\vec{u}_2$, such that

$$\frac{1}{2} <\vec{u}_1^2(\vec{x},t)>= \int_0^{+\infty} E(k,t)\ dk \qquad (XI-1-4)$$

$$E(k,t) = (D-1)\pi k^{D-1}\hat{U}(k,t) \qquad\qquad (XI-1-5)$$

where $\hat{U}(k,t)$ is the trace of the spectral tensor, spatial Fourier transform of

$$<u_{1_i}(\vec{x},t)u_{1_j}(\vec{x}+\vec{r},t)>$$

(or of the spectral tensor built with $\vec{u}_2$). D is the dimension of space (2 or 3). In the same way one can introduce the *correlated energy spectrum*

$$E_W(k,t) = (D-1)\pi k^{D-1}\hat{W}(k,t) \qquad\qquad (XI-1-6)$$

where $\hat{W}(k,t)$ is the trace of the tensor Fourier transformed of

$$<u_{1_i}(\vec{x},t)u_{2_j}(\vec{x}+\vec{r},t)> \quad ,$$

and such as

$$\frac{1}{2} <\vec{u}_1(\vec{x},t).\vec{u}_2(\vec{x},t)> = \int_0^{+\infty} E_W(k,t)\ dk \quad . \qquad (XI-1-7)$$

It is very easy to check that the spectrum of the error energy $(1/2) < (\vec{u}_1 - \vec{u}_2)^2 >$ is $2E_\Delta(k,t)$ with

$$E_\Delta(k,t) = E(k,t) - E_W(k,t) \quad . \qquad (XI-1-8)$$

$E_\Delta(k,t)$ is called the *decorrelated energy spectrum*, or *error spectrum*. The error rate defined in (XI-1-3) is thus equal to

$$r(t) = \frac{\int_0^{+\infty} E_\Delta(k,t)dk}{\int_0^{+\infty} E(k,t)dk} \quad . \qquad (XI-1-9)$$

Due to a Schwarz inequality, $|E_W(k,t)| < E(k,t)$, and $E_\Delta(k,t)$ is positive and majored by $2E(k,t)$. The case $E_\Delta(k) = E(k)$ corresponds to a complete decorrelation between the two fields. Although it is not mathematically ruled out, the situations with $E_\Delta(k) > E(k)$ are physically irrealistic, since they would correspond to a correlation of the field $\vec{u}_1$ with $-\vec{u}_2$.

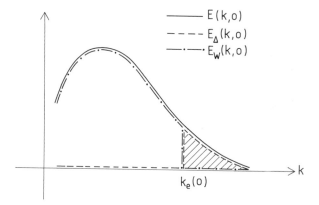

Figure XI-1: initial kinetic, correlated and decorrelated energy spectra in a typical statistical predictability study where complete uncertainty (corresponding to the shaded area) is assumed at wave numbers $k > k_e(0)$. $k_e(0)$ characterizes the initial front of the error.

Because of their meteorological motivations, the people who initiated these studies were mainly concerned with two dimensional or quasi geostrophic turbulence. But the predictability problem may be posed for three-dimensional turbulence as well, and has important implications concerning the relevance of the large-eddy simulations of these flows (see next chapter). Thus we will consider here both cases of three and two dimensions. On the other hand, the numerical prediction frame in which the predictability problem is posed usually leads one to assume that the error between the two flows $\vec{u}_1$ and $\vec{u}_2$ is confined to the small scales corresponding to the inaccuracy of the initial state or the smallest resolved scale in the numerical model: this will imply an initial error spectrum equal to zero for $k < k_e(0)$, and comprised between 0 and $E(k,0)$ for $k > k_e(0)$. The corresponding correlated energy spectrum is equal to $E(k,0)$ for $k < k_e(0)$, and comprised between $E(k,0)$ and 0 for $k > k_e(0)$. An example of an initial situation is shown on Figure XI-1. The questions which are posed concern the time evolution of the wave number $k_e(t)$ characterizing the "front" of the error spectrum, and of $r(t)$. A decrease of $k_e(t)$ as well as an increse of $r(t)$ will mean loss of predictability. Notice that in some of the situations envisaged below, the wave number $k_e(t)$ may be hard to define, and thus $r(t)$ seems to give the best measure of the error. Notice also that, even if it has been less investigated, the statistical predictability problem could also be envisaged with an initial error affecting the energy containing eddies. We will return to that point later. Notice finally that the statistical predictability problem has nothing to do with the problem where the kinetic energy spectrum of turbulence is perturbed by an energy supply: in the case where for instance a small scale kinetic energy peak is superposed at a wave number k_P upon the three dimensional Kolmogorov inertial range, it is easy to show with the aid of dimensional arguments or via the two-point closures that the peak will spread out through the whole spectrum in a few turnover times $[k_P^3 E(k_P)]^{-1/2}$ of the turbulence at the wave number k_P. In the predictability problem on the contrary, the kinetic energy spectrum $E(k,t)$ is the same for the two velocity fields considered.

In this chapter (as in the preceding ones) we will mainly be concerned with the closure results, particularly closures of the $E.D.Q.N.M.$ family, and will summarize the results obtained by Leith (1971), Leith and Kraichnan (1972), Métais (1983) and Métais and Lesieur (1986) for three and two-dimensional isotropic turbulence. Let us also mention the two-dimensional Quasi Normal study of Lorenz (1969) which concluded to an upper limit of 8 days for the deterministic forecast of the Earth's atmosphere. We must point out the pioneering study of Lilly (1973), who studied the statistical predictability problem with the aid of two-dimensional direct-numerical simulations.

The Chapter is organized as follows: Section 2 will give the $E.D.Q.$-$N.M.$ statistical predictability equations. Sections 3 and 4 will present respectively the three-dimensional turbulence and two-dimensional turbulence results.

2 - The $E.D.Q.N.M.$ predictability equations

At the present point of progress of this book, the reader might not welcome new lengthy calculations giving the $E.D.Q.N.M.$ spectral equations for $E(k,t)$ and $E_W(k,t)$. We will therefore skip such a derivation, which is at hand for any reader having mastered the $E.D.Q.N.M.$ techniques described in Chapters VII and VIII. The essence of the calculation is to write the Navier-Stokes equations in Fourier space simultaneously for $\hat{u}_1(\vec{k},t)$ and $\hat{u}_2(\vec{k},t)^*$, in order to obtain for the correlated spectral tensor an evolution equation which can be written formally as

$$(\frac{\partial}{\partial t} + 2\nu k^2) <\underline{\hat{u}}_1(\vec{k},t).\underline{\hat{u}}_2(\vec{k},t)^*> = <\hat{u}_1\hat{u}_1\hat{u}_2> + <\hat{u}_2\hat{u}_2\hat{u}_1>$$

$$(XI-2-1)$$

where the r.h.s. of (XI-2-1) involves triple moments of the two fields. Then the quasi-normal procedure applies at the level of the equation for the triple moments (see Lorenz, 1969). A new problem which arises with the eddy-damping is the choice of the triple-correlations relaxation time, which might differ from the time θ_{kpq} introduced in Chapter VII, since it now involves two *distinct* velocity fields. In fact all the studies quoted above have chosen the same time. With this choice, the resulting equations for $E(k,t)$ and $E_\Delta(k,t)$ read (see Métais and Lesieur, 1986)

$$(\frac{\partial}{\partial t} + 2\nu k^2)\, E(k,t) = \iint_{\Delta_k} dp\, dq\, S_E(k,p,q) + F(k) \quad (XI-2-2)$$

$$(\frac{\partial}{\partial t} + 2\nu k^2)\, E_W(k,t) = \iint_{\Delta_k} dp\, dq\, S_W(k,p,q) + F(k) \quad (XI-2-3)$$

$$S_E(k,p,q) = A_D(k,p,q)\, \theta_{kpq}\, a(k,p,q)\, k^{D-1}E(p)E(q)$$
$$- \frac{1}{2}b(k,p,q)\, p^{D-1}E(q)E(k) - \frac{1}{2}b(k,q,p)\, q^{D-1}E(p)E(k)$$

$$(XI-2-4)$$

$$S_W(k,p,q) = A_D(k,p,q)\, \theta_{kpq}\, a(k,p,q)\, k^{D-1}E_W(p)E_W(q)$$
$$- \frac{1}{2}b(k,p,q)p^{D-1}E(q)E_W(k) - \frac{1}{2}b(k,q,p)q^{D-1}E(p)E_W(k)$$

$$(XI-2-5)$$

$$A_3(k,p,q) = \frac{k}{pq} \; ; \; A_2(k,p,q) = \frac{2}{\pi}\frac{k^2}{pq} \quad . \qquad (\text{XI} - 2 - 6)$$

In these equations, D is still the dimension of the space (3 or 2). The coefficients $a(k,p,q)$ and $b(k,p,q)$ have already been introduced in Chapters VII (three dimensions) and IX (two dimensions). The kinetic energy transfers corresponding to (XI-2-4) are of course the same as those arising in eqs (VII-3-9) (three dimensions) and (IX-3-19) (two-dimensions), but put under a slightly different form closer to the expression of the correlated energy transfer (XI-2-5). $F(k)$ comes from a possible forcing term $\hat{\underline{f}}(\vec{k},t)$ in the r.h.s. of the Navier-Stokes equations written in Fourier space: the modelling of this forcing through the closure has been already envisaged in Chapter VII-4, and we recall the main result: when a spectral tensor trace evolution equation of the form (VI-2-1) is written, a force-velocity correlation $< \hat{\underline{f}}(\vec{k},t).\hat{u}(\vec{k},t)^* >$ appears. If the random function $\vec{f}(\vec{x},t)$ is a gaussian white noise with respect to the time, such that

$$< \hat{\underline{f}}(\vec{k}',t).\hat{\underline{f}}(\vec{k},t) >= \hat{F}(k) \; \delta(\vec{k}+\vec{k}') \; \delta(t-t') \qquad (\text{XI} - 2 - 7)$$

and statistically independent of the initial velocity field, then it may be shown (Lesieur et al., 1971)[1] that the forcing term in the r.h.s. of (XI-2-2) is

$$F(k) = 2\pi k^2 \; \hat{F}(k) \; (D=3); \; F(k) = \pi k \; \hat{F}(k) \; (D=2) \quad . \quad (\text{XI} - 2 - 8)$$

This analysis is also valid for the correlations $< f \; u_2 >$ and $< f \; u_1 >$ if the same forcing functions are taken for the two fields $\vec{u}_1$ and $\vec{u}_2$. This is what we will call *correlated forcing*, which yields the same forcing term $F(k)$ in the r.h.s. of eqs (XI-2-2) and (XI-2-3). As in the above chapters, emphasis will be given to the decaying cases where $F(k)$ is zero.

The above equations can be solved numerically with the same techniques that are described in Chapter VII. This was done in Leith (1971) and Leith and Kraichnan (1972) assuming a stationary kinetic energy spectrum, and in Métais and Lesieur (1986) in a decaying turbulence: most of the results quoted in the next two sections come from these references. Let us also mention the *E.D.Q.N.M.* predictability study performed for a quasi-geostrophic two-layer model by Herring (1984).

---

[1] The model used in this reference was the *R.C.M.*; but the analysis can be generalized to Markovian models.

3 - Predictability of three-dimensional turbulence

Let us assume that the Reynolds number of the turbulence is high enough so that a $k^{-5/3}$ Kolmogorov inertial range exists. Let $k_e(t)$ be a wave number over which most of the error is confined, and such that complete uncertainty exists for $k > k_e(t)$ $(E_\Delta(k) = E(k))$. A non-local analysis of the equation for $E_\Delta(k)$ derived from (XI-2-2) and (XI-2-3) leads, for $k < k_e(t)$ to (Leith and Kraichnan, 1972, Métais and Lesieur, 1986):

$$\frac{\partial}{\partial t} E_\Delta(k,t) = \frac{14}{15} k^4 \int_{k_e(t)}^{\infty} \theta_{kpp} \frac{E^2(p,t)}{p^2} dp \qquad (XI-3-1)$$

which reduces, due to the inertial range expression of the energy spectrum, to

$$\frac{\partial}{\partial t} E_\Delta(k,t) \sim \epsilon \, k_e^{-5}(t) \, k^4 \quad . \qquad (XI-3-2)$$

This shows in particular that the error spectrum will be $\propto k^4$ for $k < k_e$. It is then possible to develop an analysis resembling the one we did in Chapter VIII for the passive scalar when it was injected in the small scales of turbulence: assuming that

$$E_\Delta(k,t) = C_\Delta(t) \, k^4; \quad k \to 0 \qquad (XI-3-3)$$

and matching the error and energy spectra at k_e yields

$$C_\Delta(t) \sim \epsilon^{2/3} k_e^{-17/3} \qquad (XI-3-4)$$

while (XI-3-2) writes

$$\frac{dC_\Delta(t)}{dt} \sim \epsilon \, k_e^{-5} \quad . \qquad (XI-3-5)$$

For a stationary kinetic energy spectrum where ϵ is a constant, it is then easy to show that the scale k_e^{-1} follows an analogous Richardson law

$$\sigma_\Delta = \frac{1}{2} \frac{d}{dt} k_e^{-2}(t) \sim \epsilon^{1/3} k_e^{-4/3} \qquad (XI-3-6)$$

which leads to

$$k_e^{-2/3}(t) \sim \epsilon^{1/3} t + k_e^{-2/3}(t_0)$$

where t_0 is the time at which the error has been injected in the small scales of the inertial range. This shows that k_e is going to decrease following a $(t - t_0)^{-3/2}$ law, as soon as $k_e(t)$ will be sufficiently small compared with $k_e(0)$. The initial location of this wave number can then be forgotten, and the necessary time for the error, starting from very high wave numbers, to reach a given wave number k, is proportional to $\epsilon^{-1/3} k^{-2/3}$, that is the local turnover time of turbulence at k. In an infinite Reynolds number turbulence, it would therefore take a finite time for the error, injected at infinitely large wave numbers, to reach a given mode k.

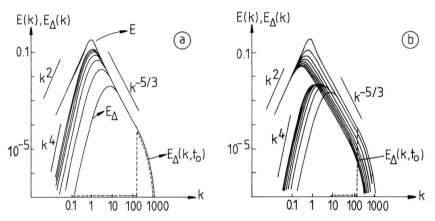

Figure XI-2: inverse cascade of the error spectrum in isotropic turbulence for:
a) a stationary three-dimensional turbulence; b) a decaying three-dimensional
turbulence with $E(k) \propto k^2$, $k \to 0$

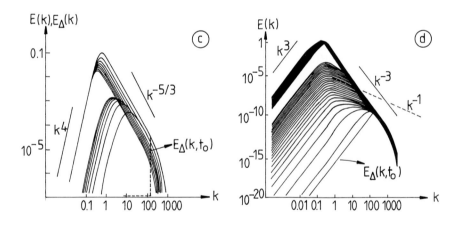

Figure XI-3: inverse cascade of the error spectrum for: c) a decaying three-
dimensional turbulence with $E(k) \propto k^4$, $k \to 0$; d) a decaying two-dimensional
turbulence (from Métais and Lesieur, 1986, courtesy J. Atmos. Sci.)

Figures XI-2-a)b) abd XI-3-c, taken from Métais and Lesieur (1986)
show an *E.D.Q.N.M.* calculation of the inverse error cascade for three
cases: a stationary turbulence forced at a mode $k_i(0)$ (and thus developing a k^2 equipartition energy spectrum at low k), and a decaying energy
spectrum in both situations of a k^2 or a k^4 infrared energy spectrum. The
initial (at $t = t_0$) energy spectrum in the decaying situation (case b) is

identical to the stationary energy spectrum of case a). In the three cases the initial error spectrum is the same, and corresponds to the typical situation described in Figure XI-1. One clearly sees the inverse cascade of error which gradually contaminates larger and larger scales. The time evolution of the error rate $r(t)$ can be understood from the approximate relation

$$r(t) \approx \frac{\int_{k_e}^{\infty} E(k)dk}{\int_{k_i}^{\infty} E(k)dk} \approx (\frac{k_e}{k_i})^{-2/3} . \qquad (XI-3-7)$$

Thus, in the stationary turbulence case, and because of the analogous Richardson law satisfied by $k_e(t)$, $r(t) \propto t$. In decaying turbulence, analytical expressions for k_e (and hence for $r(t)$) can be found in Métais and Lesieur (1986): in this case, $k_e(t)/k_i(t)$ is initially (for t close to t_0) equivalent to $(t-t_0)^{-3/2}$ and $r(t)$ also starts growing like $t-t_0$, then slows down when k_e reaches the vicinity of k_i. The *predictability time*, defined such that $r(t) = 1/2$, is equal respectively to

$$T_{r=1/2} = 17 \ \tau_0; \ \text{case } a$$
$$T_{r=1/2} = 33 \ \tau_0; \ \text{cases } b \text{ and } c \qquad (XI-3-8)$$

where τ_0 is the large eddy turnover time $(<\bar{u}^2>^{1/2} k_i)^{-1}$ of the initial (in cases b and c) energy spectrum. This predictability time is, as already stressed, independent of the initial position of $k_e(0)$ if far enough in the inertial range. This result has important implications for the large eddy simulations of three dimensional turbulence, since it shows in principle that no deterministic numerical simulation is then possible at times greater than $20 \approx 30 \ \tau_0$. We will come back to that in the next chapter on large-eddy simulations. Let us also mention that the statistical predictability of three-dimensional turbulence has been studied by Chollet and Métais (1989) using large-eddy simulations with a spectral eddy-viscosity (see next chapter). The calculations confirm that $k_e(t)$ decreases and $r(t)$ increases with time, but the resolution of the calculation (32^3) modes is too low to confirm the above theoretical predictions. In the same paper, the authors look also at the predictability of a passive scalar.

4 - Predictability of two-dimensional turbulence

Since Chapter IX, we are used to the fact that the dynamics of two-dimensional turbulence may be more intricate than in three dimensions, and this will also be the case for the predictability problem. In particular, even though some of the analytical derivations of Section 3 can be applied in the inverse energy cascade, they have no validity at all in the enstrophy cascade because of the strong non-localness of the transfers. Let us first suppose, on a dimensional ground, that in the enstrophy cascade dk_e/dt is function of β (the enstrophy dissipation rate) and k_e only. This yields

$$\frac{d}{dt}\ln k_e \sim \beta^{\frac{1}{3}} \qquad (XI - 4 - 1)$$

and k_e would then decrease exponentially. The error rate, approximated by $(k_e/k_i)^{-2}$ in a k^{-3} energy spectrum, would increase exponentially as

$$r(t) = r(t_0)\exp\ \frac{t - t_0}{\Lambda \tau_0} \qquad (XI - 4 - 2)$$

where the large eddy turnover time τ_0 is of the order of $\beta^{-1/3}$ and Λ is a constant. Since $r(t_0) \approx [k_e(t_0)/k_i(t_0)]^{-2}$, the predictability time tend to infinity with $k_e(t_0)$ in an infinite k^{-3} enstrophy cascade inertial range.

In fact, $E.D.Q.N.M.$ calculations made in Métais and Lesieur (1986) have shown that, due to the strong non-local error transfers, the exponential decrease of k_e is only valid for the early stages of the evolution, where the error is very quickly transferred from the high wave numbers to the energy containing eddies. This is shown in Figure XI-3-d: in about 5 turnover times of the turbulence, the error spreads out through the whole spectrum, and the situation is as if the error had been injected directly in the energy containing eddies. Then the error rate grows exponentially, still following approximately the law (XI-4-2) (though eq.(XI-4-1) is no more valid)[2]. The value of the constant Λ is equal to 2.6 for a stationary kinetic energy spectrum, and to 4.8 in the decaying turbulence case. The predictability time $T_{r=1/2}$ is now given by

$$T_{r=1/2} = \Lambda\ \tau_0\ \ln[1/2r(t_0)]\ . \qquad (XI - 4 - 3)$$

As shown in Métais and Lesieur (1986), this can be used to estimate the predictability times in the atmosphere and the ocean. In the atmosphere τ_0 is about 1 day: if one takes

$$\frac{k_e(t_0)}{k_i(t_0)} = \frac{100\ km^{-1}}{1000\ km^{-1}} = 10$$

$$r(t_0) = (\frac{k_e}{k_i})^{-2} \approx 10^{-2}\ ;$$

---

[2] The same behaviour for the error spectrum has been found qualitatively by Lilly (1973) in two-dimensional isotropic direct-numerical simulations already quoted above, and in the direct-numerical simulations of the temporal mixing layer done by Lesieur et al. (1988).

then the predictability time would be of 10 or 18 days whether we consider the atmosphere as a stationary turbulence or as a decaying turbulence. These values are in good agreement with predictions based on other predictability theories. In the ocean, where the large-eddy turnover time is about 20 days, the predictability times are then 20 times greater. Notice that the propagation of Rossby waves in the case of turbulence on a β-plane or on a sphere seems to slightly increase the predictability of the flow, as shown by Basdevant et al. (1981).

To finish this Chapter, it is of interest to discuss the signification of these predictability times, based on statistical estimations, with respect to the dynamics of the flow itself when only one realization is considered. It could be an error to consider that, for times greater than the predictability time, the flow is going to look different, and in particular will lose any kind of spatial organization in the case where it would possess some well identified *coherent structures*. In fact, the predictability time could in some occasions (for two-dimensional turbulence for instance) be much shorter than the life time of the organized eddies: this is in no way a contradiction, but simply means that the two flows $\vec{u}_1$ and $\vec{u}_2$ studied in the predictability problem both possess the same sort of spatially organized structures, whose spatial location ("phase") could differ appreciably from one flow to the other. Similar situations occur in meteorology at planetary scales, where one often encounters the same "mushroom" shaped [cyclonic-anticyclonic] dipole structures, whose exact location cannot be predicted accurately, in such a manner that one will never know in advance whether we will fall inside the anticyclonic part (which means fair weather) or the cyclonic one (and then bad weather). Another question which is still widely brought up concerns the role for the predictability played by the actual external forces applied to a real flow, and by the boundaries. It is obvious of course that the models of isotropic three and two-dimensional homogeneous turbulence are so idealized that they cannot pretend to represent the whole reality. Nevertheless these models show that the non-linearities of the Navier Stokes equations may be a source of unpredictability for high Reynolds number flows, and this mechanism has to be included (among others) in the physical principles and processes gear we use to try to understand Fluid Dynamics.

Chapter XII

LARGE EDDY SIMULATIONS

1 - The direct-numerical simulation of turbulence

As already stressed in the previous chapters, there is *a priori* no difficulty in envisaging a numerical solution of the unstationary Navier Stokes equations for rotational flows: the various operators are represented by discrete systems relating the values taken by the velocity or vorticity components, pressure, density, temperature, etc... on a space time grid. This grid may be spatially regular or irregular, with finite differences or finite elements methods[1]. Often an orthogonal decomposition of the flow allows a spectral method to be used (see e.g. Canuto et al., 1987). In two dimensions the use of the stream function permits the elimination of the pressure. It is not the aim of the present monograph to describe the various numerical methods used in the so-called *Numerical Fluid Mechanics*. We will insist rather on the physical limitations which arise when such a simulation is performed on a turbulent flow.

A direct-numerical simulation of a turbulent flow has to take into account explicitly all scales of motion, from the largest (imposed by the existence of boundaries or the periodicities) to the smallest (the Kolmogorov dissipation scale).

We have already seen that the total number of degrees of freedom necessary to represent a turbulent flow through this whole span of scales is of the order of $R_l^{9/4}$ in three dimensions, and R_l^2 in two dimensions, where R_l is the turbulent Reynolds number based on the integral scale of the turbulence. The calculations done in 1986 (time of the completion of the first edition of this book) could very seldom go beyond a resolu-

---

[1] On the latter method applied to fluid dynamics, see Pironneau (1989).

tion of 128^3 degrees of freedom[2] (unless particular symmetry conditions are imposed upon the flow), for a Reynolds number $R_\lambda \approx 80$. Now (in May 1990), the maximum resolution attained in three dimensions is of the order of 300^3 on a CRAY 2 machine. With symmetries, it can reach 864^3, as in the numerical simulation of the Taylor-Green vortex done by Brachet (1990). In general, the Reynolds numbers encountered in natural situations are several orders of magnitude larger: there is no hope in the near future, even with the present unprecedented computer revolution, to envisage for instance a direct numerical simulation of the atmosphere from the planetary scales (several thousands of kilometers horizontally) to the dissipation scale ($1mm$), since it would require about 10^{20} degrees of freedom to put on the computer, all these modes interacting nonlinearly. Things get even worse in a star like the sun. Even in a wind tunnel turbulence of integral scale of $5cm$ and Kolmogorov scale of $0.1mm$, about 10^8 degrees of freedom are needed (5.10^2 in each direction), still beyond the capacity of a CRAY 2 machine. The situation should improve with the development of parallel computers using a large number of processors simultaneously. However, the massively parallel computing techniques applied to fluid dynamics are only in the preliminary stage of their development. We must also mention again that the simulation of the Kolmogorov cascade in the limit of zero viscosity requires an infinite number of degrees of freedom, and is out of reach of any direct-numerical simulation.

The conclusion we draw is that, for a weakly viscous fluid, it is not possible in the near future (and perhaps not in the distant future either) to simulate explicitly all the scales of motion from the smallest to the largest. Generally, scientists or engineers are more interested in the description of the large scales of the flow, which often contain the desired information about turbulent transfers of momentum or heat for example: it is these large scales which will be simulated on the computer. The problem is no longer that of a direct-numerical simulation of turbulence (D.N.S.), but of a large-eddy simulation of turbulence (L.E.S.). These L.E.S., as will be seen, need the representation in some way (at least statistically) of the energy exchanges with the small scales which are not explicitly simulated [3]

---

[2] This corresponds to 128 modes in each direction of the physical space for a three-dimensional flow, or 64 Fourier modes in Fourier space. For a two-dimensional flow, the maximum resolution is of about 1024^2.

[3] It has to be stressed however that in combusting or reacting flows, some important phenomena or reactions take place at the molecular-diffusion scales level. This is a further difficulty for the large-eddy simulations of these flows.

2 - The Large Eddy Simulations

2.1 Large and subgrid scales

Let us first look at the philosophy of large-eddy simulations in the physical space: suppose for the sake of simplification that the numerical
method chosen involves a discretization of the fields on a regular cubic
array of points, Δx being the grid mesh. To the fields defined in the
continuous space $\vec{x}$, one will associate filtered fields (*large-scale fields*),
with the aid of a filter $H_{\Delta x}$ which can be (for instance) a gaussian of
width Δx. The filtered velocity and temperature are thus given by

$$\bar{u}(\vec{x},t) = \vec{u}(\vec{x},t) * H_{\Delta x} \qquad (XII-2-1)$$

$$\bar{T}(\vec{x},t) = T(\vec{x},t) * H_{\Delta x} \qquad (XII-2-2)$$

where $*$ is the convolution product. Let $\vec{u}'$ and T' be the fluctuations
of the actual fields with respect to the filtered fields (density is taken
constant and equal to ρ_0)

$$\vec{u} = \bar{u} + \vec{u}' \; ; \; T = \bar{T} + T' \qquad (XII-2-3)$$

The fields $\vec{u}'$ and T' concern fluctuations at scales smaller than Δx (the
"grid scale"), and will then be referred to as *subgrid-scale fields*. The
application of the filter to the dynamic equations (the temperature being
a passive scalar) yields the equations satisfied by the filtered fields

$$\frac{\partial \bar{u}_i}{\partial t} + \bar{u}_j \frac{\partial \bar{u}_i}{\partial x_j} = -\frac{1}{\rho_0} \frac{\partial \bar{p}}{\partial x_i} + \nu \nabla^2 \bar{u}_i - [\frac{\partial}{\partial x_j}\overline{u_i' u_j'} + ...] \qquad (XII-2-4)$$

$$\frac{\partial \bar{T}}{\partial t} + \bar{u}_j \frac{\partial \bar{T}}{\partial x_j} = \kappa \nabla^2 \bar{T} - [\frac{\partial}{\partial x_j}\overline{T' u_j'} + ...] \qquad (XII-2-5)$$

where it is easy to check that $\bar{u}$ (and hence $\vec{u}'$) are still nondivergent.
These filtered fields do not need to be resolved at scales smaller than
Δx, since they have been constructed in a way to eliminate all the fluctuations under this scale. They can then be properly represented by the
computer. But a new problem arises, since the averaging procedure has
produced in the equations of motion new terms involving the averages
(with respect to the filter chosen) of the subgrid-scale fields. Only the
first of these terms, analogous to the Reynolds stresses (eq. (XII-2-4))
or the turbulent heat fluxes (eq. (XII-2-5)) in non-homogeneous turbulence have been written in eqs (XII-2-4) and (XII-2-5). The other terms
(represented by the dots in these equations) arise from the fact that the
operator "$\bar{\cdot}$" is not idempotent, that is to say $\bar{\bar{f}} \neq \bar{f}$. In any case we

are faced with Navier-Stokes equations for the filtered field (large scales)
modified by supplementary subgrid-scale terms which we do not know.
The reader is referred to Rogallo and Moin (1984) for a review. The
problem of the *subgrid-scale modelling*[4] is then to express the subgrid-
scale terms as functions of the large-scale field. This is in no way an
academic problem: indeed, at least for three-dimensional developed turbulence, we absolutely need in the equations some terms allowing the
kinetic energy in the large scales and to be transferred it towards the
subgrid scales where it will eventually be dissipated by molecular viscosity. In the absence of any subgrid-scale transfer, the energy of the large
scales would tend to equipartition, from the results of Chapter X.

This kind of problem (subgrid-scale modelling) is sometimes referred
to, in mathematics, as a problem of *homogenization*, where the laws gov-
erning a medium are known at a microscopic level, and one seeks evolution laws at a macroscopic level. Here in turbulence, the "microscopic"
level corresponds to the individual fluid particle for which we were able to
write the Navier-Stokes equations (see Chapter II). The "macroscopic"
level corresponds to the filtered field (large scales, or *super-grid scales*.
Let us mention that homogenization methods applied to turbulence have
been developed (see Bègue et al., 1987). They give a more rigorous basis to the concept of eddy-viscosity and diffusivity when there exists a
separation of scales between the large and subgrid scales. But they are
not, up to now, applicable to a turbulence having a continuous distribution of energy from the large to the small scales.

2.2 L.E.S. and the predictability problem

Mathematically, the subgridscale modelling problem is not a well posed
problem, due to propagation to the large scales of the uncertainty contained initially in the subgrid-scales: we suppose that at the initial time
t_0 of the large eddy simulation, the initial flow possesses fluctuations in
the subgrid scales[5]. The large-scale fields initializing this computation ignore totally these small-scale fluctuations, and we are then in a situation
of complete uncertainty at wave numbers greater than $\sim (\Delta x)^{-1}$. Let
us now consider two possible realizations of the flow, identical in the large
scales $> \Delta x$, and completely decorrelated in the subgrid scales $< \Delta x$.
If one accepts the results on the propagation of unpredictability given in
Chapter XI, the difference between the two fields will contaminate the
large scales, and the two flows will in reality separate in these scales.
Now let us assume that we have been able to solve the subgrid-scale
modelling problem, and dispose of "closed" large-scale equations where

---

[4] also called *parameterization of the subgrid scales*

[5] This is not the case if one considers initial energy spectra sharply
peaked at a given mode k_i corresponding to the large scales

everything is expressed in terms of these scales. Then the large-eddy simulation performed on the two fields will be unable to propagate any kind of difference between them. This implies that a large-eddy simulation of turbulence, however good the subgrid-scale modelling may be, will not describe exactly the large-scale evolution from a deterministic point of view, at last for times greater than the predictability time. This point was noted in particular by Herring (1979).

Does that mean that the Large Eddy Simulations are useless at large times? Certainly not: one might interpret then the calculated flow as a different realization of the actual flow, and hope that it could possess at least the same statistical properties (in space or time), and may be the same spatially organized structures, though at a different location from the reality: this last point is of course very frustrating in meteorology, as mentioned already in Chapter XI, but might not be too crucial in a lot of engineering flows, as soon as the statistics resulting from the computed flow[6] are correct.

From these remarks we could try to propose criteria of what one could call a "good" large eddy simulation of turbulence, according to the specifications we require:

●**low grade definition**: the simulation must predict correctly the statistical properties of turbulence (spectral distributions, turbulent exchange coefficients, etc.)

●**high grade definition**: moreover, the simulation must be able to predict the shape and topology (but not the phase) of the organized vortex structures existing in the flow at the scales of the simulation.

There could have been a third definition proposed, namely the objective of predicting exactly the actual flow (in the super-grid scales). This is certainly not possible at large times, due to the inverse cascade of unpredictability mentioned above. Another remark is that the notion of *actual flow* is highly questionable for a turbulent flow in a laboratory or in a wide range of engineering situations, where no repeatable (from a deterministic point of view) realization can be obtained.

3 - The Smagorinsky model

The Smagorinsky model is the one more commonly used in non-homogeneous turbulence. The whole unknown subgrid-scale strain arising in eq. (XII-2-4) is modelled with the aid of an eddy-viscosity, exactly as was done for the Reynolds stresses in non-homogeneous turbulence

---

[6] These statistics will generally be evaluated in the computation with the aid of a spatial averaging.

(see eq. (VI-1-2)). Eqs (XII-2-4) and (XII-2-5) are thus changed into

$$\frac{\partial \bar{u}_i}{\partial t} + \bar{u}_j \frac{\partial \bar{u}_i}{\partial x_j} = -\frac{1}{\rho_0} \frac{\partial \bar{P}}{\partial x_i} + \frac{\partial}{\partial x_j} \left\{ (\nu + \nu_t) \left(\frac{\partial \bar{u}_i}{\partial x_j} + \frac{\partial \bar{u}_j}{\partial x_i} \right) \right\}$$

$$\text{(XII} - 3 - 1)$$

$$\frac{\partial \bar{T}}{\partial t} + \bar{u}_j \frac{\partial \bar{T}}{\partial x_j} = \frac{\partial}{\partial x_j} \left\{ (\kappa + \kappa_t) \left(\frac{\partial \bar{T}}{\partial x_j} \right) \right\} \qquad \text{(XII} - 3 - 2)$$

with the incompressibility condition

$$\frac{\partial \bar{u}_j}{\partial x_j} = 0 \quad . \qquad \text{(XII} - 3 - 3)$$

In eq. (XII-3-1), $\bar{P}$ is a mean pressure modified by a term proportional to the trace of the subgrid-scale strain, and whose exact form is not important since it may be eliminated with the aid of the mean incompressibility condition (XII-3-3). Smagorinsky (1963) proposes to take ν_t and κ_t proportional to

$$\nu_t \sim \kappa_t \sim \Delta x^2 \left[\left(\frac{\partial \bar{u}_i}{\partial x_j} + \frac{\partial \bar{u}_j}{\partial x_i} \right) \left(\frac{\partial \bar{u}_i}{\partial x_j} + \frac{\partial \bar{u}_j}{\partial x_i} \right) \right]^{1/2} \qquad \text{(XII} - 3 - 4)$$

with summation on the repeated indices. Smagorinsky's eddy viscosity has been used with some success for various flows such as the channel flow (Deardorff, 1970, Moin and Kim, 1982), for which it gives qualitatively good results in terms of hairpin coherent structures and alternate longitudinal streaks of slow and fast fluid. However, the validity of Smagorinsky's eddy-viscosity has been doubted by Kim et al. (1987). This might be due to the fact that Smagorinsky's eddy viscosity, built with the aid of the mean deformation, does not take into account local shears in the smallest scales in order to define the subgrid-scale dissipation. More elaborate subgrid-scale models, adopting the second-order one-point closure modelling point of view[7], have been developed by Deardorff (1974, see also Somméria, G., 1976, and Schmidt and Schumann, 1989).

4 - L.E.S. of 3-D isotropic turbulence

This chapter does not claim to give a complete review of the numerous works done on this matter, but rather to present a personal point of view on this question, based on the concepts of spectral eddy-viscosity

---

[7] with evolution equations for the subgrid-scale stresses

and spectral eddy-diffusivity introduced in Chapters VII and VIII. Related to the two definitions given in Section 2, we will try to propose a subgrid-scale procedure for the kinetic energy and temperature fluctuations predicting at least correct spectra and decay laws.

As stressed above, we will not try to distinguish, among the various subgrid-scale terms arising in (XII-2-4) and (XII-2-5), the respective contributions coming from the "Reynolds stress like" term, the so-called "Léonard term", and the "cross term", following the terminology given by Clark et al. (1979). The latter reference is an attempt to evaluate these respective terms by a comparison between a coarse mesh and a fine mesh calculation. Here we prefer to model the subgrid scales as a whole. We will work in the Fourier space, introduce a cutoff[8] wave number $k_C \sim \Delta x^{-1}$ and define the filtered field in Fourier space as

$$\bar{\underline{\hat{u}}}(\vec{k}, t) = \begin{cases} \hat{\underline{u}}(\vec{k}, t), & \text{if } |\vec{k}| \leq k_C; \\ 0, & \text{otherwise.} \end{cases} \qquad (XII-4-1)$$

and the same relation for the temperature $\bar{\hat{T}}$. This is a sharp filter in Fourier space[9]. Most of the following results come from work by Chollet and colleagues. See in particular Chollet and Lesieur (1981, 1982)), Chollet (1983, 1984), Pouquet et al. (1983), and Métais and Lesieur (1990). They use the *E.D.Q.N.M.* as a tool to model the subgrid-scale transfers.

4.1 Spectral eddy-viscosity and diffusivity

Let us first look back at the kinetic energy and temperature transfers given by the *E.D.Q.N.M.* theory in Chapters VII and VIII. One assumes first $k << k_C$. Then the transfers from the super-grid to the sub-grid scales are given by eqs (VII-10-1) (for the kinetic energy) and (VIII-4-6) (for the temperature), slightly modified in the following manner[10]

$$T_{sg}(k, t) = -\frac{2}{15} k^2 \bar{E}(k, t) \int_{k_C}^{\infty} \theta_{0pp} [5E(p, t) + p \frac{\partial E(p, t)}{\partial p}] \, dp \qquad (XII-4-2)$$

$$T_{sg}^T(k, t) = -\frac{4}{3} k^2 \bar{E}_T(k, t) \int_{k_C}^{\infty} \theta_{0pp}^T E(p, t) \, dp \ . \qquad (XII-4-3)$$

---

[8] This wave number has not to be confused with the conductive wave number k_c of the passive scalar problem.

[9] The effects of such a filter on the definition of the eddy viscosity in the physical space have been studied by Leslie and Quarini (1979).

[10] the $O(k^4)$ terms can be neglected if the kinetic energy contained in the subgridscales is small compared with the energy in the large scales. This is true in particular if k_c lies in the Kolmogorov cascade.

This allows us to write the spectral evolution equations for the super-grid scale spectra $\bar{E}(k,t)$ and $\bar{E}_T(k,t)$ ($k < k_C$):

$$(\frac{\partial}{\partial t} + 2\nu k^2)\bar{E}(k,t) = T_{<k_C}(k,t) + T_{sg}(k,t) \qquad (XII-4-4)$$

$$(\frac{\partial}{\partial t} + 2\kappa k^2)\bar{E}_T(k,t) = T^T_{<k_C}(k,t) + T^T_{sg}(k,t) \ , \qquad (XII-4-5)$$

with

$$T_{sg}(k,t) = -2\nu_t^\infty \ k^2 \ \bar{E}(k,t) \qquad (XII-4-6)$$

$$\nu_t^\infty = \frac{1}{15}\int_{k_C}^\infty \theta_{0pp}[5E(p,t) + p\frac{\partial E(p,t)}{\partial p}] \ dp \qquad (XII-4-7)$$

$$T^T_{sg}(k,t) = -2\kappa_t^\infty \ k^2 \ \bar{E}_T(k,t) \qquad (XII-4-8)$$

$$\kappa_t^\infty = \frac{2}{3}\int_{k_C}^\infty \theta^T_{0pp} \ E(p,t) \ dp \ . \qquad (XII-4-9)$$

The supergrid-scale transfers $T_{<k_C}(k,t)$ and $T^T_{<k_C}(k,t)$ correspond to triad interactions whose wave numbers lie in the super-grid range, and hence do not need any modelling, since they can be calculated exactly in the large-eddy simulation.

These asymptotic eddy-viscosities and diffusivities may be calculated analytically if the kinetic energy spectrum is given for $k \geq k_C$. Assuming for instance a $k^{-5/3}$ inertial range at wave numbers greater than k_C, we obtain:

$$\nu_t^\infty = 0.267\left[\frac{E(k_C)}{k_C}\right]^{1/2} \qquad (XII-4-10)$$

$$\kappa_t^\infty = 0.445\left[\frac{E(k_C)}{k_C}\right]^{1/2} \ . \qquad (XII-4-11)$$

The two numerical constants 0.267 and 0.445 were calculated by assuming a Kolmogorov constant of 1.4 in the energy cascade, and adjusting the time $\theta^T_{k,p,q}$ as shown in Chapter VIII, with a choice of the constants corresponding to a turbulent Prandtl number of 0.6. If one considers a kinetic energy spectrum $\propto k^{-m}$ for $k > k_C$, it has been shown by Métais and Lesieur (1990) that the eddy-viscosity given by eq. (XII-4-7) scales as $[E(k_C)/k_C]^{1/2}$ for $m < 3$. For $m > 3$, it scales as $[E(k_C)/k_C]$, and becomes negative for $m > 5$. This might explain the results of Domaradzki et al. (1987), who evaluate this asymptotic eddy-viscosity in a direct-numerical simulation[11], and find negative values: indeed, the

[11] by defining a fictitious cutoff k_C, and evaluating the transfers across k_C

kinetic energy spectrum in such a simulation is much steeper than $k^{-5/3}$. Further interesting information of Métais and Lesieur's (1990) study is that the spectral eddy-diffusivity remains finite and positive when the eddy-viscosity vanishes at $m = 5$.

When k is close to k_C, the above concept of spectral eddy-coefficients can be generalized for a $k^{-5/3}$ inertial range at wave numbers greater than k_C. Following Kraichnan's theory of eddy-viscosity in spectral space (1976), it is possible, with the aid of the *E.D.Q.N.M.* approximation, to calculate the subgrid-scale transfers, corresponding to triadic interactions where at least one of the wave numbers p and q is greater than k_C. This allows us to define two functions $\nu_t(k|k_C)$ and $\kappa_t(k|k_C)$, respectively the eddy-viscosity in spectral space (Kraichnan, 1976) and the eddy-diffusivity in spectral space (Chollet and Lesieur, 1982), such that

$$T_{sg}(k,t) = -2\nu_t(k|k_C) \; k^2 \; \bar{E}(k,t) \qquad (XII-4-11)$$

$$T_{sg}^T(k,t) = -2\kappa_t(k|k_C) \; k^2 \; \bar{E}(k,t) \qquad (XII-4-12)$$

with

$$\nu_t(k|k_C) = K(k/k_C) \; \nu_t^\infty \qquad (XII-4-13)$$

$$\kappa_t(k|k_C) = C(k/k_C) \; \kappa_t^\infty \; . \qquad (XII-4-14)$$

The functions $K(x)$ and $C(x)$ were determined respectively by Kraichnan (1976) and Chollet (1984). They are approximately constant and equal to 1, except in the vicinity of $k/k_C = 1$, where they display a strong overshoot (*cusp*-behaviour), due to the predominance of local transfers across k_C. ν_t^∞ in eq. (XII-4-13) is given by eq. (XII-4-10). κ_t^∞ in eq. (XII-4-14) depends upon the choice of the two *E.D.Q.N.M.* scalar adjustable constants: it is given by eq. (XII-4-11) for the *à la L.H.D.I.A.* case ($a_2 = 0$), the constant 0.445 being changed to 0.801 in the *à la D.I.A.* case ($a_2 = a_3 = a_1$). The turbulent Prandtl number $\nu_t(k|k_C)/\kappa_t(k|k_C)$ depends also on these constants: in the *à la L.H.D.I.A.* case, it remains approximately equal to 0.6, even in the vicinity of k_C; in the *à la D.I.A.* case, it has a plateau value of 1/3, and a cusp close to k_C where it rises to 0.6. We mention finally that the use of the subgrid-scale transfers (XII-4-11) and (XII-4-12) allow one to solve numerically the *E.D.Q.N.M.* kinetic energy and passive scalar equations (VII-3-9) and (VIII-3-14) at zero molecular viscosity and conductivity in the self-similar decaying regime (for $k \leq k_C$), as shown by Chollet and Lesieur (1981, 1982).

4.2 Spectral large-eddy simulations

Let us now come back to the evolution equations (in the spectral space) of the filtered field (for $|\vec{k}| < k_C$)

$$(\frac{\partial}{\partial t} + \nu k^2)\bar{\tilde{u}}_i(\vec{k},t) = t_{<k_C}(\vec{k},t) + t_{sg}(\vec{k},t) \qquad (XII-4-15)$$

$$(\frac{\partial}{\partial t} + \kappa k^2)\bar{\hat{T}}(\vec{k},t) = t^T_{<kc}(\vec{k},t) + t^T_{sg}(\vec{k},t) \qquad (XII-4-16)$$

with the usual distinction between the explicit super-grid transfers, still
calculated by a truncation for $k,p,q \leq k_C$ of the r.h.s. of eqs (IV-2-7)
(IV-2-8), and the unknown subgrid-scale transfers. We propose to model
the latter with the aid of $\nu_t(k|k_C)$ and $\kappa_t(k|k_C)$ determined in eqs
(XII-4-13) (XII-4-14), namely

$$t_{sg}(\vec{k},t) = -\nu_t(k|k_C)k^2\bar{\hat{u}}_i(\vec{k},t) \qquad (XII-4-17)$$

$$t^T_{sg}(\vec{k},t) = -\kappa_t(k|k_C)k^2\bar{\hat{T}}(\vec{k},t) \; . \qquad (XII-4-18)$$

The only justification of this subgrid-scale modelling is that, when one
writes the exact evolution equations for the spectra of $\bar{\hat{u}}$ and $\bar{\hat{T}}$ as
they arise from eqs (XII-4-15) (XII-4-16), one obtains the *E.D.Q.N.M.*
subgrid-scale transfers calculated in (XII-4-13) and (XII-4-14). It is then
natural, if we trust the spectral predictions of the *E.D.Q.N.M.* theory, to use these eddy-viscosity and eddy-diffusivity in spectral space for
subgrid-scale modelling purposes. The main criticisms which may be
brought to this modelling concerns the fact that the same results, as far
as the energetics are concerned, could be obtained by multiplying the
eddy-viscosity $\nu_t(k|k_C)$ by any complex number of modulus 1: indeed,
the interactions with the subgrid scales may affect the phase of the velocity vector $\hat{u}(\vec{k},t)$. However, when $k << k_C$, and due to the separation
of space and time scales, it is difficult (in isotropic turbulence) to believe
that the subgrid scales will instantaneously affect the phase of the velocity field at $\vec{k}$: this would require a finite time, comparable to the time
taken by an error in the small scales to contaminate the large scales in
the predictability problem. On the contrary, the validity of this spectral
eddy-viscosity is questionable if k lies in the vicinity of k_C. But this is
shared with other theories such as the Renormalization Group (Yakhot
and Orszag, 1986) or homogenization (Bègue et al., 1987) techniques.

As already stressed in Chapter VII, the R.N.G. analysis[12] developed by Yakhot and Orszag (1986) yields a k_C-dependent eddy-viscosity,
also proportional to $[E(k_C)/k_C]^{1/2}$ as in eq. (XII-4-10), and without
any cusp. As mentioned in Chapter VII, the constant ν_* in front of
$[E(k_C)/k_C]^{1/2}$ is equal to 0.388, and the Kolmogorov constant found
in the theory is 1.617. Now, let us consider the *E.D.Q.N.M.* eddy-
viscosity with no cusp, and adjust ν_* as proposed by Leslie and Quarini
(1979), by balancing[13] the subgrid-scale flux with the kinetic energy

---

[12] There was an earlier attempt made by Rose (1977) for the passive
scalar problem.

[13] in the inertial range

dissipation rate ϵ in the energy spectrum evolution equation. This yields $\nu_* = (2/3)C_K^{-3/2}$, where C_K is the Kolmogorov constant. A Kolmogorov constant of 1.4 leads to $\nu_* = 0.402$.

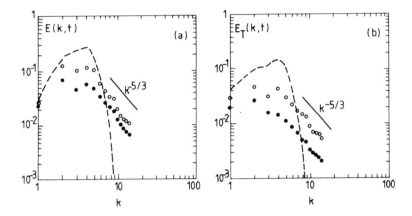

Figure XII-1: decaying kinetic energy (a) and temperature spectra (b) in a large-eddy simulation of three-dimensional isotropic turbulence using a spectral eddy-viscosity and diffusivity. The resolution is 32^3 in a spectral code (from Chollet and Lesieur, 1982, courtesy "La Météorologie")

The results of the *E.D.Q.N.M.* spectral eddy-viscosity applied to three-dimensional isotropic turbulence large-eddy simulations are satisfactory when it is implemented on a spectral code directly in Fourier space and at a low resolution: Figure XII-1, taken from Chollet and Lesieur (1982), shows the decay of the kinetic energy and temperature spectra in a 32^3 modes spectral calculation, starting from initial conditions corresponding to sharply peaked spectra. Kolmogorov and Corrsin-Oboukhov $k^{-5/3}$ cascades establish in times comparable with the critical times estimated in Chapters VII and VIII. Then the spectra decay self-similarly, with $k^{-5/3}$ slopes extending up to k_C. Notice that during the early stage of the calculation, the $\sim [E(k_C)/k_C]^{1/2}$ expression of the eddy-coefficients automatically sets their value to zero as long as the ultraviolet cascade has not reached k_C and that $E(k_C)$ is zero. This is an advantage compared to other eddy-viscosity methods which dissipate the energy more uniformly, since it permits the correct simulation of the inviscid stage preceding the establishment of the Kolmogorov energy cascade. The same calculations allow one to compute the kinetic energy and temperature variance decay, extrapolating the total energy and temperature from the supergrid scales values by assuming that infinite

$k^{-5/3}$ spectra extend beyond k_C. This yields

$$\frac{1}{2} < \vec{u}^2 > = \int_0^{k_C} \bar{E}(k,t)\, dk + \frac{3}{2}\, k_C\, E(k_C,t) \qquad (XII-4-19)$$

and the same expression for the temperature. It is found that both kinetic energy and temperature variance decay as $t^{-1.2}$, according to the Saffman law derived in Chapter VII[14]. This might be due to the coarse resolution of the large-eddy simulation in the low wave numbers, which could yield equipartition of energy in these modes, and hence a k^2 infrared spectrum.

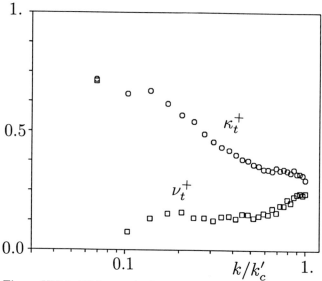

Figure XII-2: 3D isotropic decaying turbulence, resolution 128^3; spectral eddy-viscosity and diffusivity, calculated from the large-eddy simulations of Métais and Lesieur (1990)

However, we have already mentioned in Chapter VII (Figure VII-9) results concerning the same *E.D.Q.N.M.* subgrid-scale modelling[15], but with a higher resolution (128^3, see Lesieur and Rogallo, 1989, and Lesieur et al., 1990). In these calculations, the kinetic energy spectrum at the cutoff is closer to k^{-2} than to $k^{-5/3}$. At this resolution, a calculation (without cusp) with a plateau at 0.402 gives approximately the same tendency (Métais and Lesieur, 1990). We will present below a local generalization of this spectral eddy-viscosity to the physical space,

---

[14] The result is the same for the "super-grid" kinetic energy and temperature variance, corresponding only to modes $< k_C$.

[15] that is, a plateau at 0.267 followed by a cusp

which seems to improve the performances of the model. But beforehand we will look at some anomalous aspects of the spectral eddy-diffusivity, related to the large-scale intermittent character of the passive temperature already mentioned in Chapter VI and VIII.

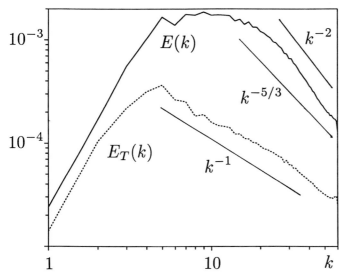

Figure XII-3: kinetic energy and temperature spectra corresponding to the calculation of Figure XII-2

4.3 The anomalous spectral eddy-diffusivity

Figure VIII-5 (from Lesieur and Rogallo, 1989), corresponding to the *E.D.Q.N.M.* large-eddy simulation of Figure VII-9 and a turbulent Prandtl number of 0.6, shows an anomalous large-scale passive scalar range close to k^{-1}, which might be due to the direct shearing of the velocity fluctuations by the large-scale velocity gradients. In these simulations, the spectral eddy-viscosity and diffusivity may be evaluated in the following manner: one defines a fictitious cutoff wave number k'_C (for instance $k'_C = k_C/2$). The transfers between $k < k'_C$ and the range $[k'_C, k_C]$ are evaluated directly in the simulation, while the transfers between k and the range $[k_C, +\infty]$ are calculated analytically using the *E.D.Q.N.M.* approximation.

Figure XII-2, taken from Métais and Lesieur (1990) shows the spectral eddy-viscosity and diffusivity normalized by $[E(k_C)/k_C]^{1/2}$, in a 128^3

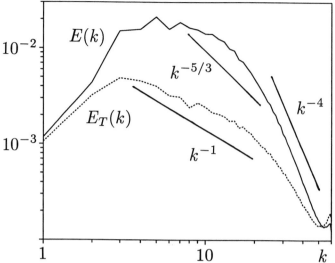

Figure XII-4: 3D isotropic decaying turbulence, resolution 128^3; kinetic energy and scalar spectra, calculated from the direct-numerical simulations of Métais and Lesieur (1990)

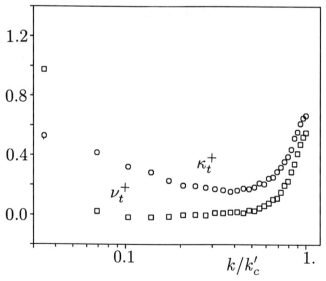

Figure XII-5: same calculation as Figure XII-4; spectral eddy-viscosity and diffusivity

spectral large-eddy simulation of decaying isotropic turbulence[16]. The eddy-viscosity displays a plateau at a value which, once corrected in or-

---

[16] $k_C = 60, k_i = 20, k'_C = 30$, same parameters as for Figure VIII-5. Only the explicit transfers are shown here.

der to take into account the transfers across k_C (see Métais and Lesieur, 1990) is very close to the 0.267 theoretical $E.D.Q.N.M.$ value. The cusp is somewhat eroded, due to the proximity of k_i with respect to k_C. The eddy-diffusivity, on the contrary, has no plateau and decreases logarithmically with k. The same behaviour had been found in the calculations of Lesieur and Rogallo (1989). Figure XII-3 shows the kinetic energy and scalar spectra corresponding to Figure XII-2. Figure XII-4 shows a direct-numerical simulation done by Métais and Lesieur (1990) in the same conditions: the k^{-1} scalar spectral range is still visible, since the shearing due to the large-scale velocity gradients may still act. Figure XII-5 shows the spectral eddy-viscosity and diffusivity corresponding to Figure XII-4: the plateau of the eddy-viscosity is now zero (and even slightly negative as in Domaradzki et al., 1987), but the eddy-diffusivity is non zero with the same logarithmic behaviour.

4.4 Alternative approaches

One may question the necessity of the presence of the "cusp" in functions $K(k/k_C)$ and $C(k/k_C)$ used in the definition of the eddy coefficients. Indeed, as stressed above, it has been checked by Métais and Lesieur (1990) for the velocity that taking $K(x) = 1$ (with a plateau at 0.402) does not modify appreciably the evolution of the energy spectrum. Then eddy viscosity and diffusivity independent of k and scaling on $[E(k_C)/k_C]^{1/2}$ could be sufficient to model this isotropic turbulence at a high Reynolds number. The advantage is that they could also be used in a large-eddy simulation performed in the physical space, provided the energy spectrum at k_C could be determined[17].

Let us also stress that this eddy-viscosity subgrid scale model is completely compatible with the energy conservation constraints arising in the $E.D.Q.N.M.$ equations, since it represents the whole transfers across k_C. These transfers reduce finally to an energy flux from the large eddies to the subgrid scales, which is quite expected in three dimensional isotropic turbulence, and, as already stressed, necessary to simulate the dissipation of the energy by molecular viscosity in the dissipative scales.

The above subgrid-scale modelling (with or without a "cusp") assumes that an inertial range at modes greater than k_C exists. However, this imposes a quite high k_C, and therefore costly calculations. When k_C lies in the energy containing eddies range, the method is certainly no more valid. Other more sophisticated methods, still based on

---

[17] It has been shown by Chollet (1983) that the *cusp* part of the eddy coefficients was close to a k^4 contribution, and would then, back in the physical space, correspond to a dissipation term $\sim \nabla^6 \vec{u}$ in the momentum equation. The latter is anyhow small compared with the $\nu_t \nabla^2 \vec{u}$ contribution of (XII-4-10).

an $E.D.Q.N.M.$ calculation of the transfers, have been developed (see Aupoix and Cousteix, 1982, and Chollet, 1984): it is a coupled method which solves simultaneously the Navier Stokes equations in the large scales, and the $E.D.Q.N.M.$ evolution equation for $E(k, t)$. This enables to recalculate at each step a spectral eddy-viscosity based on the actual flow spectrum, instead of assuming a fictitious Kolmogorov spectrum. The method is nevertheless quite heavy to implement, and it is perhaps not a safe horse to bet on.

An alternative formalism to derive a subgrid-scale modelling in spectral space exists, utilizing the stochastic Langevin model point of view (see e.g. Herring and Kraichnan, 1972). This has been done by Bertoglio (1986) who retrieves the same results as Chollet (1984) in the isotropic case, and shows in the case of a homogeneous turbulence submitted to a constant shear (and thus anisotropic) the appearance of a new class of interactions sending back energy from the subgrid to the super-grid scales. This could then render the use of the isotropic spectral eddy-viscosity questionable in non isotropic situations. Further comparisons between both methods have to be made in order to reach a decision. Let us mention finally the work of Yoshizawa (1982), based on $D.I.A.$ approximation, which allows also an extension of the isotropic eddy-viscosity to anisotropic situations.

These large-eddy simulations allow predictability studies in the isotropic case, in the following manner: Chollet (1984) has considered the large-eddy simulations of two flows $\vec{u}_1$ and $\vec{u}_2$, differing either by the resolution of the calculation (16^3 and 32^3) or by a perturbation involving modes close to k_C. Bertoglio (1986) was able, with the Langevin model method, to let the error come from the subgrid scales to the large scales. Both studies show an inverse cascade of error qualitatively analogue to the closure predictions of Chapter XI, with a k^4 infrared ($k \to 0$) error spectrum. This result is very encouraging, displaying both the performances of the large-eddy simulations and the validity of the closures.

The three-dimensional numerical simulations modelling the interaction of vortex filaments via Biot and Savart laws and mentioned in Chapter II (Léonard, 1985, Coüet, 1985) need also some sort of subgrid-scale modelling, necessary when a vortex tube is too much distorted and develops oscillations of length scale smaller than the spatial resolution of the calculation: these small scales are removed by a smoothing, whose dynamical significance in terms of subgrid-scale parameterization is not completely clear.

4.5 A local formulation of the spectral eddy-viscosity

To finish this section, let us consider how the spectral eddy-viscosity (without cusp) can be employed in the physical space. The inconvenient of an eddy-viscosity of the type

$$\nu_t(k_C) = 0.402 \left[\frac{E(k_C)}{k_C} \right]^{1/2} \quad , \tag{XII $-$ 4 $-$ 20}$$

is that it is uniform in space when used in physical space. Obviously, the eddy-viscosity should take into account the intermittency and inhomogeneity of turbulence: there is no need for any subgrid-scale modelling in regions of space where the flow is calm or transitional. On the other hand, it is essential to dissipate in the subgrid scales the local bursts of turbulence if they become too intense. Considering also that turbulence in the small scales may not be too far from isotropy, we propose to come back to the classical formulation (XII-3-1) in the physical space, the eddy-viscosity being determined with the aid of (XII-4-20), where $E(k_C, \vec{x})$ is now a local kinetic energy spectrum: this spectrum corresponds to a fictitious homogeneous turbulence which would be obtained by filling the whole space periodically with small boxes containing the flow in the neighbourhood of $\vec{x}$. How can we determine such a spectrum? We have seen in Chapter VI-4 that, for isotropic turbulence with an infinite $k^{-5/3}$ inertial range, the second-order velocity structure function $F_2(r, t)$ is equal to $4.82 \, C_K \, (\epsilon r)^{2/3}$. Taking $k_C = \pi / \Delta x$, the eddy-viscosity (XII-4-20) writes

$$\nu_t(\Delta x, t) = 0.0396 \, \Delta x \, \sqrt{F_2(\Delta x, t)} \quad . \tag{XII $-$ 4 $-$ 21}$$

This is now an expression which may be evaluated locally in physical space, provided the velocity structure function should be calculated locally, for instance by an averaging over the 6 points[18] a distance Δx apart from $\vec{x}$. A last difficulty arises from the fact that the structure function of the filtered field does not take into account the velocity fluctuations in the subgrid scales. Métais and Lesieur (1990) propose a relation allowing to express the structure function of the whole signal in terms of the filtered one. With this correction, the local eddy-viscosity may be written as

$$\nu_t(\Delta x, t) = 0.063 \, \Delta x \, \sqrt{\bar{F}_2(\vec{x}, \Delta x, t)} \quad , \tag{XII $-$ 4 $-$ 22}$$

where $\bar{F}_2(\vec{x}, \Delta x, t)$ is the locally averaged second-order velocity structure function in the large-eddy simulation. Preliminary calculations done by Métais and Lesieur (1990) in isotropic turbulence show that it yields a spectrum closer to the Kolmogorov spectrum at the cutoff. However, the validity of this local eddy-viscosity has to be tested for non-homogeneous

---

[18] for a regular grid of mesh Δx

flows. It must be remarked that, although it is based on an isotropic analysis, eq. (XII-4-22) takes into account inhomogeneities of turbulence, via the local structure function: in a turbulent boundary-layer calculation for instance, the velocity fluctuations in the vicinity of the boundary will be damped, which will diminish the local eddy-viscosity.

5 - L.E.S. of two-dimensional turbulence

Since the *Numerical Fluid Mechanics* was initiated by the meteorologists in quasi two-dimensional situations, it is in this context that the subgrid-scale modelling problem was posed, by Smagorinsky (1963) for instance. The problem is extremely far from being solved, but empirical recipes have been developed, which seem to work quite well in comparison with the other physical processes to be parameterized in the global atmospheric prediction models.

Let us consider a flow which, for some reason, is quasi two-dimensional in the large scales and three-dimensional in the small scales, and assume that the cutoff wave number k_C in the large-eddy simulation corresponds to quasi two-dimensional scales. Then the subgrid scale modelling to be developed has to take into account the two-dimensional dynamics of the large scales (conservation of kinetic energy and enstrophy) on the one hand, and also possible interactions with small-scale three-dimensional turbulence on the other hand. It is then difficult at this level to know whether the subgrid-scale modelling to use has to consider the small-scale three-dimensional point of view (and then employ methods presented in Section 4), or the two-dimensional point of view. Smagorinsky's example illustrates this point, and it has been remarked by Herring (1979) that *"N.A. Phillips (1956), in one of the first numerical experiments treating the general circulation of the atmosphere, introduced an eddy viscosity A whose value was chosen in accordance with Richardson's empirical law $A = 0.2 \ (\Delta x)^{4/3}$ "*. This law[19] concerns three-dimensional turbulence, and it is clear that such a subgrid-scale model was not much concerned with the still unknown enstrophy cascade dynamics.

From the point of view of strictly two-dimensional turbulence in the context of two-dimensional Navier-Stokes equations, the problem of subgrid-scale modelling is far from being solved: if k_C lies in the (possible) enstrophy cascade, the parameterization of the small scales needs to ensure both a constant enstrophy flux and a zero kinetic energy flux through k_C. The two constraints are very difficult to satisfy:

---

[19] which is not far from the ideas expressed in Section 4

a method calculating the fluxes with the aid of the *E.D.Q.N.M.* approximation, and paralleling the one developed in Section 4, was proposed by Basdevant et al. (1978). The resulting large-eddy simulation gave "satisfactory" spectral results[20], but appeared to strongly affect the shape of the spatially organized eddies characteristic of two-dimensional turbulence (Basdevant and Sadourny, 1983). More empirical techniques were then developed, generalizing the *biharmonic dissipation* operator $-\nu_1 \nabla^2 \omega$ used by the oceanographers in the r.h.s. of the vorticity equation (Holland, 1978). Thus Basdevant and Sadourny (1983) performed a systematic study of large-eddy simulations with a subgrid-scale diffusion operator (in the vorticity equation) proportional to $-(-\nabla^2)^\alpha \omega$, and came to the conclusion that the "optimal value" of α was of the order of $4 \sim 8$ (the value of 2 already giving much better results than the "viscous" value $\alpha = 1$)[21]. The effect of these modified dissipativities is, with an adjustment of the numerical constant, to push the effects of dissipation in the neighbourhood of k_C, as can be understood easily when looking at the expression $\sim -k^{2\alpha}\hat{\omega}$ of the dissipative term in Fourier space. It seems difficult for numerical reasons at these quite low resolution calculations to increase the value of α above 8. Let us mention also another new method, called the *anticipated potential vorticity method* (Sadourny and Basdevant, 1985), which could prove to be promising in quasi-geostrophic turbulence when the mesh size Δx is larger than the internal Rossby radius of deformation.

As already mentioned in Chapter II, numerical methods based on point vortex dynamics have been developed for two dimensional turbulence (see e.g. Zabusky and Deem, 1971). Another method, called the *random vortex method*, has been proposed by Chorin (1973): it solves the vorticity advection equation by a Lagrangian method, and models the small-scale effects assuming a random Brownian-like distribution of small-scale vortices.

For flows of great engineering interest such as two-dimensional wakes behind airfoils, new methods have recently been developed, based on the statistical mechanics of a *lattice gas* (see e.g. Frisch et al., 1986, d'Humières et al., 1986): instead of discretizing the real continuous flow equations on a grid, as was done in the classical methods of direct or large-eddy simulations, the method calculates explicitly the motions of the particles of a fictitious gas, these particles being allowed to travel on a regular grid from one point to one of its neighbours with the same velocity modulus[22]V. The above quoted two-dimensional calculations

---

[20] satisfactory, for those who believe in the enstrophy-cascade concept

[21] An application to the simulation of a mixing layer will be presented in the next chapter.

[22] and sometimes also to remain stationary on the same site

are made with a triangular array of points. The collision laws between the particles are chosen in such a way that a "macroscopic velocity" $\bar{u}$, calculated as an average on all the particles lying at a given time inside a " macroscopic site" including a given number of individual grid points, should follow the Navier-Stokes equations[23]. When programmed on classical vectorial computers, these methods allow one to reproduce Karman vortex streets in times of the order[24] of the finite differences or spectral methods computational times. The future of the lattice-gas methods could come from the possibility of designing massively parallel computers devoted to these cellular automata. The application to three-dimensional flows, however, is to date an open question.

To conclude this section, it seems presently[25] that the development of computers will soon render possible the direct-numerical simulation of all the two-dimensional flows, at least when one is only interested by the coherent structures dynamics. In this case, no subgrid-scale modelling is needed if the resolution is high enough, and a regular molecular viscosity should be sufficient. But the problem of subgrid-scale modelling is certainly one of the highest difficulties in the computations of three-dimensional turbulence.

---

[23] This result supposes however certain validity conditions, as the assumption of a "Mach number" of the theory $\bar{u}/V$ small compared with one, which are not fulfilled in the numerical applications. Then, and since the "macroscopic velocities" do not in practice follow exactly the Navier-Stokes equations, the philosophy of the lattice-gas method could be closer to the large-eddy simulations one than to the direct-numerical simulations.

[24] larger, however

[25] in 1990, at the time of the second edition

Chapter XIII

TOWARDS "REAL WORLD TURBULENCE"

1 - Introduction

Up to now, the analytical statistical theories we have considered concerned mainly[1] what one could call *ideal turbulence*, in the sense that isotropy was nearly always assumed, either in three or in two dimensions. We have, however, also discussed in Chapter III the dynamics of more realistic flows, in the context of transition and coherent structures in particular. We have chosen in this last chapter to consider two examples of turbulent flows which both have a great practical importance, and which may be considered as prototypes for the applications of certain concepts, analysis and techniques developed in the preceding chapters.

The first problem envisaged here will be that of a three-dimensional homogeneous turbulence submitted to a stable stratification. This is no longer an isotropic turbulence, and we will mainly focus on what happens if an isotropic turbulence is created in a stably-stratified fluid, through some external device which afterwards will be suddenly turned off. This problem has important applications in meteorology and oceanography. The study of this question will permit the use of some of the isotropic phenomenological results derived above, and also enable us to see how three-dimensional isotropic dynamics is modified by the buoyant forces. In particular we will show how the spectral eddy-viscosity and eddy-diffusivity derived in Chapter XII can be used during the initial stage of decay.

The second problem will be two- and three-dimensional turbulent phenomena resulting from the development of the inflexional shear in-

---

[1] except for the mixing-length theory of turbulent shear flows considered in Chapter VI

stability[2], in the case of the mixing layer. This question has already been mentioned at several occasions in this monograph, and it will be looked at in detail here, mainly from the point of view of the direct and large-eddy simulations. We will in particular examine up to which point it may be studied within the framework of two-dimensional turbulence, and what is the importance of three-dimensional instabilities and structures which develop within such a flow. This is the first step towards the understanding of complex flows like the axisymmetric jet or the plane wake or jet, or even the boundary-layer.

The chapter will be divided into five sections: section 2 will deal with stably-stratified turbulence, section 3 with the mixing layer considered from the point of view of two-dimensional turbulence, and section 4 will report various three-dimensional direct or large-eddy simulations of the mixing layer. Section 5 will be a tentative conclusion of this monograph.

2 - Stably-stratified turbulence

2.1 The so-called "collapse" problem

The problem of the evolution of an initially three-dimensional isotropic turbulence submitted to the action of a stable stratification has already been introduced in Chapter III-6. This problem is of essential concern for the *mesoscale atmosphere*[3], and the ocean, where it governs the vertical fluxes of temperature for instance.

In the atmosphere, small-scale three-dimensional turbulence, produced e.g. by the breaking of lee waves behind the mountains or by convective storms, may find itself afterwards imbedded in a stably-stratified density field. The question is then to know how the turbulence will reorganize under the action of gravity. A conversion of three-dimensional into quasi-horizontal turbulence has been proposed by Riley et al. (1981) on the basis of an expansion of the Navier-Stokes equations within the Boussinesq approximation, properly normalized, with respect to the Froude number. If the latter is sufficiently small, this procedure may lead, in certain conditions, to two-dimensional Navier-Stokes equations for a given vertical level. Physically, this may be understood by the existence of a vertical scale l_B characterizing the buoyancy and preventing the larger amplitude vertical motions from developing. This scale will be calculated below. It would lead to the "collapse" of turbulence into horizontal "pancake-shaped" eddies[4] of thickness l_B and having a

---

[2] also called barotropic instability

[3] that is, the atmospheric motions between one to a few hundred kilometers

[4] Maxworthy, private communication

two-dimensional dynamics[5]. This idea has been used by Gage (1979) and Lilly (1983) to envisage the existence of a small-scale[6] two-dimensional forcing in the atmosphere, which would feed an inverse horizontal $k^{-5/3}$ energy cascade responsible for the atmospheric mesoscale energy spectrum and extending up to several hundred kilometers. It is claimed that such a spectrum has been measured experimentally[7], which would contradict the EOLE[8] experiment conclusions of a k^{-3} spectrum extending from 1000 to 100 km (Morel and Larchevêque, 1974). This question of the structure of atmospheric mesoscale turbulence is extremely open: vertical soundings of the atmosphere seem to confirm the fact that the turbulence is concentrated into multiple extremely thin horizontal layers, which could be an argument favouring the inverse cascade theory. Notice however that such a structure is sometimes interpreted by the physicists of the "middle atmosphere" as due to the vertical propagation of gravity waves.

In the ocean, the interaction of gravity waves with three-dimensional turbulence involves various complex phenomena which contribute to the formation of the mixed layer and are responsible for vertical exchanges arising in this layer and at the level of the thermocline. Among these mechanisms are again the breaking of internal waves into small-scale turbulence, and the reorganization of this turbulence into waves and possibly horizontal motions. Notice that in the experiment of Gargett et al. (1984), already quoted as an evidence of a small-scale Kolmogorov energy spectrum in the ocean, this turbulence was actually the result of the breaking of internal waves. The evolution of this turbulence under the stratification was then examined, in such a way that these measurements are an oceanic counterpart of the grid-stratified turbulence experiments done by Stillinger et al. (1983), Itsweire et al. (1986), Lienhard and Van Atta (1990) and Thoroddsen and Van Atta (1990).

A number of laboratory experiments have been carried out on this problem (see e.g. Hopfinger, 1987, for a review). A particularly impres-

---

[5] This turbulence is not, properly speaking, a two-dimensional turbulence as was defined in Chapter IX, to be, in particular, independent of z. Here, the correct term should be *horizontal turbulence*, since these horizontal layers are more or less vertically decorrelated, and hence depend upon the z direction. Some authors call this *quasi-two-dimensional turbulence*

[6] a few kilometers

[7] Gage, private communication

[8] In this experiment, 500 constant level 200 mb balloons were released in the southern hemisphere, and tracked with the aid of the EOLE satellite: statistics on their relative dispersion then allowed measurement of the energy spectrum.

sive result was obtained by Pao (1973), who showed that the wake of a spherical obstacle in a stably-stratified flow, initially three-dimensional behind the obstacle, would suddenly collapse into a thin horizontal vortex street. This is a purely experimental result, which has not yet been theoretically explained. It is of course an argument in favour of the theory of collapse to quasi two-dimensional turbulence, but corresponds to initial conditions different from the atmospheric or oceanic conditions, or the grid turbulence experiments: in some of the latter for instance, an apparently abrupt change occurs in the dissipation of kinetic energy, leading to a much less dissipative state. But it could be attributed either to a transition to an internal-wave field or to two-dimensional turbulence, since both states do not dissipate the kinetic energy.

2.2 A numerical approach to the collapse

We consider a fluid satisfying the Boussinesq approximation in its simplest form, with a constant Brunt-Vaisala frequency N, and neglect rotation. We assume statistical homogeneity. The force of gravity is such that $\vec{g} = -g\vec{z}$. It is easy to check from (II-8-8) that the fluid satisfies

$$\frac{D\vec{u}}{Dt} = -\nabla p + \vartheta \vec{z} + \nu \nabla^2 \vec{u}$$
$$\nabla.\vec{u} = 0 \qquad\qquad\qquad\qquad (XIII - 2 - 1)$$
$$\frac{D\vartheta}{Dt} + N^2 w = \kappa \nabla^2 \vartheta$$

where $\vartheta = -\tilde{\rho}g$. The assumption that N is a constant is certainly not fulfilled in the atmospheric or oceanic situations mentioned above, but is a very useful step in order to understand the physics of these complex interactions between turbulence and waves. The characteristic scales of the problem are the integral scale $l = v^3/\epsilon$, and a characteristic scale of stratification l_B. The latter is obtained by defining a turbulent velocity $(\epsilon l_B)^{1/3}$ such that the associated Froude number $(\epsilon l_B)^{1/3}/N l_B = 1$. This yields

$$l_B = (\frac{\epsilon}{N^3})^{1/2} \qquad\qquad\qquad (XIII - 2 - 2)$$

This scale is generally referred to as the Osmidov scale (Osmidov, 1975), and corresponds to a balance between the inertial and buoyant effects. Since the Froude number of turbulence $F = u/Nl$ is equal to $\epsilon^{1/3}/N l^{2/3}$, one obtains the important relation

$$F = (\frac{l_B}{l})^{2/3} \qquad\qquad\qquad (XIII - 2 - 3)$$

which shows that stratification has a negligible effect on turbulence if $l << l_B$, and becomes dynamically important when the eddies are of the order of or greater than the Osmidov scale. It justifies the phenomenology discussed in Chapter III-6, where, if the initial Froude number is large, the turbulence will approximately decay as in the isotropic case, with $v/l \sim t^{-1}$, and the turbulent Froude number will be proportional to $(Nt)^{-1}$. It will then, in a period of time of the order of N^{-1}, reach values of the order of one, where it will no longer be possible to neglect the gravity. During this evolution, the integral scale of turbulence will grow, according to the laws derived in Chapter VII ($t^{0.3} \approx t^{0.5}$); the Osmidov scale will decay like $\epsilon^{1/2}$. Both scales will collapse when the Froude number is one. Further evolution is difficult to predict by the phenomenology.

Within the Boussinesq approximation with constant N, the potential energy is

$$E_p = \frac{1}{2} N^{-2} < \vartheta^2 >$$

and it is very easy to check from (XIII-2-1) that the total energy

$$\frac{1}{2} < \vec{u}^2 > + E_p$$

is conserved by the non-linear terms of the equations. But, as in three-dimensional isotropic turbulence, these quantities might be dissipated at a finite rate by viscosity and conductivity. Notice in particular that when the Froude number is large and that buoyancy may be neglected, the potential energy is proportional to the variance of the passive scalar ϑ, and decays according to the laws derived in Chapter VIII.

This problem has been studied, at a Prandtl number ν/κ of one, with the aid of direct[9] numerical simulations by Riley et al. (1981), with a spectral code of 32^3 modes, and starting initially with an isotropic velocity field. The Reynolds number was of course much smaller than in the laboratory experiments, but an extremely interesting behaviour was obtained, namely a finite dissipation of kinetic energy for times greater than N^{-1}, without any marked transition (or "collapse") at N^{-1}. More specifically, the kinetic energy decay curve also exhibited a wavy tendency, which could be attributed to the appearance of waves. The same calculation was performed by Métais (1985) within a large-eddy simulation in the spectral space using the spectral eddy-viscosity and diffusivity defined in (XII-4-11, 4-12), and starting initially from a turbulent Froude number of 3, close to the experimental value of Itsweire et al. (1986). The use of such an isotropic subgrid scale parameterization can be justified in the early stage of the evolution[10], since, as recalled in Chapter

---

[9] that is taking into account all the scales of motion, including the dissipative and conductive ones

[10] when the integral scale is smaller than the Osmidov scale

XII, it allows the kinetic energy and the integral scale to evolve in such a way that the condition $v/l \sim t^{-1}$ is still fulfilled. The results thus obtained have been interpreted in terms of the vortex-wave decomposition (or Craya decomposition) presented in Chapters IV and V, and in particular the "vortex kinetic energy" and the "wave kinetic energy", respectively

$$\bar{\Phi}_1(t) = \int (e - Z_1) d\vec{k} \quad ; \quad \bar{\Phi}_2(t) = \int (e + Z_1) d\vec{k} \ ,$$

introduced in eq. (V-7-1) for axisymmetric turbulence[11]. We recall that the total kinetic energy is equal to $\bar{\Phi}_1(t) + \bar{\Phi}_2(t)$. Figure XIII-1 shows in this calculation the evolution of these normalized energies and of the potential energy $<\vartheta^2>$ both in an isotropic calculation (no stratification) and in the stratified case: in the isotropic case, the various kinetic energies (horizontal kinetic $<\bar{u}_H^2>/2$, vertical kinetic $<w^2>/2$, vortex kinetic, wave kinetic) properly normalized are all equal, and decay within the large-eddy simulation like $t^{-1.2}$, as stressed in Chapter XII. This is also true for the temperature variance[12] $<\vartheta^2>$, which represents an analogous "potential" energy. All these isotropic energies (plus the total [13] energy) are represented by curve A. In the stratified case, the vortex kinetic energy (curve B) differs negligibly from the isotropic case. The wave kinetic energy (C) presents oscillations of period $\approx \pi/N$. The potential energy (D), initially weak, starts building up, then decays with oscillations of the same period, in phase opposition with the wave kinetic energy. It has been checked that the total energy (kinetic plus potential) presents no oscillations and decays following the $t^{-1.2}$ law, exactly as in the non-stratified case. In this particular calculation, it seems that the vortex and wave kinetic energies are very close to respectively the horizontal and vertical kinetic energies.

The wavy behaviour of the "wave kinetic" and "potential" components of the energy in the stratified case is the same as found by Riley et al. (1981), even if the initial Froude number is now higher. The period π/N found numerically seems to characterize the horizontal propagation of gravity waves: indeed, a wave of energy

$$|\phi|^2 \propto \sin(2Nt + \alpha)$$

corresponds to a fluctuation $\phi \propto \sin(Nt + \gamma)$ (α and γ are arbitrary phases), and hence a frequency $\varpi = N$ in eq. (II-9-10). This implies

---

[11] Since the initial conditions are axisymmetric, this condition will be preserved with time.

[12] at this low resolution of 32^3

[13] kinetic + potential

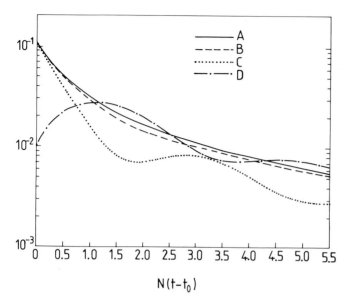

Figure XIII-1: time evolution of the normalized energies in a non-stratified and stratified spectral large-eddy simulation (resolution 32^3). A: kinetic or potential energy (isotropic case). B: "vortex" kinetic energy (stratified case). C: "wave" kinetic energy (stratified case). D: potential energy (stratified case) (from Métais, 1985).

$k_3 = 0$. These waves affect the large scales of the flow, as shown by the spectra of these various energies (averaged on a sphere of radius $k = |\vec{k}|$) presented on Figure XIII-2: the vortex spectrum in the stratified case (B) does not differ very much from the isotropic one (A); the waves are clearly displayed on the spectra C and D for modes $k <\sim k_i$, the small scales $k > k_i$ behaving in the same way as in A and B with a $k^{-5/3}$ spectrum extending up to the cutoff k_C. It is not clear whether these waves are physically realistic, or are due to the artificial periodic boundary conditions[14] of the *turbulence within a box* which has been considered. But even if these waves are strongly amplified by the numerical model, they do not seem to significantly affect the dynamics of the "vortex component", nor to reduce the total energy dissipation rate.

These calculations show also that the integral scale and the Osmidov scale cross at about $t = N^{-1}$, as predicted by the phenomenology. At that time, the turbulent Froude number has decreased from its initial value of 3 to 1. Afterwards it continues to decrease and is about 0.15 after

---

[14] which indeed put in phase the fluctuations at the boundaries, thus forcing waves at the scale of the computing box

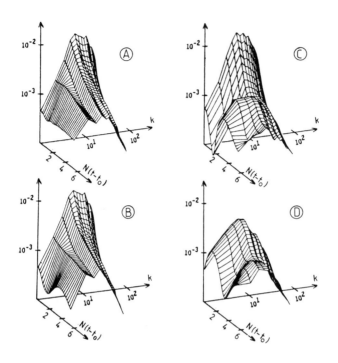

Figure XIII-2: time evolution of the energy spectra in the same conditions as in Figure XIII-1 (from Métais, 1985).

$5\,N^{-1}$. This is quite a low value, but there is not yet any two-dimensional turbulence tendency. This is clear in particular when looking at the kinetic energy spectra, which do not show any significant inverse transfer characteristic of two-dimensional turbulence.

Later on, direct-numerical simulations at a higher resolution (64^3) and for longer times ($50\,N^{-1}$) were made by Métais and Herring (1989). Their results are qualitatively similar to those of Riley et al. (1981) and Métais (1985) for $t < 5\,N^{-1}$. However, they find a clear slowdown in the decay of the vortex kinetic energy at $t \approx 6\,N^{-1}$, while the other energies (wave and potential) continue to decay at the same rate. Before this time[15], the vortex and wave spectra are identical in the small scales. Afterwards, the wave energy cascades faster to high k than the vortex

---

[15] The existence of this collapse time was confirmed simultaneously in a spectral large-eddy simulation (at a resolution of 32^3) done by Métais and Chollet (1989). This calculation shows also that the vertical diffusion of a thin initially horizontal layer of passive scalar is strongly inhibitted by stratification after the collapse, as shown on Figure XIII-3.

energy, and it seems that there is a tendency for the vortex energy to be trapped in the large scales. It is found that the "collapse time" $6\ N^{-1}$ is such that the Osmidov scale and the Kolmogorov dissipative scale are equal, which implies that all the scales of motion are influenced by buoyancy, including the smallest ones. But the vortex kinetic energy dissipation rate is still important, even though it has decreased, and the situation is in no way that of a two-dimensional turbulence, although Métais and Herring (1989) find that the isopycnal surfaces become quasi two-dimensional after the collapse.

Other initial situations can be considered: for instance a completely horizontal field where all the energy is under its vortex form (Métais and Herring, 1989). Such a state implies important vertical gradients of the horizontal velocity. In this case, two possibilities exist: if the stratification is sufficiently high, the local Richardson number will be everywhere greater than its local critical value 0.25, and no instability will develop: the flow remains in horizontal layers, and is not affected by gravity waves which propagate; but again the vortex kinetic energy dissipation is important, due to the strong vertical gradients of the horizontal velocity. Otherwise, internal Kelvin-Helmholtz waves will emerge, and possibly break up into small-scale three dimensional turbulence. This is a mechanism which converts vortex energy into wave energy[16].

Returning to the above-considered problem of the collapse of initially isotropic turbulence under stratification, we have to mention the recent spectral large-eddy simulation performed by Métais and Lesieur (1990) at a resolution of 128^3 and up to $t = 7.5\ N^{-1}$: it confirms the existence of the collapse time at $6N^{-1}$, by considering the instantaneous time-decay exponent α_V of the vortex kinetic energy, which starts decreasing appreciably after the collapse. The exponent α_W corresponding to the potential and wave energy seems, after some adjustment time, to become of the same order as α_V, which shows a strong difference with the anomalous behaviour of the passive temperature found in Chapter XII for the same calculation. Here, in the stratified case, the potential energy spectrum loses the k^{-1} large-scale range displayed in the non-stratified case of Figure XII-3.

This is shown in Figure XIII-4, presenting respectively the vortex, wave and potential energy spectra at $t = 1.48\ N^{-1}$. Figure XIII-5, to be compared with Figure XII-2 in the isotropic case, shows that, now, the

---

[16] Though there does not seem to be any doubt about the fact that the "vortex" energy is a good indicator of horizontal kinetic energy, it is on the contrary more difficult to distinguish, in the "wave" kinetic energy, the gravity wave contribution from the small-scale three-dimensional turbulence contribution.

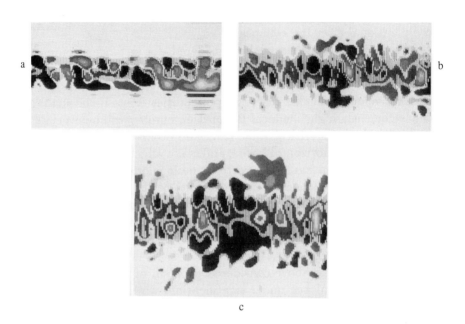

Figure XIII-3: numerical simulation of the vertical diffusion of passive scalar done by Métais and Chollet (1989). A vertical section is shown: initially (a); at $t = 12.85\ N^{-1}$ in the stratified case (b); at the same time in the isotropic case (c).

eddy-diffusivity has also lost its logarithmic k dependency, and is approximately constant, with an eddy-Prandtl number of the order of 0.45. At the same time, the p.d.f. of the temperature is now gaussian, as was the velocity in the isotropic case (see Chapter VI). Hence, the coupling between temperature and velocity due to buoyancy seems to have suppressed the large-scale intermittency of the temperature which we had mentioned in isotropic turbulence. This is confirmed by Figure XIII-6, taken from Métais and Lesieur (1990), which shows the temperature p.d.f. in a spectral large-eddy simulation and direct-numerical simulation[17], determined respectively in the isotropic and stratified case. The strong temperature intermittency in the isotropic case is illustrated in Plate 15.

---
[17] see also Métais and Lesieur (1989)

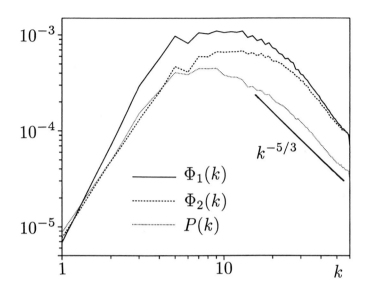

Figure XIII-4: vortex kinetic energy spectrum $\Phi_1(k)$, wave kinetic energy spectrum $\Phi_2(k)$ and potential energy spectrum $P(k)$ at $t = 1.48\ N^{-1}$ in the stratified large-eddy simulation of Métais and Lesieur (1990).

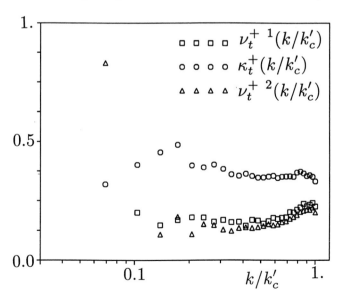

Figure XIII-5: spectral vortex and wave eddy-viscosity, and eddy-diffusivity, corresponding to the calculation of Figure XIII-4.

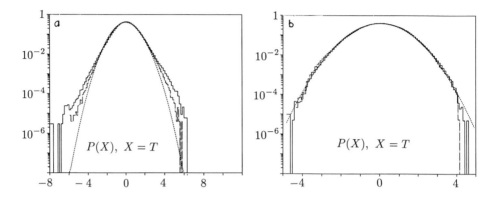

Figure XIII-6: temperature p.d.f. determined at $t = 1.48 \ N^{-1}$ in the D.N.S. (continuous line) or L.E.S. by Métais and Lesieur (1990); (a) isotropic calculation; (b) stratified calculation. The dashed line indicate a gaussian distribution.

At the later times of the stratified L.E.S. and D.N.S. calculation of Métais and Lesieur (1990), various spectra confirm that the vortex kinetic energy tends to be trapped in the large scales.

New grid-generated stratified turbulence experiments carried out by Thoroddsen and Van Atta (1990) show that the stratification induces horizontal anisotropy also, but without any evidence of two-dimensional turbulence development. Forced stratified direct-numerical simulations have been carried out by Herring and Métais (1989), in order to check the conjectures concerning the possibility of an inverse energy cascade of two-dimensional turbulence driven by the collapse of three-dimensional turbulence under stratification. Although there is a significant transfer to large scales in the horizontal, no $k^{-5/3}$ inverse energy cascade is found in these spectral calculations (at a resolution of 64^3). However, higher resolutions and different types of forcing and dissipation are needed in order to settle this point.

Let us mention that stratified-shear flow three-dimensional direct-numerical simulations have been performed by Gerz et al. (1989) in the case of a constant shear, and by Staquet and Riley (1989) for a mixing layer. In the latter case, stratification produces a sort of horizontal collapse of the Kelvin-Helmholtz billows, producing vertical vorticity. Experiments (see e.g. Browand and Winant, 1973) show this collapse of the mixing layer under stratification. Finally, Plate 16 shows a gravity wave developing in a strongly-stratified flow behind an obstacle, in a three-dimensional finite-volume calculation.

3 - The two-dimensional mixing layer

3.1 Generalities

This monograph has already discussed at several occasions the mixing layer between two flows of different velocities, when mentioning for instance the existence of coherent structures, or considering the stability of an inflexional velocity shear. In addition to the observations on the downstream persistance of the large structures made by Brown and Roshko (1974), one must quote the evidence on the pairing of these structures displayed by Winant and Browand (1974) (see also Ho and Huerre, 1984, for a review). One often encounters mixing-layer type structures in the atmosphere or in the oceans[18], and in many separated flows. As seen in Chapter III, such a flow is extremely important as a prototype of transition to turbulence. An important question is to what extent the large quasi-two-dimensional scales have two-dimensional dynamics (i.e. obey two-dimensional Navier-Stokes or Euler equations) and what is their interaction with small-scale three-dimensional turbulence. This is important in particular for numerical modelling purposes since it is a great advantage when only two spatial dimensions need to be considered.

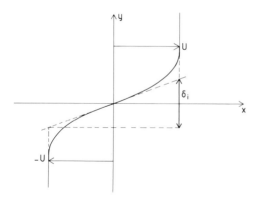

Figure XIII-7: basic inflexional velocity shear in the mixing layer.

Concerning the three-dimensionalization processes of the layer, we have already mentioned in Chapter III some of the numerous experiments concerning the appearance of hairpin-vortex filaments longitudinally strained

---

[18] They may be either of vertical or horizontal axis

by the flow[19] (see e.g. Konrad, 1976, Breidenthal, 1981, Bernal, 1981, Jimenez et al., 1985, Bernal and Roshko, 1986, and Lasheras and Choi, 1988). From the discussion initiated in Chapter III, and the stability analysis performed on a spanwise vortex filament in III-3 − 3 − 1, thin spanwise vortices of low vorticity (and located preferentially in the stagnation region), might[20] first be lifted by self-induction into the direction of larger velocities, then be strained longitudinally by the basic shear. Simultaneously, a sine-like undulation of the two-dimensional Kelvin-Helmholtz billows[21] might develop as the result of the translative instability investigated by Pierrehumbert and Widnall (1982). This instability could possibly impose its most-amplified spanwise wave length to the longitudinal hairpin vortices. Some of these three-dimensional instabilities have been studied numerically by Corcos and Lin (1984).

Although we already know that the two-dimensional approximation is not a very good one for the mixing layer, it is of interest to look at it from a purely two-dimensional point of view: this allows one to identify (in the experiments for instance) the mechanisms which are two-dimensional; furthermore, experiments on two-dimensional mixing layers can now be envisaged, in liquid films (Gharib and Derango, 1989) for instance[22]. Computations are of course much less costly in two dimensions[23]. In the present section, one will approach the mixing layer in two distinct manners: we will first consider it from the two-dimensional turbulence point of view[24], and will look at the spatial statistics of such a turbulence during the downstream evolution of the layer. Then we will present a particular spanwise one-mode spectral truncation allow-

---

[19] These filaments, when considered in the y (transverse)-z (spanwise) direction display a "mushroom" structure corresponding to two eddies of opposite longitudinal vorticity.

[20] if they are slightly perturbed in the spanwise direction

[21] created by the instability of the basic inflexional velocity profile or resulting from various two-dimensional interactions (pairings for instance) of fundamental eddies

[22] A rotation with axis parallel to the spanwise direction may, in certain conditions, have a two-dimensionalization action. Three-dimensional direct-numerical simulations of rotating mixing layers will be presented in the following section.

[23] See e.g. Riley and Metcalfe (1980), Corcos and Sherman (1984), and Lesieur et al. (1988), for the temporal problem; Ashurst (1979), for the spatially-growing problem. The calculation of Ashurst is done via vortex dynamics methods.

[24] Evidence for the sensitivity of the flow to initial infinitesimal perturbations, and hence of its unpredictability, has been given in Chapter I.

ing to relate some mechanisms of three-dimensionality growth to a two-dimensional predictability problem.

3.2 Two-dimensional turbulence in the mixing layer

The simpler case to study is the *temporal mixing layer*, where the basic flow is independent of x, equal respectively to $U\vec{x}$ and $-U\vec{x}$ for $y = \pm\infty$, and generally chosen as[25]

$$\bar{u}(y) = U \tanh \frac{2y}{\delta_i} \qquad\qquad (\text{XIII} - 3 - 1)$$

where

$$\delta_i = 2U / \left[\frac{d\bar{u}}{dy}\right]_{y=0} . \qquad\qquad (\text{XIII} - 3 - 2)$$

The velocity profile given by eq. (XIII-3-1) is shown on Figure XIII-7. δ_i is the initial vorticity thickness. The flow is assumed to be periodic in the x direction, of period[26] D_N. This is an approximation of the more realistic[27] *spatial mixing layer*, where the two parallel flows which mix have two different velocities U_1 and U_2 of same sign: if one takes a reference frame moving with the average velocity $(U_1 + U_2)/2$ and neglects locally the lateral spreading of the layer in the y direction, one recovers a temporal mixing layer such that $U_1 - U_2 = 2U$; therefore, the distance x downstream of the splitter plate behind which the flows are in contact in the spatial mixing layer will correspond to $t = 2x/(U_1 + U_2)$ in the temporal mixing layer.

When the time evolves, instabilities and eddies will develop at the interface, and the layer thickness will grow with time: this can be measured by a mean velocity profile $\bar{u}(y,t)$, defined in the temporal mixing layer by averaging[28] in the x direction for a given y. It can be checked (see for instance Lesieur et al., 1988, and the calculations of Plate 7) that this mean velocity profile keeps the form (XIII-3-1), δ_i being replaced by a mean vorticity thickness $\delta(t)$ still given by (XIII-3-2)[29].

---

[25] Notice, however, that the laminar or turbulent velocity profile of the temporal mixing layer is an error function, as stressed in Chapter VI-1 − 2 − 5. In fact, this is very close to a hyperbolic-tangent velocity profile.

[26] N is the number of fundamental eddies which will first appear, due to the linear instability.

[27] However, the temporal mixing layer may be more relevant to geophysical situations.

[28] In the experiments of the spatial mixing layer, the mean velocity profile can be measured by a time averaging at a given position.

[29] The momentum thickness can also be defined. It is approximately one fifth of the vorticity thickness.

Spatial mixing-layer experiments show that $\delta(x)$ increases linearly with x, which may be understood on the basis of the mixing-length theory considered in Chapter VI. The same theory predicts that $\delta(t) \propto t$ in the temporal mixing layer. In the latter case, the characteristic turn-over time is δ_i/U, and one expects to have

$$\frac{d}{dt}\delta(t) = \alpha_1 \; \delta_i/U \quad , \qquad\qquad (XIII - 3 - 3)$$

where α_1 is a universal constant. Associating a spatial mixing layer to this temporal mixing layer yields

$$\frac{d}{dx}\delta(x) = \frac{d}{dt}\delta(t) \; \frac{dx}{dt}$$

which leads to

$$\frac{U_1 + U_2}{U_1 - U_2} \frac{d}{dx}\delta(x) = \alpha_2 \quad , \qquad\qquad (XIII - 3 - 4)$$

where $\alpha_2 = \alpha_1$ if the two problems (temporal/vs spatial) were exactly equivalent, which is not the case. In the experiments of Liepmann and Laufer (1949), Wygnanski and Fiedler (1970), Brown and Roshko (1974) and Bernal (1981), the average growth law corresponds to $\alpha_2 \approx 0.18$. The numerical simulations of the spatially-growing mixing layer forced upstream by a white-noise perturbation of small amplitude, done by Normand (1989) for various velocity parameters[30] confirm this law. In the numerical simulations of the temporal mixing layer[31] done by Lesieur et al. (1988), it is found that $\alpha_1 = 0.10$ for a calculation involving several pairings. This is a significant difference (compared with α_2) which needs however to be confirmed[32]. In this temporal calculation, a layer of size δ will double its size in a time of $10 \; \delta/U$. These calculations also show that the first fundamental eddies which appear have a vorticity thickness of $2 \; \delta_i$, and are formed at a time of $15 \; \delta_i/U$. Then the first pairing will occur approximately at $35 \; \delta_i/U$, and the subsequent[33]

---

[30] See also Normand et al. (1988): in these calculations, it may be checked also that the coherent eddies travel downstream with approximately the velocity $(U_1 + U_2)/2$, as stressed in Chapter 3.

[31] still forced initially by a white-noise

[32] It must be stressed, however, that manipulations of the inflow perturbations in the mixing-layer experiments may lead to substantial variations of the spreading rate, as shown by Dziomba and Fiedler (1985).

[33] This reasoning is simply based on the assumption that the vorticity thickness will have doubled after each pairing.

pairings at $75\,\delta_i/U$ and $155\,\delta_i/U$. Here, the white-noise perturbation injects energy in all the longitudinal spatial modes, and might approximate the physical case of a real mixing layer which is naturally submitted to a broad-band disturbance spectrum. From the linear-stability analysis considered in Figure III-1, all the modes corresponding (for a given Reynolds number) to the unstable region will amplify exponentially, and Michalke's most-amplified wave number, given by eq. (III-3-2) (with $\delta_i = 2\delta_0$) will emerge the first as long as the Reynolds number is high enough. Thus, the length of the computational domain has to be taken equal to

$$D_N = 7N\delta_i \quad , \qquad\qquad (XIII - 3 - 5)$$

in order to satisfy the periodicity condition in the $x-$ direction. One obtains thus N Kelvin-Helmholtz fundamental vortices: this is what we will call an "N-eddy calculation". As already mentioned above, these calculations of Lesieur et al. (1988) display successive pairings. This is due to the existence of initial *subharmonic modes*[34], as discussed in Chapter III-3 $-$ 1 $-$ 1. Furthermore, Corcos and Sherman (1984) have checked with a direct-numerical simulation that there is no pairing in the absence of subharmonic perturbations. Note that a non-linear stability analysis of the pairing has been developed by Kelly (1967).

Figure XIII-8, taken from Lesieur et al. (1988), shows the evolution of the 4-eddy mixing layer in the physical space represented by the vorticity contours, compared with the corresponding longitudinal spatial energy spectra: at $t = 20\,\delta_i/U$, the spectrum corresponds to a sharp peak at the fundamental mode ($k_4 = 2\pi/\lambda_a$), which is responsible for smaller peaks at its harmonics $2k_4$ and $3k_4$. The background spectrum corresponds to more complex interactions involving the first and second subharmonic $k_4/2$ and $k_4/4$. At $t = 40\,\delta_i/U$, that is at the end of the first pairing, the two distinct parts of the spectrum collapse to form a continuous range of exponent close to k^{-4}. Then the flow can really be called "turbulent", since it possesses a broad band spectrum. If one accepts the association of this range to an enstrophy-cascading inertial range of two-dimensional turbulence, this indicates clearly the importance of the pairing interactions for the establishment of the enstrophy cascade, as was already mentioned in Chapter IX. At time $t = 80\,\delta_i/U$, at the end of the second pairing, the second subharmonic $k_1 = \pi/2\lambda_a$ has grown and rejoins the inertial range.

Another result of these calculations is that the vorticity thickness stops growing when there remains approximately two eddies in the computational square domain, due to the longitudinal periodicity[35]. This

---

[34] of wave lengths $\lambda_a, 2\lambda_a, 4\lambda_a, etc.$

[35] and not to the parallel boundaries, as erroneously stressed in the

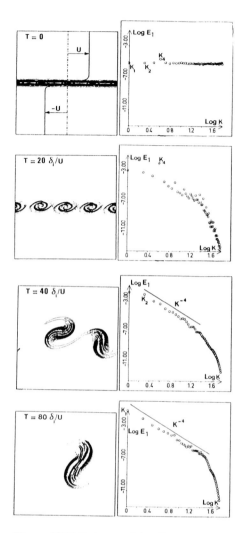

Figure XIII-8: isovorticity lines and longitudinal kinetic energy spectra in the temporal mixing layer at times $0, 20, 40$ and 80 δ_i/U (from Lesieur et al., 1988, courtesy the Journal of Fluid Mechanics).

does not prevent the last pairing to occur, but it does so without any vorticity thickness growth.

3.3 Two-dimensional unpredictability

Such a predictability study has been performed numerically in the case of the temporal mixing layer by Lesieur et al. (1988). Two velocity fields $\vec{u}_1(x, y, t)$ and $\vec{u}_2(x, y, t)$ are considered: they are supposed to be

first edition of this monograph; this point has been verified in Lesieur et al. (1988).

two solutions of the Navier-Stokes equation in the same domain D_N, with the boundary conditions of the temporal mixing layer. The error rate $r(t)$ is defined as a properly spatially averaged[36] kinetic energy of the difference between the two flows, divided by the average kinetic energy of one of the flows. The two fields are prepared in such a way that they initially (at time 0) differ only by a white-noise perturbation, superposed in both cases to the same basic hyperbolic-tangent velocity shear to which a small deterministic longitudinal sine perturbation at the fundamental wave length has been added: at the time t_0 of the rollup, the difference existing between $\vec{u}_1$ and $\vec{u}_2$ will therefore have lost the artificial character it had initially due to the white noise, and the two fields thus generated at t_0 will be mainly composed of coherent structures extremely close from one field to the other. The simultaneous numerical calculation of the evolution of $\vec{u}_1$ and $\vec{u}_2$ allows one to determine $r(t)$, which has roughly an exponential growth of characteristic time $15\, \delta_0/U$. The error saturates together with the vorticity thickness, due to the longitudinal periodicity constraints already mentioned above. Hence, it can be concluded that the error grows as long as the vorticity thickness grows.

3.4 Two-dimensional unpredictability and 3D growth

Here we will develop a two-mode[37] non-linear approach allowing one to relate the above two-dimensional predictability study to the three-dimensionalization of the layer. This analysis is taken from Lesieur et al. (1988), and has some analogies with the linear studies of Pierrehumbert and Widnall (1982), Corcos and Lin (1984) and Herbert (1988) mentioned in Chapter III: let us consider a quasi-two-dimensional flow, whose velocity and pressure fields $\vec{u}(\vec{x},t)$ and $p(\vec{x},t)$ are expanded as

$$\vec{u}(x,y,z,t) = \vec{u}_{2D}(x,y,t) + \sqrt{2}\, \vec{u}_{3D}(x,y,t)\sin k_z z \qquad (XIII-3-6)$$

$$p(x,y,z,t) = p_{2D}(x,y,t) + \sqrt{2}\, p_{3D}(x,y,t)\sin k_z z \qquad (XIII-3-7)$$

where $\vec{u}_{2D}$ and $\vec{u}_{3D}$ are two horizontal (in the $x-y$ direction) non-divergent velocity fields, and k_z the wave-number of the spanwise perturbation. The flow is non-divergent, since the spanwise velocity is zero. $\vec{u}_{2D}$ corresponds more or less to a basic Kelvin-Helmholtz billow, while $\vec{u}_{3D}$ is the amplitude of a spanwise oscillation of the billow, initially small. The following decomposition is analogous to the two-mode spectral vertical decomposition of the quasi-geostrophic potential-vorticity

---

[36] in a test domain

[37] one mode is the two-dimensional flow, and the second is obtained through a one-mode spanwise truncation of the flow

equation presented in Chapter IX-2. Projection of the Navier-Stokes equation (with constant density) on these two modes may be performed either as was done in Chapter IX-2, or in the following manner: one substitutes the particular expansions (XIII-3-6) and (XIII-3-7) into the three-dimensional Navier-Stokes equations (without rotation). One obtains

$$\frac{\partial \vec{u}_{2D}}{\partial t} + \vec{u}_{2D}.\vec{\nabla}\vec{u}_{2D} + 2\sin^2 k_z z \;\; \vec{u}_{3D}.\vec{\nabla}\vec{u}_{3D} +$$

$$\sqrt{2}\sin k_z z \; (\frac{\partial \vec{u}_{3D}}{\partial t} + \vec{u}_{2D}.\vec{\nabla}\vec{u}_{3D} + \vec{u}_{3D}.\vec{\nabla}\vec{u}_{2D}) = -\frac{1}{\rho}\vec{\nabla}p_{2D}$$

$$- \sqrt{2}\sin k_z z \; \frac{1}{\rho}\vec{\nabla}p_{3D} + \nu[\nabla^2\vec{u}_{2D} + \sqrt{2}\sin k_z z \; (\nabla^2\vec{u}_{3D} - k_z^2\vec{u}_{3D})].$$

$$(XIII - 3 - 8)$$

Since $2\sin^2 k_z z = 1 - \cos 2k_z z$, and eliminating the $\cos 2k_z z$ term (which corresponds to the truncation), we obtain after identification of the $\sin k_z z$ terms

$$\frac{\partial \vec{u}_{2D}}{\partial t} + \vec{u}_{2D}.\vec{\nabla}\vec{u}_{2D} + \vec{u}_{3D}.\vec{\nabla}\vec{u}_{3D} = -\frac{1}{\rho}\vec{\nabla}p_{2D} + \nu\nabla^2\vec{u}_{2D} \quad (XIII - 3 - 9)$$

$$\frac{\partial \vec{u}_{3D}}{\partial t} + \vec{u}_{2D}.\vec{\nabla}\vec{u}_{3D} + \vec{u}_{3D}.\vec{\nabla}\vec{u}_{2D} = -\frac{1}{\rho}\vec{\nabla}p_{3D} + \nu\nabla^2\vec{u}_{3D} \quad (XIII - 3 - 10)$$

Notice that, in eq. (XIII-3-10), the $\sim \nu k_z^2\vec{u}_{3D}$ term arising in (XIII-3-8) has been neglected in the dissipation term; we think this term would have only a negligible influence on the following results.

Now let

$$\vec{u}_1(x,y,t) = \vec{u}_{2D} + \vec{u}_{3D}, \quad \vec{u}_2(x,y,t) = \vec{u}_{2D} - \vec{u}_{3D} \quad (XIII - 3 - 11)$$

and similarly for p_1 and p_2. It can be easily checked from eqs (XIII-3-9) and (XIII-3-10) that $\vec{u}_1$ and $\vec{u}_2$ both satisfy independent two-dimensional Navier-Stokes equations, with the same boundary conditions as $\vec{u}(x,y,z,t)$ for $y = \pm\infty$. Therefore, and since initially $\vec{u}_1$ and $\vec{u}_2$ are very close, the growth of this particular three-dimensional perturbation can be expressed in terms of the predictability problem studied above: the study of $\frac{1}{2}(\vec{u}_1 - \vec{u}_2) = \vec{u}_{3D}$ in the predictability problem will give access to the three-dimensional perturbation amplitude, the error rate being identified with the kinetic energy of the three-dimensional perturbation, averaged over a wave length $2\pi/k_z$ in the spanwise direction (see Lesieur et al., 1988).

Such an analysis is of course subject to criticism, since there is no spanwise velocity in the velocity field. The discarding of the $\cos 2k_z z$ term eliminates the harmonic spanwise wave length π/k_z which contains certainly some important three-dimensional characteristics of the flow, such as the existence of secondary streamwise vortices mentioned in Chapter III, or small-scale three-dimensional turbulence. However, this analysis could take into account certain three-dimensional characteristics of the large coherent eddies. At any rate, the present two-mode expansion provides a systematic formalism relating a wave-like tilting of the Kelvin-Helmholtz billows to the two-dimensional predictability. This gives a mathematical justification to and sheds a new light on an analogy already proposed by Lesieur (1984) and Staquet et al. (1985), and discussed from the point of view of the statistical theory of turbulence by Métais and Lesieur (1986): within this formalism, the two fields $\vec{u}_1$ and $\vec{u}_2$ were two cross-sections of the flow in the $x - y$ plane at two distinct spanwise locations, which were assumed to be decorrelated. This could be valid only for spanwise wave lengths much longer than the vorticity thickness. Here, on the contrary, the spanwise wavenumber k_z may be arbitrary.

Due to this analogy, the growth of unpredictability displayed above may be associated with a growth of three-dimensionality. One of the main reasons for the two-dimensional unpredictability between $\vec{u}_1$ and $\vec{u}_2$ introduced in eq. (XIII-3-11) is here the randomness of the initial subharmonic perturbation, which will be responsible for an important decorrelation between the two fields at the time of the first pairing. The maximum decorrelation would be obtained if $\vec{u}_1$ and $\vec{u}_2$ are two rows of vortices modulated by a subharmonic perturbation of wave length $2\lambda_a$, as indicated in Figure XIII-9-a: in this case, the resulting fields after the pairing will consist in two rows of Kelvin-Helmholtz vortices of wave length $2\lambda_a$ in phase opposition (Fig XIII-9-b).

The three-dimensional field given by eq. (XIII-3-6) and associated with Figure XIII-9-a consists in a row of Kelvin-Helmholtz billows oscillating (with a spanwise wave length k_z) in a vertical plane (y, z), in phase opposition from one billow to the other, as shown on Figure XIII-10-a. This corresponds to a subharmonic[38] longitudinal mode $k_a/2$ in $\vec{u}_{3D}(x, y, t)$. Hence the basic flow $\vec{u}_{2D}$ in eq. (XIII-3-6) is here perturbed by a subharmonic excitation modulated by a sine wave in the spanwise direction: in this case, the "peaks" P will tend to move to the right[39], entrained by the upper flow. The "valleys" will move to the left, and there will be a tendency for the pairings to occur as indicated on Figure XIII-10-b: the resulting topology will be that of a *vortex lattice*, as shown on the figure. It must be stressed[40] that the same vortex-lattice topology has also been proposed by Pierrehumbert and Widnall (1982) as the result of a *helical-*

---

[38] with the proper phase for the pairing

[39] that is, in the direction of positive x

[40] I am indebted to P. Comte and J. Riley for drawing my attention to the following points.

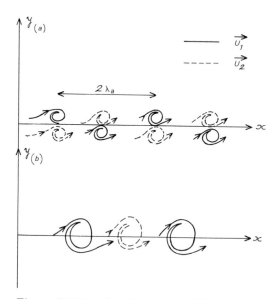

Figure XIII-9: a) initial fields $\vec{u}_1(x,y,0)$ and $\vec{u}_2(x,y,0)$, modulated by two subharmonic perturbations in phase opposition; b) the velocity fields after the pairing.

pairing instability, corresponding to the growth of the same perturbation (longitudinal subharmonic modulated by a spanwise sine wave) through the secondary-instability analysis which leads also to the translative instability already mentioned. Within this stability study, the translative instability is is more amplified than the helical-pairing one[41]. Our result is independent of k_z , which has been eliminated from the equations, and hence the three-dimensional amplitude will grow at the same rate whatever the spanwise wave length of the perturbation. This is a difference with respect to the *helical-pairing instability* of Pierrehumbert and Widnall (1982). Similar branchings of vortices have also been observed by Browand and Troutt (1980). Even in a plane wake, Cimbala, Nagib and Roshko (1988) observe a three-dimensional vortex structure resembling very much the above-discussed one, and attribute it to *a secondary instability in the form of a parametric subharmonic resonance similar to that analysed by Pierrehumbert and Widnall (1982) in free shear layers, and more extensively for boundary layers and channel flows by Herbert (1988).*

---

[41] thus called after the experimental observations of Chandrsuda et al. (1978) who discovered such a structure in a plane mixing layer with a high level of turbulence in the free stream

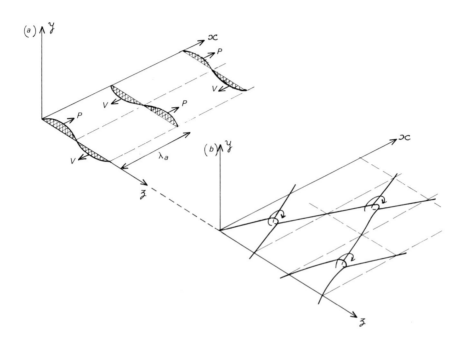

Figure XIII-10: a) three-dimensional field $\vec{u}(x, y, z, 0)$ given by eq. (XIII-3-6) and corresponding to Figure XIII-9-a; b) the vortex lattice structure resulting from the evolution of the initial field shown in (a).

4 - 3D numerical simulations of the mixing layer

4.1 Direct-numerical simulations

We have already mentioned in Chapter III the work of Metcalfe et al. (1988) and Sandham and Reynolds (1990) for respectively an incompressible and compressible mixing layer. Now we present some results which appear to validate (for particular initial conditions) the vortex lattice model of three-dimensionality growth proposed above: the same vortex lattice structure has been recently found by Comte and Lesieur

(1990) in a direct-numerical simulation of a three-dimensional temporal mixing layer submitted initially to a small three-dimensional random perturbation. This calculation, using pseudo-spectral methods, is done at a resolution of 128^3 Fourier wave vectors, at a Reynolds number $U\delta_i/\nu = 100$. The vortex lattice structure which develops is shown on Plate 17-a, and resembles strikingly the schematic lattice[42] of Figure 10 b. On Plate 17 b are shown the corresponding vortex lines, which display also the existence of thinner longitudinal vortex filaments strained by the flow from the zone of merging between the fundamental billows. These filaments are also visible on Plate 17 a. Plate 17 c shows the interface between the two layers, visualized with the aid of a *numerical dye*. Plate 18 shows a side view (in the $x - y$ plane) of this interface, indicating apparently regular Kelvin-Helmholtz vortices[43].

On the contrary, the same calculation done with an initial quasi-two-dimensional perturbation shows quasi-two-dimensional billows, between which thin longitudinal vortex filaments are strained (see Plate 19). In fact, it is not so easy to discriminate experimentally on the basis of velocity measurements between the *longitudinal vortex filament structure* and the vortex-lattice structure: Huang and Ho (1990) show the existence of spanwise velocity fluctuations, and attribute it to *the development of streamwise vortices*. However, a vortex-lattice structure of the billows would also be responsible for spanwise velocity.

In Plate 20 is shown the direct-numerical simulation of a mixing layer behind a backwards-facing step, done by Silveira (1990). The longitudinal and spanwise vorticity components show very clearly the formation of thin vortices strained longitudinally, and whose spanwise wave length increases downstream with the layer thickness. It would be interesting to see whether these structures exist in the laboratory experiments.

4.2 Large-eddy simulations of mixing layers

A three-dimensional spectral calculation was carried out by Comte (1989), using the spectral subgrid-scale modelling defined by eq. (II-4-13). The computational box contained two fundamental longitudinal wave lengths, and a three-dimensional random perturbation was superimposed onto

---

[42] It is interesting to note that the same three-dimensional tendency was found by Sandham and Reynolds (1990) for the numerical simulation of a compressible mixing layer. Let us mention also the kinematic model of three-dimensional vortex lattice proposed by Chorin (1986) in order to describe the inertial range of turbulence.

[43] It is possible that some mixing-layer experiments showing coherent structures when visualizing cross sections of the flow could correspond to a high degree of three-dimensionalization.

the basic velocity profile. The calculation evolved into two big rollers, straining vortex filaments longitudinally (see Plate 21). When the turbulence in the small scales has developed (during the pairing), the variances of the three velocity components are found to be:

$$<u'^2>= 0.11 \ U^2 \ , \ <v'^2>= 0.07 \ U^2 \ , \ <w'^2>= 0.08 \ U^2 \ ,$$

which is in fairly good agreement with the experiments (see e.g. Browand and Troutt, 1980).

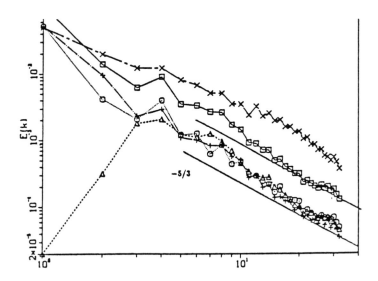

Figure XIII-11: large-eddy simulation of the mixing layer; three-dimensional kinetic energy spectra of (from top to bottom) the passive temperature, the kinetic energy and the three velocity components (longitudinal, transverse and spanwise). From Comte (1989).

Figure XIII-11 shows the three-dimensional spatial spectra[44] of the passive temperature, the kinetic energy, and the three components of velocity, all taken at the same point in time. It indicates how, starting from initial spectra exponentially decaying at high wavenumbers, a cascade has developed towards the small scales, with a slope which is fairly close to the Kolmogorov spectrum. This is again in good agreement with the experimental measurements of these spectra (see e.g. Browand and Ho, 1983).

4.3 Recreation of the coherent structures

We have proposed above some mechanisms which may[45] contribute to the three-dimensionalization of the mixing layer and, in certain condi-

---

[44] that is, integrated on a sphere of radius k
[45] for some particular initial conditions

tions, to a three-dimensional disruption of the coherent structures. This corresponds to the phenomenon of intermittency well known by experimentalists. If this is the case, one can then propose a mechanism of recreation, after the possible destruction due to the three-dimensionality growth: let us assume now that we are in a situation of complete three-dimensionalization of the layer, that is of three-dimensional turbulence superposed to a mean inflexional velocity shear[46]. Let δ be the mean vorticity thickness. We assume also, from Chapter XII, that the action of three-dimensional turbulence on the two-dimensional large-scale motions can be modelled with the aid of an eddy-viscosity ν_t

$$[\frac{\partial}{\partial t} + J(., \psi)]\nabla^2 \psi = \nu_t \nabla^2 (\nabla^2 \psi) \qquad (XIII - 4 - 1)$$

where $\psi(x, y, t)$ is a stream function representing the two-dimensional large scales (mean shear + a two-dimensional broad-band spectrum perturbation $\tilde{\psi}(x, y, t)$). The value of ν_t could simply be evaluated by the experimental measurements of the Reynolds stresses (see Liepmann and Laufer, 1949; Wygnanski and Fiedler, 1970). This leads to an eddy-Reynolds number

$$R_t = \frac{U\delta}{\nu_t} = 30 \approx 40 \qquad (XIII - 4 - 2)$$

which is high enough for the inviscid linear stability theory to be applied (see Fig III-1). Then the perturbation $\tilde{\psi}$, initially small, will satisfy an Orr-Sommerfeld equation and the most amplified coherent structure $\lambda_a = 7\,\delta$ will appear in a characteristic time $\approx 10 \sim 15\,\delta/U$: the coherent structure would then emerge from the smaller-scale turbulence. This fact has been verified in three-dimensional direct-numerical simulations done by Metcalfe et al. (1988).

We have then proposed two competing mechanisms, one of destruction and one of recreation of the coherent structures. They are certainly extremely schematic, but could shed some light on the dynamics of the mixing layer. It is also possible that some of the conclusions relating to the three-dimensional instability growth and the recreation of coherent structures could be applied to other large Reynolds number shear flows, even if the shear is not inflexional. The dynamics of the turbulent boundary layer in particular could be envisaged in this way.

4.4 Rotating mixing layers

To finish with this section, we would like to give some details on the three-dimensional rotating mixing-layer simulations done by Yanase et

---

[46] This is for instance the case when a grid is placed in a mixing layer behind the splitter plate.

al. (1990, see also Lesieur et al., 1990) and already mentioned in Chapter III. The solid-body rotation vector is parallel to the z axis. These direct-numerical simulations are performed using pseudo-spectral methods (with lateral boundaries to infinity treated using a coordinate mapping) at a resolution of $48 \times 48 \times 24$ (two longitudinal fundamental modes) starting initially with a three-dimensional random perturbation superposed on the basic flow $U \tanh 2y/\delta_i$. The Rossby number is now defined as $R_O = U/\Omega \delta_i$. Figure XIII-12 shows the growth of the spanwise kinetic energy[47] ($< w'^2 >$), for values of the Rossby number ranging from infinity (no rotation) to 0.25. In the cyclonic case (where the solid-body vorticity is of same sign as the coherent vortices), rotation is always stabilizing with respect to the non-rotating case: an interesting feature may be noticed, that is, the initial decay of the three-dimensional perturbation, followed by a growth which is delayed as the rotation increases. In the anticyclonic case, and at $t = 25 \, \delta_i/U$ (slightly before the pairing), the highest three-dimensional energy is for $R_O = 4 \sim 5$. Visualizations of the pressure field for $R_O = \infty$ (non-rotating case) show that, at $t = 25$, the flow possesses strong quasi-two-dimensional billows. For $R_O = 11$ (anticyclonic case), the billows are three-dimensionally distorted, as in the secondary instabilities[48] discussed above, but remain coherent. On the other hand, the $R_O = 5$ case shows a totally three-dimensional and chaotic pressure field, which seems to confirm the ideas of explosive stretching of absolute vorticity proposed in Chapter III-5. Still in the anticyclonic case, rotation is stabilizing again for $R_O < \approx 2$. This confirms the qualitative theory of Chapter III, and suggests that an anticyclonic solid-body rotation of spanwise axis may be a very efficient mechanism to destroy the coherent structures, as already noticed experimentally.

5 - Conclusion

It is of course not possible to give a definitive conclusion to this monograph, which poses certainly more questions than it provides answers. Therefore I will leave the reader to draw his own conclusions, hoping that I have been able to propose some new trails to his imagination.

The world of turbulence is so wide that we have chosen to present only a limited number of topics and techniques, which are not usually encountered in textbooks, and which could be of great practical use: the

---

[47] which is a good indicator of three-dimensionality
[48] vortex lattice, or helical pairing

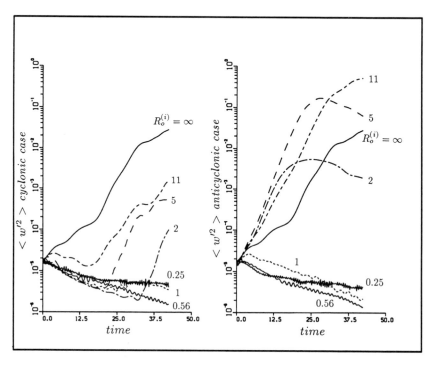

Figure XIII-12: growth of the three-dimensional kinetic energy in the rotating mixing layer; a) cyclonic case; b) anticyclonic case. From Yanase et al. (1990).

stochastic modelling tools are extremely effective to understand the energy transfers between different scales of motion. We have in this context proposed a synthetic presentation of the *non-local expansion techniques.* The stochastic models allow one to calculate the time evolution of various quantities related to turbulence. They also permit us to introduce new concepts such as the growth of unpredictability. They predict[49] the behaviour of scalars diffused by turbulence in a wide range of situations, and show in particular that the classical Reynolds analogy between the momentum and scalar diffusion has to be reconsidered. These turbulent diffusion results could be of great help in combustion or pollution problems for instance. The limitation of the theories presented here to isotropic situations is acceptable to describe the small scales of turbulence. Thus they can provide useful information for more applied theories such as the "$K - \epsilon$" method.

We also have put emphasis on two-dimensional turbulence dynamics, both for its geophysical importance, and because the severe cons-

---

[49] at least qualitatively

traint imposed by two-dimensionality marks the limit of validity of such an approximation.

We have placed all these theories in a double context: the basic principles of fluid dynamics, and the spectral formalism of homogeneous turbulence. The latter has been presented by referral to new orthogonal expansions of the velocity field, such as the *wave-vortex* or *complex helical waves* decomposition.

The Direct and Large-eddy simulations of turbulence have allowed us to study the dynamics of isotropic turbulence, for which they provide a new tool allowing one to assess the statistical theories[50]. Applied to stably-stratified turbulence, they permit one to study the interaction between internal gravity waves and turbulence. This is of great importance for the mesoscale atmosphere and the vertical mixing processes in the ocean. Finally, these numerical-simulation techniques[51] are becoming an exceptional tool to study *transition* and *vortex dynamics*.

A continuous preoccupation throughout the book concerned the question of the existence of *coherent structures* as a part of the turbulence itself, and how they interact with smaller scales in the flow. We have looked at the turbulent mixing layer within this context.

What will be the near future of Fluid Mechanics? It seems that a period has come in which three-dimensional large-eddy simulations of flows of reasonable complexity [52] are not only at hand, but are even less expensive than laboratory experiments[53]. These numerical simulations need efficient subgrid-scale modelling in order to take into account the small-scale turbulence. They need also to be devised in such a way as to reproduce perfectly the results of the linear instability theory [54]. The numerical codes have to be validated on simpler cases such as the isotropic turbulence, and must be consistent with the physical principles which have emerged from the theory, like the existence of inertial cascades, the correct decay laws or the unpredictability. It is

---

[50] However, the numerical simulation of a Kolmogorov cascade of several decades does not appear to be at hand.

[51] allied with three-dimensional visualizations

[52] "reasonable", in the sense of the large eddies around a wing or in the jet of an engine, or the planetary scales of atmospheres and oceans, or the convective cells in a heated flow; but certain extremely complex industrial flows (like within the core of a fast-breeder reactor for instance) do not fall into this category.

[53] In some situations such as in very high speed or high temperature flows, the experiment (or measurements) may even be impossible (or too dangerous).

[54] This point is extremely important in the context of the study of the large coherent structures.

then important in this context to maintain and develop the stochastic modelling tools, and make them as simple as possible to handle.

Such a trend in Fluid Dynamics research will in no way exclude the recourse to theory. Laboratory experiments will of course be necessary, either for a better understanding of fundamental physics, or in complex situations where developing a calculation is not realistic. They are also needed in order to validate the numerical codes.

Of course, all these ambitious projects will be rendered possible only with a proper development of the computing, data processing and visualization facilities. Parallel calculations performed simultaneously on a large number of processors will certainly play an important role in this development.

It might finally happen that this would be only a necessary transition stage towards the definition of new fluid dynamical concepts which would render obsolete and useless the complicated analytical and numerical techniques which helped create them.

REFERENCES

Allègre, C., 1983, *L'écume de la Terre*, Fayard.

André, J.C., 1974, "Irreversible interaction between cumulants in homogeneous, isotropic, two-dimensional turbulence theory", Phys.Fluids, **17**, pp 15-21.

André, J.C., 1988, private communication.

André, J.C. and Lesieur, M., 1977, "Influence of helicity on high Reynolds number isotropic turbulence", J. Fluid Mech., **81** , pp 187-207.

Anselmet, F., Gagne, Y., Hopfinger, E.J. and Antonia, R.A., 1984, "High-order velocity structure functions in turbulent shear flows", J. Fluid Mech., **140** , pp 63-89.

Antonia, R.A., Hopfinger, E..J., Gagne, Y. and Anselmet, F., 1984, "Temperature structure functions in turbulent shear flows", Phys. Rev. A, **30**, pp 2704-2707.

Aref, H., 1983, "Integrable, chaotic and turbulent vortex motion in two-dimensional flows", Ann. Rev. Fluid Mech., **15**, pp 345-389.

Aref, H., 1985, "Chaos in the dynamics of few vortices - fundamentals and applications", in *Theoretical and Applied Mechanics*, F.L. Niordson and N. Olhoff eds., Elsevier Science Publishers, pp 43-68.

Argoul, F., Arnéodo, A., Grasseau, G., Gagne, Y. Hopfinger, E.J. and Frisch, U., 1989, "Wavelet analysis of turbulence reveals the multifractal nature of the turbulent cascade", Nature, **338**, pp 51-53.

Arnal, D., 1988, "Stability and transition of two-dimensional laminar boundary layers in compressible flow over an adiabatic wall", Rech. Aérosp., **1988-4**, pp 15-32.

Ashurst, W.T., 1979, "Numerical simulation of turbulent mixing layers via vortex dynamics", in *Turbulent Shear Flows I*, Springer-Verlag, pp 402-413.

Atten, P., Caputo, J.G., Malraison, B. and Gagne, Y., 1984, "Determination of attractor dimension of various flows", in *Bifurcations and chaotic behaviours*, J. Mec. Theor. Appl., Suppl., G. Iooss and R. Peyret ed., pp 133-156.

Aubry, N., Holmes, P., Lumley, J.L. and Stone, E., 1988, "The dynamics of coherent structures in the wall region of a turbulent-boundary layer", J. Fluid Mech., **192**, pp 115-173.

Aupoix, B. and Cousteix, J., 1982, "Modèles simples de tension sous-maille en turbulence homogène et isotrope", La Recherche Aérospatiale, **4** , pp 273-283.

Babiano, A., Basdevant, C. and Sadourny, R., 1985, "Structure functions and dispersion laws in two-dimensional turbulence", J. Atmos. Sci., **42** , pp 941-949.

Babiano, A., Basdevant, C., Legras, B. and Sadourny, R., 1987, "Vorticity and passive-scalar dynamics in two-dimensional turbulence", J. Fluid Mech., **183**, pp 379-397.

Balint, J.L., Vukoslavcevic, P. and Wallace, J.M., 1986, "A study of the vortical structure of the turbulent-boundary layer", in *Advances in turbulence*, G. Comte-Bellot and J. Mathieu, eds, Springer-Verlag, pp 456-464.

Bartello, P. and Holloway, C., "Passive scalar transport in β-plane turbulence", J. Fluid Mech., in press.

Basdevant, C., 1981, "Contribution à l'étude numérique et théorique de la turbulence bidimensionnelle", Thèse de doctorat d'Etat, Université Pierre et Marie Curie, Paris.

Basdevant, C., Legras, B., Sadourny, R. and Beland, B., 1981, "A study of barotropic model flows: intermittency, waves and predictability", J. Atmos. Sci., **38** , pp 2305-2326.

Basdevant, C., Lesieur, M. and Sadourny, R., 1978, "Subgrid-scale modelling of enstrophy transfer in two dimensional turbulence", J. Atmos. Sci., **35** , pp 1028-1042.

Basdevant, C. and Sadourny, R., 1975,"Ergodic properties of inviscid truncated models of two dimensional incompressible flows", J. Fluid Mech., **69** , pp 673-688.

Basdevant, C. and Sadourny, R., 1983, "Parameterization of virtual scale in numerical simulation of two-dimensional turbulent flows", in *Two-dimensional turbulence*, J. Mec. Theor. Appl., Suppl. , R. Moreau ed., pp 243-270.

Batchelor, G.K., 1949, "Diffusion in a field of homogeneous turbulence: I. Eulerian analysis", Austral. J. Sci. Res., **2**, pp 437-450.

Batchelor, G.K., 1952, "Diffusion in a field of homogeneous turbulence: II. The relative motion of particles", Proc. Camb. Phil. Soc., **48** , pp 345-362.

Batchelor, G.K., 1953, "The theory of homogeneous turbulence", Cambridge University Press.

Batchelor, G.K., 1959, "Small scale variation of convected quantities like temperature in turbulent fluid. Part 1. General discussion and the case of small conductivity". J. Fluid Mech., **5** , p 113-134.

Batchelor, G.K., 1967, *An introduction to fluid dynamics*, Cambridge University Press.

Batchelor, G.K., 1969, "Computation of the energy spectrum in homogeneous two-dimensional turbulence", Phys. Fluids Suppl., II **12** , pp 233-239.

Batchelor, G.K., Howells, I.D. and Townsend, A., 1959, "Small scale variation of convected quantities like temperature in turbulent fluid. Part 2. The case of large conductivity", J. Fluid Mech., **5** , pp 134-139.

Batchelor, G.K. and Melb, M. 1946, Proc. Roy. Soc. A, **186** , pp 840-902.

Batchelor, G.K. and Townsend, A.A., 1947, "Decay of vorticity in isotropic turbulence", Proc. Roy. Soc., Vol A **191** , pp 534-550.

Batchelor, G.K. and Townsend, A.A., 1949, Proc. Roy. Soc., Vol A **199**, p 238

Bègue, C., Chacón, T., Ortegón, F. and Pironneau, O., 1987, "3D simulation of two length scales turbulent flows by homogenization, in *Advances in turbulence*, G. Comte-Bellot and J. Mathieu, eds, pp 135-142, Springer-Verlag.

Bergé, P., Pomeau, Y. and Vidal, C., 1984, *L'ordre dans le chaos*, Hermann, Paris.

Bernal, L.P., 1981, "The coherent structure of turbulent mixing layers" Ph.D. Thesis. Calif. Inst. Technol. Pasadena.

Bernal, L.P. and Roshko, A., 1986, "Streamwise vortex structure in plane mixing layer", J. Fluid Mech., **170**, pp 499-525.

Bertoglio, J.P., 1986, Thèse de Doctorat d'Etat, Université Claude Bernard, Lyon.

Betchov, R. and Szewczyk, G. , 1963, "Stability of a shear layer between parallel streams", Phys. Fluids, **6** , pp 1391-1396.

Blackwelder, R.F., 1979, "Boundary-layer transition", Phys. Fluids, **22**, pp 583-584.

Blanc-Lapierre, A. and Picinbono, B., 1981, "Fonctions Aléatoires", Masson.

Blumen, W., 1970, "Shear-layer instability of an inviscid compressible fluid", J. Fluid Mech., **40**, pp 769-781.

Bogdanoff, D.W., 1983, "Compressibility effects in turbulent shear layers", AIAA J., **21**, pp 926-927.

Brachet, M.E., 1982, "Intégration numérique des équations de Navier-Stokes en régime de turbulence développée", C. R. Acad. Sci. Paris, Ser. B, **294**, pp 537-540.

Brachet, M.E., 1990, "Géométrie des structures à petite échelle dans le vortex de Taylor-Green", C. R. Acad. Sci. Paris, Ser. B, to appear.

Brachet, M.E., Meiron, D.I., Orszag, S.A., Nickel, B.G., Morf, R.H. and Frisch, U., 1983, "Small scale structure of Taylor Green vortex", J. Fluid Mech., **130** , pp 411-452.

Brachet, M.E., Méneguzzi, M. and Sulem, P.L., 1986, "Small scale dynamics of high Reynolds number two-dimensional turbulence", Phys. Rev. Lett., **57**, p 683.

Brachet, M.E., Méneguzzi, M., Politano, H. and Sulem, P.L., 1988, "The dynamics of freely decaying two-dimensional turbulence", J. Fluid Mech., **194**, pp 333-349.

Breidenthal, R., 1981, "Structure in turbulent mixing layers and wakes using a chemical reaction", J. Fluid Mech., **109**, pp 1-24.

Bretherton, F.P. and Haidvogel, D.B., 1976, "Two- dimensional turbulence above topography", J. Fluid Mech., **78** , pp 129-154.

Brissaud, A., Frisch, U., Léorat, J., Lesieur, M., Mazure, A., Pouquet, A., Sadourny, R. and Sulem, P.L., 1973 a, "Catastrophe énergétique et nature de la turbulence", Annales de Géophysique (Paris), **29**, pp 539-546.

Brissaud, A., Frisch, U., Léorat, J., Lesieur, M., and Mazure, A., 1973 b, "Helicity cascades in fully developed isotropic turbulence", Phys. Fluids, **16** , pp 1366-1367.

Browand, F.K. and Troutt, T.R., 1980, "A note on spanwise structure in the two-dimensional mixing layer" J. Fluid Mech., **93**, pp 325-336.

Browand, F.K. and Ho, C.M., 1983, "The mixing layer: an example of quasi two-dimensional turbulence", in *Two-dimensional turbulence*, J. Mec. Theor. Appl., Suppl. , R. Moreau ed., pp 99-120.

Brown, G.L. and Roshko, A., 1974, "On density effects and large structure in turbulent mixing layers", J. Fluid Mech., **64** , pp 775-816.

Busse, F.H., 1981, "Transition to turbulence in Rayleigh-Bénard convection", in *Hydrodynamic instabilities and the transition to turbulence*, edited by H.L. Swinney and J.P. Gollub, Springer-Verlag, pp 97-137.

Busse, F.H., 1983, "A model of mean zonal flows in the major planets", Geophys. Astrophys. Fluid Dynamics, **23** , pp 153-174.

Cambon, C., 1983, "Etude spectrale de la turbulence: rotation pure et introduction à la décomposition modale", Contract DRET-Ecole Centrale de Lyon 83225.

Cambon, C., Jeandel, D. and Mathieu, J., 1981, "Spectral modelling of homogeneous non isotropic turbulence", J. Fluid Mech., **104** , pp 247-262.

Cambon, C., Tesseidre, C. and Jeandel, D., 1985, "Effets couplés de déformation et de rotation sur une turbulence homogène", J. Mec. Theor. Appl., **4** , pp 629-657.

Canuto, C., Hussaini, M.Y., Quarteroni, A. and Zang, T.A., 1987, "Spectral methods in fluid dynamics", Springer.

Capéran, P., 1982, "Contribution à l'étude expérimentale de la turbulence homogène *M.H.D.*. Première caractérisation de son anisotropie", Thèse, INPG and USMG, Grenoble.

Carnevale, G.F., 1982, " Statistical features of the evolution of two dimensional turbulence", J. Fluid Mech., **122** , pp 143-153.

Carrigan, C.R. and Busse, F.H., 1983, "An experimental and experimental investigation of the onset of convection in rotating spherical shells", J. Fluid Mech., **126**, pp 287-305.

Castaing, B., Gunaratne, G., Heslot, F., Kadanoff, L., Libchaber, A., Thomae, S., Wu, X.Z., Zaleski, S. and Zanetti, G., 1989, "Scaling of hard thermal turbulence in Rayleigh-Bénard convection", J. Fluid Mech., 204, pp 1-30.

Chabert d'Hières, G., Davies, P.A. and Didelle, H., 1988, "Laboratory studies of pseudo-periodic forcing due to vortex shedding from an isolated solid obstacle in a homogeneous rotating fluid", 20 th Int. Liège Colloquium on Ocean Hydrodynamics, 2-6 May 1988, Elsevier.

Champagne, F.H., Friehe, C.A., LaRue, J.C., Wyngaard, J.C., 1977, J. Atmos. Sci., **34**, p 515

Chandrasekhar, S., 1950, "The theory of axisymmetric turbulence", Phil. Trans. A, **242**, pp 557-577.

Chandrsuda, C., Mehta, R.D., Weir, A.D. and Bradshaw, P., 1978, "Effect of free-stream turbulence on large structure in turbulent mixing layers", J. Fluid Mech., **85**, pp 693-704.

Charney, J.G., 1947, "The dynamics of long waves in a baroclinic westerly current", J. Meteor., **4**, pp 135-163.

Charney, J.G., 1971, "Geostrophic turbulence", J. Atmos. Sci., **28**, pp 1087-1095.

Chasnov, J., Canuto, V.M. and Rogallo, R.S., 1988, "Turbulence spectrum of a passive temperature field: results of a numerical simulation", Phys. Fluids, **31**, pp 2065-2067.

Chen, J.H., Cantwell, B.J. and Mansour, N.N., 1990, "The effect of Mach number on the stability of a plane supersonic wake", Sandia Report SAND90-8423, to appear in Phys. Fluids.

Chollet, J.P., 1983, "Turbulence tridimensionnelle isotrope: modélisation statistique des petites échelles et simulation numérique des grandes échelles", Thèse de Doctorat d'Etat, Grenoble.

Chollet, J.P., 1984, "Two-point closures as a subgrid scale modelling for large eddy simulations", in "Turbulent Shear Flows IV", edited by F. Durst and B. Launder, Lecture Notes in Physics, Springer-Verlag.

Chollet, J.P. and Lesieur, M., 1981, "Parameterization of small scales of three-dimensional isotropic turbulence utilizing spectral closures", J. Atmos. Sci., **38** , pp 2747-2757.

Chollet, J.P. and Lesieur, M., 1982, "Modélisation sous maille des flux de quantité de mouvement et de chaleur en turbulence tridimensionnelle isotrope", La Météorologie, **29-30** , p 183-191.

Chollet, J.P. and Métais, O., 1989, "Predictability of three-dimensional turbulence in large-eddy simulations", Eur. J. Mech., B/Fluids, 8, pp 523-548.

Chorin, A.J., 1973, "Numerical study of slightly viscous flow", J. Fluid. Mech., **57**, pp 785-796.

Chorin, A.J., 1986, "Turbulence and vortex stretching on a lattice", Comm. Pure and Appl. Maths, **XXXIX**, pp S47-S65.

Chou, P.Y., 1940, "On an extension of Reynolds' method of finding apparent stress and the nature of turbulence". Chin. J. Phys. 4, pp 1-33.

Cimbala, J.M., Nagib, H.M. and Roshko, A., 1988, "Large structure in the far wakes of two-dimensional bluff bodies", J. Fluid Mech., **190**, pp 265-298.

Clark, R.A., Ferziger, J.H. and Reynolds, W.C., 1979, "Evaluation of subgrid scale models using an accurately simulated turbulent flow", J. Fluid Mech., **91** , pp 1-16.

Colin de Verdière, A., 1979, "Mean flow generation by topographic Rossby waves", J. Fluid Mech., **94** , pp 39-64.

Colin de Verdière, A., 1980, "Quasi geostrophic turbulence in a rotating homogeneous fluid", Geophys. Astrophys. Fluid Dynamics, **15**, pp 213-251.

Comte, P., 1989, "Etude par simulation numérique de la transition à la turbulence en écoulement cisaillé libre", Thèse, Institut National Polytechnique de Grenoble.

Comte, P. and Lesieur, M., 1990, "Large and small-scale stirring of vorticity and passive scalar in a 3D temporal mixing layer", IUTAM Symposium on *Fluid Mechanics of Stirring and Mixing*, La Jolla, August 1990. To appear in the Physics of Fluids.

Comte, P., Lesieur, M. and Chollet, J.P., 1987, "Simulation numérique d'un jet plan turbulent", C. R. Acad. Sci., ser. II, **305**, pp 1037-1044.

Comte, P., Lesieur, M. Laroche, H. and Normand, X., 1989, "Numerical simulations of turbulent plane shear layers", in *Turbulent Shear Flows 6*, Springer-Verlag, pp 360-380.

Comte-Bellot, G. and Corrsin, S., 1966, "The use of a contraction to improve the isotropy of a grid generated turbulence", J. Fluid Mech., **25** , pp 657-682.

Constantin, P., Foias, C., Manley, O.P. and Temam, R., 1985, "Determining modes and fractal dimension of turbulent flows", J. Fluid Mech., **150** , pp 427-440.

Corcos, G.M. and Sherman, F.S., 1984, "The mixing layer: deterministic models of a turbulent flow. Part 1: Introduction and the two-dimensional flow" J. Fluid Mech., **139**, pp 29-65.

Corcos, G.M. and Lin, S.J., 1984, "The mixing layer: deterministic models of a turbulent flow. Part 2. The origin of the three-dimensional motion", J. Fluid Mech., **139**, pp 67-95.

Corrsin, S., 1951, J. Appl. Phys., **22**, p 469

Corrsin, S., 1962, "Turbulent dissipation fluctuations", Phys. Fluids, **5**, p 1301.

Corrsin, S., 1964, "The isotropic turbulent mixer: Part II. Arbitrary Schmidt number". A.I.Ch.E.J. **10** , pp 870-877.

Couder, Y., 1984, "Two-dimensional grid turbulence in a thin liquid film", J. Phys. Lett., **45**, pp 353-360.

Coüet, B., 1985, "Simulation tridimensionnelle vortex in cell: méthode et application", in *Société Française de Physique, Nice 1985*, Les Editions de Physique, pp 5-14.

Craya, A., 1958, "Contribution à l'analyse de la turbulence associée à des vitesses moyennes", P.S.T. Ministère de l'Air, **345** .

Crow, S.C. and Champagne, F.H., 1971, "Orderly structure in jet turbulence", J. Fluid Mech., **48**, pp 547-591.

Dannevik, W.P., Yakhot, V. and Orszag, S.A., 1988, "Analytical theories of turbulence and the ϵ expansion", Phys. Fluids, **30**, pp 2021-2029.

Deardorff, J.W., 1970, "A numerical study of three-dimensional turbulent channel flow at large Reynolds number", J. Fluid Mech., **41**, pp 453-480.

d'Humières, D., Lallemand, P. and Frisch, U., 1986, "Simulations d'allées de Von Karman bidimensionnelles à l'aide d'un gaz de réseau", C. R. Acad. Sci. Paris, B **301**, p 11

Domaradzki, J.A., Metcalfe, R.W., Rogallo, R.S. and Riley, J., 1987, "Analysis of subgrid-scale eddy viscosity with the use of the results from direct numerical simulations", Phys. Rev. Lett. **58**, pp 547-550.

Drazin, P.G. and Howard, L.N., 1966, "Hydrodynamic stability of parallel flow of inviscid fluid", Adv. Appl. Mech., **7**, pp 142-250.

Drazin, P.G. and Reid, W.H., 1981, "Hydrodynamic stability", Cambridge University Press.

Dumas, R., 1962, "Contribution à l'étude des spectres de turbulence", Thèse de Doctorat d'Etat, Université d'Aix-Marseille.

Dziomba, B. and Fiedler, H.E., 1985, "Effect of initial conditions on two-dimensional free shear layers", J. Fluid Mech., **152**, pp 419-442.

Eswaran, V. and O'Brien, E.E., 1989, "Simulations of scalar mixing in grid turbulence using an eddy-damped closure model", Phys. Fluids A, **1** , pp 537-548.

Fal'kovich, G.E. and Shafarenko, A.V., 1988, "Blow-up in weak-acoustic turbulence", Preprint N 393, Institute of Aut. and Elec., Sib. USSR Acad. Sci.

Farge, M. and Sadourny, R., 1989, "Wave-vortex dynamics in rotating shallow water", J. Fluid Mech., **206**, pp 433-462.

Favre, A., Guitton, H. and J., Lichnerowicz, A. and Wolf, E., 1988, *De la causalité à la finalité: à propos de la turbulence*, Maloine, Paris.

Favre-Marinet, M. and Binder, G., 1989, "Coherent structures and the turbulent Prandtl number", in *Advances in Turbulence 2*, H.H. Fernholtz and H.E. Fiedler eds., Springer-Verlag.

Fjortoft, R., 1953, "On the changes in the spectral distribution of kinetic energy for two-dimensional non-divergent flow", Tellus, **5**, pp 225-230.

Foias, C. and Penel, P., 1975, "Dissipation totale de l'énergie dans une équation non linéaire liée à la théorie de la turbulence", C. R. Acad. Sci., Paris, A **280**, pp 629-632.

Fornberg, B., 1977, "A numerical study of two-dimensional turbulence", J. Comp. Phys., **25**, p 1

Forster, D., Nelson, D.R. and Stephen, M.J., 1977, "Large distance and long time properties of a randomly stirred field", Phys. Rev. A, **16**, pp 732-749.

Fournier, J. D., 1977, "Quelques méthodes systématiques de développement en turbulence homogène", Thèse, Université de Nice.

Fournier, J.D. and Frisch, U., 1983 a, "L'équation de Burgers déterministe et statistique", J. Mec. Theor. Appl., **2**, pp 699-750.

Fournier, J.D. and Frisch, U., 1983 b, "Remarks on the renormalization group in statistical fluid dynamics", Phys. Rev. A, **28**, pp 1000-1002.

Fox, D.G. and Orszag, S.A., 1973, "Inviscid dynamics of two dimensional turbulence", Phys. Fluids, **16** , pp 167-171.

Friehe, C.A., Van Atta, C.W. and Gibson, C.H., 1971, "Jet turbulence: dissipation rate measurements and correlations". Proc. A.G.A.R.D. Spec. Meeting on Turbulent Shear Flow, London.

Frisch, U., 1974, private communication.

Frisch, U. and Fournier, J.D., 1978, in Nice Winter School on *Turbulence and non-linear Physics*, Nice Observatory, January 1978 (unpublished).

Frisch, U., Hasslacher, B. and Pomeau, Y., 1986, "Lattice gas automata for the Navier Stokes equation", Phys. Rev. Lett., **56**, p 1505

Frisch, U., Lesieur, M. and Brissaud, A., 1974, "A Markovian Random Coupling Model for turbulence" J. Fluid Mech., **65** , pp 145-152.

Frisch, U., Lesieur, M. and Schertzer, D., 1980, "Comment on the Quasi Normal markovian approximation for fully developed turbulence", J. Fluid Mech., **97** , pp 181-192.

Frisch, U., Pouquet, A., Léorat, J. and Mazure, A., 1975, "Possibility of an inverse cascade of magnetic helicity in magneto hydrodynamic turbulence" J. Fluid Mech., **68** , pp 769-778.

Frisch, U. and Sulem, P.L., 1984, "Numerical simulation of the inverse cascade in two-dimensional turbulence", Phys. Fluids, **27** , pp 1921-1923.

Frisch, U., Sulem, P.L. and Nelkin, M., 1978, "A simple dynamical model of intermittent fully developed turbulence", J. Fluid Mech., **87** , pp 719-736.

Frisch, U. and Orszag, S.A., 1990, "Turbulence: challenges for theory and experiment", Physics Today, January, pp 24-32.

Fulachier, L. and Dumas, R., 1976, "Spectral analogy between temperature and velocity fluctuations in a turbulent boundary layer", J. Fluid Mech., **77** , pp 257-277.

Fyfe, D., Montgomery, D. and Joyce, G., 1977, Phys. Fluids, **117**, p 369

Gage, K.S., 1979, "Evidence for a $k^{-5/3}$ law inertial range in meso scale two dimensional turbulence", J. Atmos. Sci., **36** , pp 1950-1954.

Gagne, Y., 1978, "Contribution à l'étude expérimentale de l'intermittence de la turbulence à petite échelle", Thèse de Docteur-Ingénieur, Grenoble.

Gagne, Y., 1987, "Etude expérimentale de l'intermittence et des singularités dans le plan complexe en turbulence développée", Thèse de Doctorat d'Etat, Grenoble.

Gargett, A.E., Osborn, T.R. and Nasmyth, T.W., 1984, "Local isotropy and the decay of turbulence in a stratified fluid", J. Fluid Mech., **144** , pp 231-280.

Gerz, T., Schumann, U. and Elghobashi, S.E., 1989, "Direct numerical simulation of stratified homogeneous turbulent shear flows", J. Fluid Mech., **200**, pp 563-594.

Gharib, M. and Derango, P., 1989, "A liquid film (soap film) tunnel to study laminar and turbulent shear flows", Physica D, **37**, pp 406-416.

Gibson, M.M., 1963, "Spectra of turbulence in a round jet", J. Fluid Mech., **15**, pp 161-173.

Giger, M., Jirka, G.H. and Dracos, Th., 1985, "Meandering jets in a shallow fluid layer", "Turbulent Shear Flows V", Cornell University, edited by J.L. Lumley.

Gilbert, A.D., 1988, "Spiral structures and spectra in two-dimensional turbulence", J. Fluid Mech., **193**, pp 475-498.

Gill, A., E., 1982, *Atmosphere-Ocean dynamics*, Academic Press.

Gonze, M.A., Métais, O.M. and Lesieur, M., 1990, "Direct-numerical simulations of periodic wakes", preprint, Institut de Mécanique de Grenoble.

Grant, H.L., Stewart, R.W. and Moilliet, A., 1962., "Turbulent spectra from a tidal channel", J. Fluid Mech.,**12** , pp 241-268.

Greenspan, H.P., 1968, "The theory of rotating fluids", Cambridge University Press.

Herbert, T., 1984, "Stability and transition of laminar flows", Von Karman Institute for Fluid Dynamics, Special Course on *Stability and transition of laminar flows*, March 1984, Brussels.

Herbert, T., 1988, "Secondary instability of boundary layers", Ann. Rev. Fluid Mech., **20**, pp 487-526.

Herring, J.R., 1974, "Approach of an axisymmetric turbulence to isotropy", Phys. Fluids, **17**, pp 859-872.

Herring, J.R., 1979, "Subgrid scale modeling - An introduction and overview", Turbulent Shear Flows I, edited by F. Durst et al., Springer Verlag, pp 347-352.

Herring, J.R., 1980, "Statistical theory of quasi-geostrophic turbulence", J. Atmos. Sci., **37** , pp 969-977.

Lasheras, J.C. and Choi, H., 1988, "Three-dimensional instability of a plane free shear layer: an experimental study of the formation and evolution of streamwise vortices", J. Fluid Mech., **189**, pp 53-86.

Lasheras, J.C. and Meiburg, E., 1990, "Three-dimensional vorticity modes in the wake of a flat plate", Phys. Fluids A, **3**, pp 371-380.

Laufer, J., Kaplan, R.E. and Chu, W.T., 1973, "On the generation of jet noise", AGARD Specialists Meeting *Noise Mechanisms*, Brussels.

Launder, B.E. and Spalding, D.B., 1972, *Mathematical models of turbulence*, Academic Press.

Lee, T.D., 1952, "On some statistical properties of hydrodynamical and magneto hydrodynamical fields", Q. Appl. Math., **10** , pp 69-74.

Lee, J., 1979, "Dynamical behaviour of the fundamental triad-interaction system in three-dimensional homogeneous turbulence", Phys. Fluids, **22** , pp 40-53.

Lee, J., 1982, "Development of mixing and isotropy in inviscid homogeneous turbulence", J. Fluid Mech., **120** , pp 155-183.

Legras, B., 1978, "Turbulence bidimensionnelle sur une sphère en rotation, simulations directes et modèles stochastiques", Thèse, Université Pierre et Marie Curie", Paris.

Legras, B., 1980, "Turbulent phase shift of Rossby waves", Geophys. Astrophys. Fluid Dyn., **15** , pp 253-281.

Leith, C.E., 1968, "Diffusion approximation for two-dimensional turbulence", Phys. Fluids, **11**, pp 671-673.

Leith, C.E., 1971, "Atmospheric predictability and two-dimensional turbulence", J. Atmos. Sci., **28** , pp 145-161.

Leith, C.E., 1984, "Minimum enstrophy vortices", Phys. Fluids, **27**, pp 1388-1395.

Leith, C.E. and Kraichnan, R.H., 1972, "Predictability of turbulent flows", J. Atmos. Sci., **29** , pp 1041-1058.

Lele, S.K., 1989, "Direct-numerical simulation of compressible free-shear flows", A.I.A.A. Paper, n⁰ 89-0374.

Léonard, A., 1985, "Computing three dimensional incompressible flows with vortex elements", Ann. Rev. Fluid Mech. **17** , pp 523-559.

Léorat, J., Pouquet, A. and Frisch, U., 1981, "Fully-developed MHD turbulence near critical magnetic Reynolds number", J. Fluid Mech., **104**, pp 419-443.

Leray, J., 1934, "Sur le mouvement d'un fluide visqueux emplissant l'espace", J. Acta. Math., **63** , pp 193-248.

Lesieur, M.,1972,"Décomposition d'un champ de vitesse non divergent en ondes d'hélicité", Turbulence (Observatoire de Nice).

Lesieur, M., 1973, "Contribution à l'étude de quelques problèmes en turbulence pleinement développée", Thèse de Doctorat d'Etat, Université de Nice.

Lesieur, M., 1982, "La turbulence développée", La Recherche, **139** , pp 1412-1425.

Lesieur, M., 1983, "Introduction to two-dimensional turbulence", in *Two-dimensional turbulence*, J. Mec. Th. Appl., Suppl. , R. Moreau ed., pp 5-20.

Lesieur, M., 1984, "Intermittency of coherent structures, an approach using statistical theories of isotropic turbulence", In *Turbulence and chaotic phenomena in fluids*, edited by T. Tatsumi, North-Holland, pp 339-350

Lesieur, M., Frisch, U. and Brissaud, A., 1971, "Théorie de Kraichnan de la turbulence, application à l'étude d'une turbulence possédant de l'hélicité", Ann. Geophys., **27** , pp 151-165.

Lesieur, M. and Herring, J.R., 1985, "Diffusion of a passive scalar in two-dimensional turbulence", J. Fluid Mech., **161** , pp 77-95.

Lesieur, M. and Rogallo, R., 1989, "Large-eddy simulation of passive scalar diffusion in isotropic turbulence", Phys. Fluids A, **1**, pp 718-722.

Lesieur, M., Métais, O. and Rogallo, R., 1989, "Etude de la diffusion turbulente par simulation des grandes échelles", C.R. Acad. Sci. Paris, Ser. II, **308**, pp 1395-1400.

Lesieur, M., Montmory, C. and Chollet, J.P., 1986, "The decay of kinetic energy and temperature variance in three-dimensional isotropic turbulence", Phys. Fluids, **30**, pp 1278-1286.

Lesieur, M. and Schertzer, D., 1978, "Amortissement auto similaire d'une turbulence à grand nombre de Reynolds," Journal de Mécanique, **17** , pp 609-646.

Lesieur, M., Somméria, J. and Holloway, G., 1981, "Zones inertielles du spectre d'un contaminant passif en turbulence bidimensionnelle", C.R. Acad. Sci. Paris, Ser II, **292** , pp 271-274.

Lesieur, M., Staquet, C., Le Roy, P. and Comte, P., 1988, "The mixing layer and its coherence examined from the point of view of two-dimensional turbulence", J. Fluid Mech., **192**, pp 511-534.

Leslie, D.C., 1973, *Developments in the theory of turbulence*, Clarendon Press, Oxford.

Leslie, D.C. and Quarini, G.L., 1979, "The application of turbulence theory to the formulation of subgrid modelling procedures", J. Fluid Mech., **91** , pp 65-91.

Lessen, M., Fox, J.A. and Zien, H.M., 1965, "On the inviscid stability of the laminar mixing of two parallel streams of a compressible fluid", J. Fluid Mech., **23**, pp 355-367.

Lessen, M., Fox, J.A. and Zien, H.M., 1966, "Stability of the laminar mixing of two parallel streams with respect to supersonic disturbances", J. Fluid Mech., **25**, pp 737-742.

Lienhard, V.J. and Van Atta, C.W., 1990, "The decay of turbulence in thermally stratified flow, J. Fluid Mech., **210**, pp 57-112.

Liepmann, H.W. and Laufer, J., 1949, "Investigation of free turbulent mixing", N.A.C.A. Tech. Note 1257.

Lighthill, J., 1978, *Waves in fluids*, Cambridge University Press.

Lilly, D.K., 1969, "Numerical simulation of two-dimensional turbulence", Phys. Fluids Suppl. II **12**, pp 240-249.

Lilly, D.K., 1971, "Numerical simulation of developing and decaying two-dimensional turbulence", J. Fluid Mech., **45**, p 395.

Lilly, D.K., 1973, "Lectures in sub-synoptic scales of motion and two-dimensional turbulence", in *Dynamic Meteorology*, Proceedings of the Lannion 1971 C.N.E.S. Summer School of Space Physics, P. Morel ed., Reidel Publishing Company.

Lilly, D.K., 1983, "Stratified turbulence and the mesoscale variability of the atmosphere", J. Atmos. Sci., **40**, pp 749-761.

Lilly, D.K., 1984, "Some facets of the predictability problem for atmospheric mesoscales", in *Predictability of fluid motions*, La Jolla Institute 1983, AIP Conference Proceedings **106**, edited by G. Holloway and B.J. West, pp 287-294.

Lilly, D.K., 1986, "The structure, energetics and propagation of rotating convective storms. Part II: helicity and storm stabilization", J. Atmos. Sci., **43**, pp 126-140.

Lin, J.T., 1972, "Relative dispersion in the enstrophy cascading inertial range of homogeneous two-dimensional turbulence", J. Atmos. Sci., **29** , pp 394-396.

Lions, J.L., 1989, "Remarks on turbulence", in *Turbulence and coherent structures*, M. Lesieur and O. Métais eds., Kluwer.

Lorenz, E.N., 1963, "Deterministic nonperiodic flow", J. Atmos. Sci., **20**, pp 130-141.

Lorenz, E.N., 1969, "The predictability of a flow which possesses many scales of motion", Tellus, **21** , pp 289-307.

Lowery, P.S. and Reynolds, W.C., 1986, "Numerical simulation of a spatially-developing, forced, plane mixing layer", Report TF-26, Stanford University.

Lucretius, "De Natura Rerum ".

Lumley, J.L., 1967, "The structure of inhomogeneous turbulent flows", in *Atmospheric turbulence and radio wave propagation*, ed. A.M. Yaglom and V.I. Tatarski, pp 166-176, NAUKA, Moscow.

Lumley, J.L., 1978, Advances in Appl. Mech., **18**, pp 123-175.

Lundgren, T.S., 1981, "Turbulent pair dispersion and scalar diffusion", J. Fluid Mech., **111** , pp 27-57.

Mandelbrot, B., 1975, *Les objets fractals: forme, hasard et dimension*, Flammarion, Paris.

Mandelbrot, B., 1976, "Intermittent turbulence and fractal dimension", in "Turbulence and Navier-Stokes equations", R. Temam ed., Lecture Notes in Mathematics, **565**, Springer-Verlag, pp 121-145.

Mandelbrot, B., 1977, *The fractal geometry of nature*, W.H. Freeman and Company, New York.

Mack, L.M., 1969, "Boundary-layer stability theory" (2 volumes), Jet Propulsion Laboratory.

Maréchal, J., 1972, "Etude expérimentale de la déformation plane d'une turbulence homogène", J. Mécanique, **11**, pp 263-294.

Maslowe, S.A., 1981, "Shear-flow instabilities and transition", in *Hydrodynamic instabilities and the transition to turbulence*, edited by H.L. Swinney and J.P. Gollub, Springer-Verlag, pp 181-228.

Maslowe, S.A., Laroche, H. and Lesieur, M., 1989, "Some direct numerical simulations of stable and unstable waves on a zonal shear flow", preprint, Institut de Mécanique de Grenoble.

Merilees, P.E. and Warn, H., "On energy and enstrophy exchanges in two-dimensional non-divergent flow", J. Fluid Mech., **69**, pp 625-630.

Meshalkin, L.D. and Sinai, Y.G., 1961, J. Appl. Math. Mech., **25**, p 1700

Métais, O., 1983, "La prédicibilité des écoulements turbulents évoluant librement", Thèse de Docteur Ingénieur, Grenoble.

Métais, O., 1985, "Evolution of three-dimensional turbulence under stratification", Proceedings of "Turbulent Shear Flows V", Cornell University, edited by J.L. Lumley.

Métais, O. and Herring, J.R., 1989, "Numerical simulations of freely-evolving turbulence in stably-stratified fluids", J. Fluid Mech., **202**, pp 117-148.

Métais, O. and Chollet, J.P., 1989, "Turbulence submitted to a stable density stratification: large-eddy simulation and statistical theory", in *Turbulent Shear Flows VI*, Springer-Verlag, pp 398-416.

Métais, O. and Lesieur, M., 1986, "Statistical predictability of decaying turbulence", J. Atmos. Sci., **43**, pp 857-870.

Métais, O. and Lesieur, M., 1989, "Large-eddy simulation of isotropic and stably-stratified turbulence", in *Advances in Turbulence 2*, H.H. Fernholz and H.E. Fiedler eds, Springer-Verlag, pp 371-376.

Métais, O. and Lesieur, M., 1990, "Spectral large-eddy simulations of isotropic and stably-stratified turbulence", preprint, Institut de Mécanique de Grenoble.

Metcalfe, R.W., Orszag, S.A., Brachet, M.E., Menon, S. and Riley, J., 1987, "Secondary instability of a temporally growing mixing layer", J. Fluid Mech., **184**, pp 207-243.

Michalke, A., 1964, "On the inviscid instability of the hyperbolic tangent velocity profile", J. Fluid Mech., **19**, pp 543-556.

Millionshtchikov, M., 1941, "On the theory of homogeneous isotropic turbulence", Dokl. Akad. Nauk. SSSR, **32** , pp 615-618.

Mirabel, A.P. and Monin, C.A., 1982, Physics of the atmosphere and oceans, **19** , p 902

Moffatt, H.K, 1970, "Dynamo action associated with random inertial waves in a rotating conducting fluid", J. Fluid Mech., **44** , pp 705-719.

Moffatt, H. K., 1978, *Magnetic field generation in electrically conducting fluids*, Cambridge University Press.

Moffatt, H.K., 1985 "Magnetostatic equilibria and analogous Euler flows of arbitrarily complex topology. Part 1. Fundamentals", J. Fluid. Mech., **159**, pp 359-378.

Moffatt, H.K., 1986, "Geophysical and astrophysical turbulence", in *Advances in turbulence*, G. Comte-Bellot and J. Mathieu, eds, Springer-Verlag, pp 228-244.

Moin, P. and Kim. P., 1982, "Numerical investigation of turbulent channel flow", J. Fluid Mech., **118** , pp 341-377.

Monin, A.S. and Yaglom, A.M., 1975, "Statistical Fluid Mechanics: Mechanics of Turbulence", Vol. 2, M.I.T. Press, Cambridge.

Montmory, C., 1986, private communication.

Moreau, R, 1990, *Magneto-Hydrodynamic Turbulence*, Kluwer, Dordrecht (to appear).

Morel, P. and Larchevêque, M., 1974, "Relative dispersion of constant level balloons in the 200 mb general circulation", J. Atmos. Sci., **31**, pp 2189-2196.

Morf, R.H., Orszag, S.A. and Frisch, U., 1980, "Spontaneous singularity in three-dimensional inviscid, incompressible flow", Phys. Rev. Letters.,**44** , p 572

Morkovin, M., 1988, private communication

Mory, M. and Capéran, P., 1987, "On the genesis of quasi-steady vortices in a rotating turbulent flow", J. Fluid Mech., **185**, pp 121-136.

Moses, H. E., 1971, "Eigenfunctions of the curl operator, rotationally invariant Helmholtz theorem and applications to electromagnetic theory and fluid mechanics", SIAM J. Appl, **21** , pp 114-144.

Munk, W.H. and MacDonald, G.J.F., 1975, *The rotation of the Earth*, Cambridge University Press.

Nelkin, M. and Kerr, R.M., 1981, "Decay of scalar variance in terms of a modified Richardson law for pair dispersion", Phys. Fluids, **24**, pp 1754-1756.

Newman, G.R. and Herring, J.R., 1979, "A test field model study of a passive scalar in isotropic turbulence", **94**, pp 163-194.

Nikuradse, J, 1932, Forschungsheft, **356**

Normand, X., 1989, private communication.

Normand, X., Comte, P. and Lesieur, M., 1988, "Numerical simulation of a spatially-growing mixing layer", La Recherche Aérospatiale, **6**, pp 45-52.

Normand, X., Fouillet, Y. and Lesieur, M., 1989, "How much are the small scales of high Mach number turbulence affected by compressibility?", preprint, Institut de Mécanique de Grenoble.

Novikov, E.A., 1959, Izv. Acad. Sci. USSR, Geophys. Ser. **11**

Novikov, E.A. and Stewart, R.W., 1964, "Intermittency of turbulence and spectrum of fluctuations in energy dissipation", Izv. Akad. Nauk. SSSR, Ser. Geophys., **3** , p 408

Oboukhov, A.M., 1949, "Structure of the temperature field in turbulent flows", Isv. Geogr. Geophys. Ser., **13** , pp 58-69.

Oboukhov, A.M., 1962, "Some specific features of atmospheric turbulence", J. Fluid Mech., **12** , pp 77-81.

O'Brien, E.F. and Francis, G.C., 1963, "A consequence of the zero fourth cumulant approximation", J. Fluid Mech., **13** , pp 369-382.

Ogura, Y., 1963, "A consequence of the zero fourth cumulant approximation in the decay of isotropic turbulence", J. Fluid Mech., **16** , pp 33-40.

Orszag, S.A. 1970 a, *The statistical theory of turbulence* (this monograph was unfortunately never published, though excerpts are contained in Orszag (1977). I am greatly indebted to the author for having let copies of his manuscript available).

Orszag, S. A., 1970 b, "Analytical theories of turbulence", J. Fluid Mech., **41** , pp 363-386.

Orszag, S.A., 1977, "Statistical theory of turbulence", in *Fluid Dynamics 1973*, Les Houches Summer School of Theoretical Physics, R. Balian and J.L. Peube eds, Gordon and Breach, pp 237-374.

Orszag, S.A. and Patterson, G.S., 1972, "Numerical simulations of turbulence", in "Statistical models and turbulence", edited by M. Rosenblatt and C. Van Atta, Springer Verlag, **12** , pp 127-147.

Orszag, S.A. and Patera, A.T., 1983, "Secondary instability of wall-bounded shear flows", J. Fluid Mech., **128**, pp 347-385.

Osmidov, R.V., 1975, "On the turbulent exchange in a stably stratified ocean", Izvestiya, Academy of Sciences, USSR, Atmospheric and Oceanic Physics, **1**, pp 493-497.

Panchev, S., 1971, *Random functions and turbulence*, pp 198-203, Pergamon Press, Oxford.

Pao, Y.H., 1973, "Measurements of internal waves and turbulence in two-dimensional stratified shear flows", Boundary-Layer Met., **5**, p 177

Papamoschou, D. and Roshko, A., 1988, "The compressible turbulent shear layer: an experimental study", J. Fluid Mech., **197**, pp 453-477.

Papamoschou, D., 1989, "Stucture of the compressible turbulent shear layer", A.I.A.A. Paper, n^0 89-0126.

Papoulis, A., 1965, "Probability, random variables and stochastic processes", Mc Graw Hill.

Pedlosky, J., 1979, *Geophysical Fluid Dynamics*, Springer Verlag.

Phillips, N.A., 1954, "Energy transformations and meridional circulations associated with simple baroclinic waves in a two-level, quasigeostrophic model", Tellus, **6**, pp 273-286.

Phillips, N.A., 1956, "The general circulation of the atmosphere: a numerical experiment", Q. J. R. Meteorol. Soc., **82** , pp 123-140.

Pierrehumbert, R.T. and Widnall, S.E.,1982. "The two- and three-dimensional instabilities of a spatially periodic shear layer", J. Fluid Mech., **114**, pp 59-82.

Pironneau, O., 1989, *Finite-elements methods in fluids*, Wiley-Masson.

Pouquet, A., Frisch, U. and Chollet, J.P., 1983, "Turbulence with a spectral gap", Phys. Fluids., **26** , pp 877-880.

Pouquet, A., Frisch, U. and Léorat, J., 1976, "Strong MHD helical turbulence and the non-linear dynamo effect", J. Fluid Mech., **77** , pp 321-354.

Pouquet, A., Lesieur, M., André, J.C. and Basdevant, C., 1975, "Evolution of high Reynolds number two-dimensional turbulence", J. Fluid Mech., **72** , pp 305-319.

Prandtl, Z.A., 1925, "Bericht über Untersuchugen zur Ausgebildeten Turbulenz", Zs Angew. Math. Mech., **5** , pp 136-169.

Prigogine, I., 1980, "From being to becoming", Freeman.

Proudman, I. and Reid, W.H., 1954, "On the decay of a normally distributed and homogeneous turbulent velocity field", Phil. Trans. Roy. Soc., Vol A **247** , pp 163-189.

Reynolds, O., 1883, "An experimental investigation of the circumstances which determine whether the motion of water shall be direct and sinuous, and the law of resistance in parallel channels", Phil. Trans. Roy. Soc., pp 51-105.

Rhines, P.B., 1975, "Waves and turbulence on a β-plane", J. Fluid Mech., **69** , pp 417-443.

Rhines, P.B., 1979, "Geostrophic turbulence", Ann. Rev. Fluid Mech., **11**, pp 401-441.

Richardson, L.F., 1922, *Weather prediction by numerical process*, Cambridge University Press.

Richardson, L.F., 1926, "Atmospheric diffusion shown on a distance neighbor graph", Proc. Roy. Soc. London, Ser A **110** , pp 709-737.

Riley, J.J., Metcalfe, R.W. and Weissman, M.A., 1981, "Direct numerical simulations of homogeneous turbulence in density stratified fluids", in *Non linear properties of internal waves*, La Jolla Institute, AIP Conference Proceedings **76** , edited by B.J. West, pp 79-112.

Roberts, P.H., 1961, "Analytical theory of turbulent diffusion", J. Fluid Mech., **11** , pp 257-283.

Robertson, H.P., 1940, Proc. Cambridge Phil. Soc., **36**, p 209

Rogallo, R.S., 1981, "Numerical experiments in homogeneous turbulence", NASA TM 81315.

Rogallo, r.S. and Moin, P., 1984, "Numerical Simulation of turbulent flows", Ann. Rev. Fluid Mech., **16**, pp 99-137.

Rogers, M.M. and Moin, P., 1987 a, "The structure of the vorticity field in homogeneous turbulent flows", J. Fluid Mech., **176**, pp 33-66.

Rogers, M.M. and Moin, P., 1987 b, "Helicity fluctuations in incompressible turbulent flows", Phys. Fluids, **30**, pp 2662-2671.

Rose, H.A., 1977, "Eddy diffusivity, eddy noise and subgrid scale modelling", J. Fluid Mech., **81** , pp 719-734.

Rose, H. and Sulem, P.L., 1978, "Fully developed turbulence and statistical mechanics", Journal de Physique (Paris), **39**, pp 441-484.

Rossby, C.G. et al., 1939, "Relation between variations in the intensity of the zonal circulation of the atmosphere and the displacements of the semi permanent centers of action", J. Marine Res., **2**, pp 38-55.

Rouse, H. and Ince, S., 1957, *History of Hydraulics*, Iowa Institute of Hydraulic Research.

Sadourny, R. and Basdevant, C., 1985, "Parameterization of subgrid scale barotropic and baroclinic eddies in quasi geostrophic models: anticipated potential vorticity method", J. Atmos. Sci., **42**, pp 1353-1363.

Saffman, P.G., 1967, "Note on decay of homogeneous turbulence", Phys. Fluids, **10** , p 1349

Sandham, N.D. and Reynolds, W.C., 1989, "The compressible mixing layer: linear theory and direct simulation", AIAA Paper 89-0371.

Sandham, N.D. and Reynolds, W.C., 1990, "Three-dimensional simulations of the compressible mixing layer", J. Fluid Mech., in press.

Saffman, P.G., 1971, "On the spectrum and decay of random two-dimensional vorticity distributions of large Reynolds number", Stud. Appl. Math.,**50**, pp 377-383.

Salmon, R., 1978, "Two-layer quasi-geostrophic turbulence in a simple special case", Geophys. Astrophys. Fluid Dyn., **10** , pp 25-52.

Sato, H. and Kuriki, K., 1961, "The mechanism of transition in the wake of a thin flat plate placed parallel to a uniform flow", J. Fluid Mech., **11**, pp 321-352.

Schertzer, D. and Lovejoy, S., 1984, "The dimension and intermittency of atmospheric dynamics", in "Turbulent Shear Flows IV", edited by B. Launder, Springer-Verlag.

Schlichting, H., 1968, *Boundary-layer theory*, Pergamon Press, London.

Schmidt, H. and Schumann, U., 1989, "Coherent structure of the convective boundary layer derived from large eddy simulations", J. Fluid Mech., **200**, pp 511-562.

Schubauer, G.B. and Skramstad, H.K., 1948, "Laminar boundary layer oscillations and transition on a flat plate", N.A.C.A. Report 909.

Schubauer, G.B. and Klebanoff, P.S., 1956, "Contributions on the mechanics of boundary layer transition", N.A.C.A. Report 1289.

Schwartz, L., 1967, "Théorie des distributions", Hermann, Paris.

Serres, M., 1977, "La naissance de la Physique dans le texte de Lucrèce, fleuves et turbulences", Editions de Minuit.

Shepherd, T., 1987, "Non-ergodicity of inviscid two-dimensional flow on a beta-plane and on the surface of a rotating sphere", J. Fluid Mech., **184**, pp 289-302.

Sivashinsky, G. and Yakhot, V., 1985, "Negative viscosity effect in large-scale flows", Phys. Fluids, **28**, pp 1040-1042.

Shtilman, L., Levich, E., Orszag, S.A., Pelz, R.B. and Tsinober, A., 1985, "On the role of helicity in complex fluid flows", Phys. Let., **113 A**, pp 32-37.

Silveira, A., Grand, D. and Lesieur, M., 1989, "Simulation numérique des grandes échelles d'un écoulement turbulent bidimensionnel derrière une marche", Rapport CEA-R-5510, Commissariat à l'Energie Atomique, France.

Smagorinsky, J., 1963, "General circulation experiments with the primitive equations", Mon. Weath. Rev., **91, 3** , pp 99-164.

Soetrisno, M., Eberhardt, S., Riley, J. and McMurtry, P., 1988, "A study of inviscid, supersonic mixing layers using a second-order TVD scheme", AIAA/ASME/SIAM/APS 1st Nat. Fluid Dynamics Congress, Cincinnati, AIAA paper 88-3676-CP.

Soetrisno, M., Greenough, J.A., Eberhardt, S. and Riley, J., 1989, "Confined compressible mixing layers: part I. Three-dimensional instabilities", AIAA 20 Fluid Dynamics, Plasma Dynamics and Lasers Conference, Buffalo, pp th st Nat. Fluid Dynamics Congress, Cincinnati, AIAA paper 89-1810.

Somméria, G., 1976, "Three-dimensional simulation of turbulent processes in an undisturbed trade wind boundary layer", J. Atmos. Sci., **33**, pp 216-241.

Somméria, J., 1986, "Experimental study of the two-dimensional inverse energy cascade in a square box", J. Fluid Mech., **170**, pp 139-168.

Somméria, J., Meyers, S.D. and Swinney, H.L., 1988, "Laboratory simulation of Jupiter's Great Red Spot", Nature, **331**, pp 689-693.

Sreenivasan, K.R., Tavoularis, S., Henry, S. and Corrsin, S., 1980, "Temperature fluctuations and scales in grid-generated turbulence", J. Fluid Mech., **100** , pp 597-621.

Sreenevasan, K.R. and Strykowski, P.J., 1984 "On analogies between turbulence in open flows and chaotic dynamical systems", in "Turbulence and chaotic phenomena in fluids", edited by T. Tatsumi, North-Holland, pp 191-196.

Stapountzis, B., Sawford, B.L., Hunt, J.C. and Britter, R.E., 1986, "Structure of the temperature field downwind of a line source in grid turbulence", J. Fluid Mech., **165**, pp 401-424.

Staquet, C., 1985, "Etude numérique de la transition à la turbulence bidimensionnelle dans une couche de mélange", Thèse de l'Université de Grenoble.

Staquet, C., Métais, O. and Lesieur, M., 1985, "Etude de la couche de mélange et de sa cohérence du point de vue de la turbulence bidimensionnelle", C.R. Acad. Sci., ser II, **300** , pp 833-838.

Staquet, C. and Riley, J., 1989, "A numerical study of a stably-stratified mixing layer", in *Turbulent Shear Flows VI*, Springer-Verlag, pp 381-397.

Stillinger, D.C., Helland, K.N., and Van Atta, C.W., 1983, "Experiments on the transition of homogeneous turbulence to internal waves in a stratified fluid", J. Fluid Mech., **131**, pp 691-722.

Sulem, P.L., Lesieur, M. and Frisch, U., 1975, "Le Test Field Model interprété comme méthode de fermeture des équations de la turbulence", Ann. Géophys. (Paris), **31** , pp 487-495.

Tatsumi, T., 1957, "The theory of decay process of incompressible isotropic turbulence", Proc. Roy. Soc. London, Ser. A **239** , pp 16-45.

Tatsumi, T., 1980, "Theory of homogeneous turbulence", Advances in Applied Mechanics, **20** , pp 39-133.

Tatsumi, T., Kida, S. and Mizushima, J., 1978, "The multiple scale turbulent expansion for isotropic turbulence", J. Fluid Mech., **85** , pp 97-142.

Tatsumi, T. and Yanase, S., 1981, "The modified cumulant expansion for two-dimensional isotropic turbulence", J. Fluid. Mech., **110**, pp 475-496.

Taylor, G.I., 1921, Proc. London Math. Soc., **20**, p 126.

Taylor, G.I., 1922, "The motion of a sphere in a rotating liquid", Proc. Roy. Soc. London, Ser.A, **102**, pp 213-215.

Taylor, G.I. and Green, A.E. 1937, Proc. Roy. Soc. London, Ser.A, **158**, p 499

Temam, R., 1977, *Navier-Stokes equations*, North Holland (revised edition 1984).

Tennekes, H., 1968, "Simple model for the small-scale structure of turbulence", Phys. Fluids, **11** , pp 669-671.

Tennekees, H. and Lumley, J.L., 1972, *A first course in turbulence*, MIT Press, Cambridge, Mass.

Thompson, P.D., 1957, "Uncertainty of initial state as a factor in the predictability of large-scales atmospheric flow patterns", Tellus, **9** , pp 275-295.

Thompson, P. D., 1984, "A review of the predictability problem", in *Predictability of fluid motions*, La Jolla Institute 1983, AIP Conference Proceedings **106** , edited by G. Holloway and B.J. West, pp 1-10.

Thoroddsen, S.T. and Van Atta, C.W., 1990, "Experiments on small-scale anisotropy and dissipation in stably-stratified turbulence", preprint, University of California San Diego.

Thorpe, S.A., 1968, "A method of producing a shear-flow in a stratified fluid", J. Fluid Mech., **32**, pp 693-704.

Townsend, A.A., 1967, *The structure of turbulent shear flow*, Cambridge University Press.

Tritton, D.J. and Davies, P.A., 1985, "Instabilities in Geophysical Fluid Dynamics", in *Hydrodynamic instabilities and the transition to turbulence*, edited by H.L. Swinney and J.P. Gollub, Springer-Verlag, pp 181-228 (second edition).

Tsinober, A. and Levitch, E., 1983, "On the helical nature of three-dimensional coherent structures in turbulent flows", Phys. Letters, **99 A**, pp 321-324.

Van Atta, C.W. and Chen, W.Y., 1970, "Stucture functions of turbulence in the atmospheric boundary layer over the ocean", J. Fluid Mech., **44** , pp 145-159.

Van Dyke, M., 1982, *An album of fluid motion*, The Parabolic Press, Stanford, California.

Verron, J. and Le Provost, C., 1985, "A numerical study of quasi-geostrophic flow over isolated topography", J. Fluid Mech., **154**, pp 231-252.

Warhaft, Z. and Lumley, J.L. 1978, "An experimental study of the decay of temperature fluctuations in grid generated turbulence." J. Fluid Mech, **88** , pp 659-684.

Warhaft, Z., 1984, "The interference of thermal fields from line sources in grid turbulence", J. Fluid Mech., **144**, pp 363-387.

Werlé, H., 1974, *Courants et couleurs*, ONERA.

Williams, G.P., 1979, "Planetary circulations: 2. The Jovian quasi-geostrophic regime", J. Atmos. Sci., **36**, pp 932-968.

Winant,C.D. and Browand,F.K., 1974, "Vortex pairing, the mechanism of turbulent mixing layer growth at moderate Reynolds number", J. Fluid Mech., **63**, pp 237-255.

Wray, A., 1988, private communication.

Wygnanski, I. and Fiedler, H.E., 1970, "The two-dimensional mixing region", J. Fluid Mech. **41**, pp 327-361.

Williams, G.P., 1978, "Planetary circulations: 1. barotropic representation of Jovian and terrestrial turbulence", J. Atmos. Sci., **35**, pp 1399-1426.

Yaglom, A.M., 1966, "Effect of fluctuations in energy dissipation rate on the form of turbulence characteristics in the inertial range", Dokl. Akad. Nauk. SSSR, **166** , p 49

Yakhot, V. and Orszag, S., 1986, "Renormalization Group (RNG) methods for turbulence closure", J. Scientific Computing, **1**, pp 3-52.

Yamada, H. and Matsui, T., 1978, "Preliminary study of mutual slip-through of a pair of vortices", Phys. Fluids, **21**, pp 292-294.

Yanase, S., Métais, O. and Lesieur, M., 1990, "Direct-numerical simulations of rotating mixing layers", preprint, Institut de Mécanique de Grenoble.

Yeh, T.T. and Van Atta, C.W., 1973, "Spectral transfer of scalar and velocity fields in heated-grid turbulence", J. Fluid Mech., **48**, pp 41-71.

Yoshizawa, A., 1982, "A statistically derived subgrid model for the large eddy simulations of turbulence", Phys. Fluids, **25** , pp 1532-1538.

Zabusky, N.J. and Deem, G.S., 1971. "Dynamical evolution of two-dimensional unstable shear flows", J. Fluid Mech., **47**, pp 353-379.

index

coherence 17, 71.
coherent eddies 7, 264, 270, 283, 352.
coherent structures 6, 17, 50, 58, 127, 143, 154, 270, 275, 315, 336, 349, 355, 361, 365.
coherent vortices 49, 253, 271.
Colin de Verdière 245, 265.
collapse 345.
collapse time 345.
collapse to two-dimensional turbulence 97, 340.
combustion 29, 43, 364.
complex helical waves 97, 116, 365.
complex helical waves decomposition 192.
compressibility 18, 32.
compressible mixing layer 73.
computational-Fluid Dynamics 7.
Comte 58, 65, 357, 361.
Comte and Lesieur 63, 359.
Comte et al. 271.
Comte-Bellot and Corrsin 193, 196.
conductive wave number 200, 202, 204.
conductivity 341.
conservation of kinetic energy 334.
constant-skewness model 212.
Constantin et al. 142.
continuity equation 20, 28.
convective cells 365.
convective Mach number 73.
convective storms 338.
convergence zones 74.
Corcos and Lin 62, 72, 350, 355.
Corcos and Sherman 350, 353.
Coriolis 23, 95.
Coriolis force 21, 34, 42, 80, 243, 248, 262.
Coriolis parameter 56, 248.
correlated energy spectrum 306.
correlated forcing 310.
Corrsin 31, 202.
Corrsin-Oboukhov 226, 327.

Corrsin-Oboukhov constant 209, 220, 229.
Couder 246.
Coüet 32, 332.
Craya 95, 103, 110, 113.
Craya decomposition 95, 99, 132.
critical Mach number 75.
critical phenomena 197.
critical points 54.
critical Rayleigh number 79.
critical Reynolds number 50, 75.
critical Rossby number 83.
Crow and Champagne 66, 70.
cumulants 164, 173.
cusp 325, 331.
cutoff wave number 225.
cyclonic 9, 249, 288, 304, 315, 363.
cyclonic disturbances 253.
cyclonic perturbations 84.
cyclonic rotation 83.
d'Humières et al. 335.
Dannevik et al. 162, 202.
Deardorff 322.
decaying isotropic turbulence 159, 330.
decimation 200.
decorrelated energy spectrum 307.
deficit-velocity profile 124.
deformation tensor 21.
degrees of freedom 11, 13, 143, 270, 317.
density-velocity correlation 108.
departures from gaussianity 15.
depression 9, 24.
detailed conservation 129, 297.
deterministic 2, 14, 313, 321,.
developed turbulence 143.
$D.I.A.$ inertial-range 177.
differential heating 8.
differential rotation 46, 53, 264, 288, 302, 334.
diffusion coefficient 234.
diffusion equation 234.

dimension of the attractor 143.
dipole structures 315.
Dirac 90,110.
direct-numerical simulation 7, 14,
21, 49, 58, 69, 75, 94, 173, 192,
215, 276, 283, 288, 295, 304, 309,
317, 330, 346, 353, 360.
Direct-Interaction Approximation
161, 171.
displaced fluid particle 58.
displacement thickness 67.
dissipative range 140, 168, 179,
195, 200.
dissipative scales 10, 70,142, 295,
331.
dissipative structures 31, 86, 192.
dissipative systems 11, 17.
distributions 90.
divergence of the enstrophy 184.
Domaradzki et al. 324, 331.
drag 1.
Drazin and Howard 253.
Drazin and Reid 52, 56, 64, 78,
85.
dry adiabatic lapse rate 40.
Dumas 138.
dynamic viscosity 21.
dynamical system 11, 16, 143, 305.
dynamo effect 106, 189.
Dziomba and Fiedler 352.
Eady model 253.
Earth 1, 8, 11, 19, 34, 41, 46, 81,
250, 260, 288, 308.
Earth's atmosphere 16, 63.
Earth's magnetic field 106.
eddy-conductivity 209.
Eddy-Damped Quasi-Normal approximation 168.
Eddy-Damped Quasi-Normal Markovian approximation 169.
eddy-damping rate 167.
eddy-damping 168, 309.
eddy-diffusivity 218, 235, 285, 325,
337, 346.

eddy-dispersion coefficient 229.
eddy-Prandtl number 346.
eddy-viscosity 16, 120, 135, 193,
209, 218, 247, 278, 285, 320, 325,
333.
eddy-viscosity subgrid scale model
331.
E.D.Q.N.M. approximation 161.
E.D.Q.N.M. closure 193, 195.
E.D.Q.N.M. energy flux 176.
E.D.Q.N.M. equation for the spectral tensor 169.
E.D.Q.N.M.-type 232.
Einstein convention 93.
Ekman layer 246, 254, 260, 274,
287.
Ekman number 260.
Ekman pumping 261.
Ekman spiral 259, 260.
El Nino 263.
energy cascade 137, 176, 324.
energy containing eddies 215, 331.
energy flux 191.
energy spectrum 166, 179, 190.
energy transfers 15, 174.
energy-containing range 140, 216,
226.
ensemble average 101, 104.
enstrophy 112, 115, 130, 145, 177,
279, 300, 334.
enstrophy cascade 135, 142, 179,
269, 277, 314, 334, 353.
enstrophy divergence 211.
enstrophy flux 278.
enstrophy transfer 131.
enstrophy-dissipation rate 271.
enstrophy-divergence time 147.
enthalpy 26, 28, 43, 47.
enthalpy equation 26.
entrainment vorticity 80.
entropy 86.
EOLE experiment 287, 339.
equilibrium solutions 298.
equipartition 320.

passive tracers 12.
passive vector 32.
peak 68.
Peclet number 2, 27, 200.
Pedlosky 30, 80, 246, 251, 254, 260, 264.
perfect barotropic fluid 32.
perfect fluid 4, 26, 30, 211.
periodic channel 69, 71.
perturbed vortex filament 72.
phase space 11, 297.
phenomenological theories 17, 341.
phenomenology 341.
Phillips 257, 334.
physical space 333.
phytoplancton 23.
Pierrehumbert and Widnall 62, 350, 355, 358.
pipe flows 49.
Pironneau 317.
plane Couette flow 51, 55, 69.
plane jet 56.
plane Poiseuille flow 51, 69.
plane symmetry 106.
plane wake 65, 67.
planetary scales 1, 8, 11, 81, 246, 289, 318, 365.
planetary scale motions 143.
planetary vorticity 34.
point-vortex dynamics 335.
point vortices 12.
Poiseuille flow 50, 55, 69, 71.
Poiseuille parabolic velocity profile 50.
pollution 364.
potential energy 341, 342, 343.
potential enstrophy 290, 302.
potential enstrophy cascade 289, 291.
potential temperature 28, 34, 38.
potential vorticity 17, 33, 38, 45, 251, 255, 263, 287, 302.
potential vorticity conservation 35, 251, 261, 289.

Pouquet et al. 135, 167, 180, 189, 245, **273**, 277, 280, 299, 323.
Prandtl number 78, 199, 202, 341.
Prandtl viscous boundary-layer equations **63**.
Prandtl 16, 120, 121.
Prandtl's mixing-length 16, 64.
predictability 16, 86, 170, 189, 220, 305, 354.
predictability of a passive scalar 313.
predictability problem, studies 223, 326, 332.
predictability time 313, 315, 321.
pressure drop coefficient 50.
pressure field 363.
pressure highs 8, 24, 74.
pressure term 212.
pressure troughs 74.
pressure-velocity correlation 108, 129.
Prigogine 87.
primitive equations 292.
probability distribution functions 158.
propagator 199.
Proudman and Reid 164, 185.
Proudman-Taylor theorem 33, 80, 250, 292.
pseudo-spectral 272.
pseudo-spectral methods 43, 94, 360.
pure helicity cascade 191.
quadratic invariants of turbulence 133.
quasi-geostrophic analysis 81.
quasi-geostrophic potential vorticity conservation 287.
quasi-geostrophic potential vorticity equation 252, 355.
quasi-geostrophic theory 246.
quasi-geostrophic turbulence 287, 335.